MICROWAVE ELECTRONIC CIRCUIT TECHNOLOGY

MICROWAVE ELECTRONIC CIRCUIT TECHNOLOGY

YOSHIHIRO KONISHI

Tokyo Institute of Technology
Atsugi, Japan

MARCEL DEKKER, INC. NEW YORK • BASEL • HONG KONG

ISBN: 0-8247-0101-1

The publisher offers discounts on this book when ordered in bulk quantities. For more information, write to Special Sales/Professional Marketing at the address below.

This book is printed on acid-free paper.

Marcel Dekker, Inc.
270 Madison Avenue, New York, New York 10016
http://www.dekker.com

Current printing (last digit):
10 9 8 7 6 5 4 3 2 1

PRINTED IN THE UNITED STATES OF AMERICA

Preface

High-frequency devices are used for movable communication systems, satellite communication systems, and wireless LAN as well as other systems. New devices with new materials will be very important not only in the multimedia fields mentioned above, but also in developing areas such as energy transformation and new sensors. To create these new devices, an understanding of the physical requirements is necessary.

The aim of this book is to explain the fundamental principles and physical meanings of practical devices. A thorough foundation in basic theories and concepts is given to provide engineers with a springboard from which to create new devices and inventions.

Chapter 1 offers an introduction to guided waves. The text clearly explains not only the physical relationships but also the equivalent network concepts and their corresponding modes.

In Chapter 2, basic network concepts such as matrix and eigenvalue problems are discussed in connection with circuit energy. This will give an understanding of the physical meaning of circuit operation and help to create an intuitive feeling for possible applications.

In Chapter 3, the basic network concept for distributed lines is explained together with applications. This chapter also discusses the important area of the design of miniaturized ceramic filters used for portable movable communications. The multicoupled lines theory is explained as is the equivalent network. Multilines in inhomogeneous ceramic materials are used to make couplers between resonators. The inhomogeneous part is expressed by the minimum phase networks along with the perturbation theory, which is fairly useful in practical applications.

Chapter 4 reviews the basic concepts of resonators and filters. The coupling mechanism is physically explained, which will help the designer to consider new devices with resonators.

Several microwave devices are explained in Chapter 5. Their physical descriptions are also given.

In Chapter 6, microwave circuits with high dielectric materials are discussed. Analysis of the types of characteristics of dielectric circuits required in microwave devices is given (for example, the contribution of a magnetic wall).

After an introduction to the microwave performance of ferrite, Chapter 7 covers devices made of this material. Attention is given to recently developed magnetostatic wave circuits along with their physical description. The Appendix outlines important theories relating to group velocity, cavity and waveguide perturbation, determination of capacitances, design of power dividers, principles of directional couplers, and Faraday rotation in an infinite medium.

Yoshihiro Konishi

Contents

1

Basic Knowledge of Guided Waves

1 TYPICAL WAVEGUIDES AND MODES

The construction of typical waveguides, the applications, and the characteristics are summarized in Fig. 1. The applications of the waveguides and modes shown in Fig. 1 are as follows:

(a) The *parallel wire line* is used for power lines, telephone lines, and a VHS TV feeder. In a high frequency, transmission loss is increased by radiation.
(b) The *coaxial line* is used from low frequency to microwave because there is no radiation loss.
(c) The *strip line* is used for small devices because it has no radiation loss and thin construction.
(d) The *microstrip line* is used in low-frequency to millimeter wave components because it is easily made on a substrate as a planar circuit.
(e) The *waveguide* is used for microwave and millimeter wave devices because it has small loss and is available to high power.
(f) The *ridge waveguide* is available to the lower frequency of microwave and small-sized components.

The electric and magnetic fields of Figs. 1a–1d exist only in the plane perpendicular to the axis of wave propagation. Such a distribution of electromagnetic field is called a TEM (transverse electric magnetic) mode, and

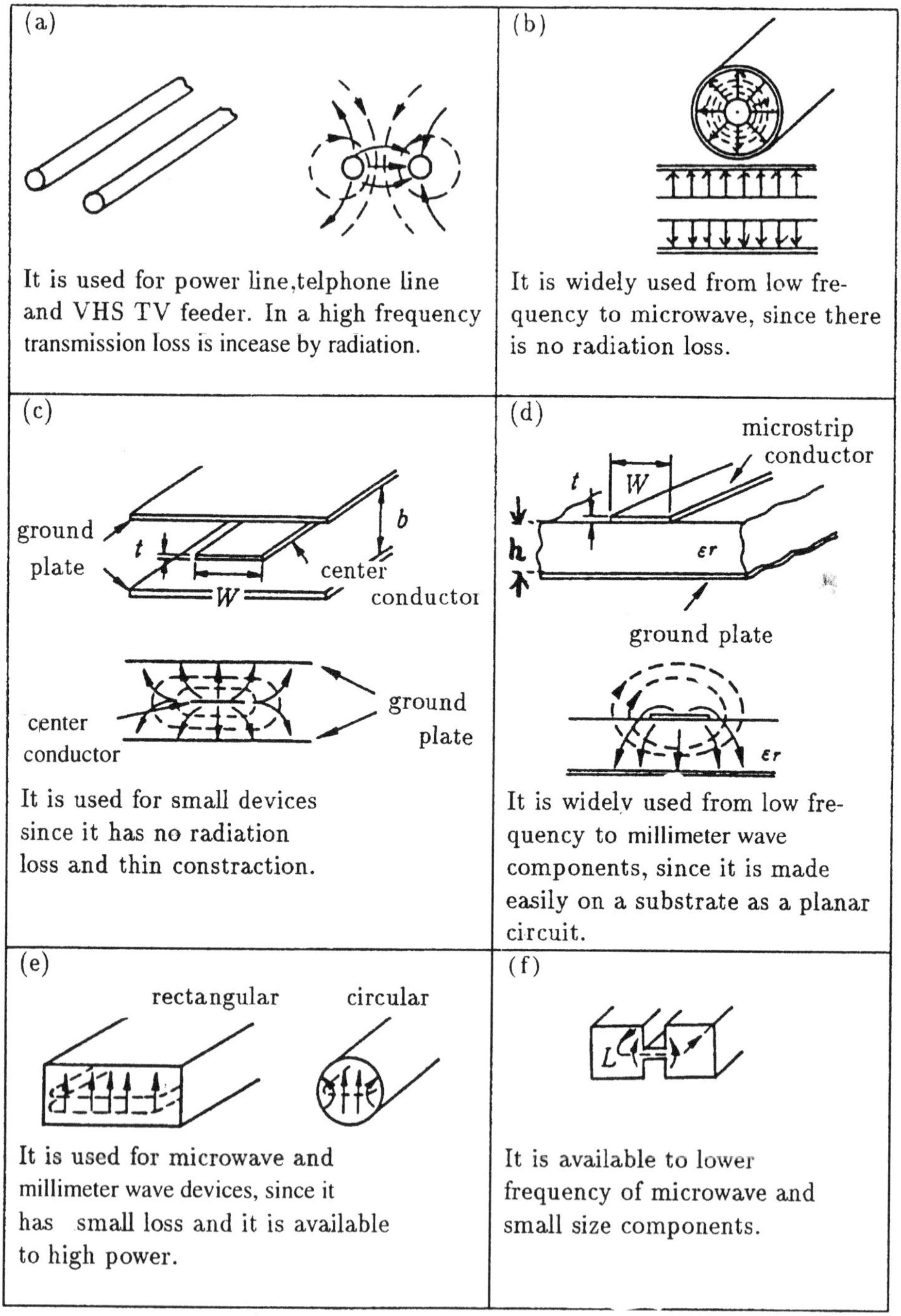
(a)
It is used for power line,telphone line and VHS TV feeder. In a high frequency transmission loss is incease by radiation.
(b)
It is widely used from low frequency to microwave, since there is no radiation loss.
(c)
ground plate
t
b
W
center conductor
center conductor
ground plate
It is used for small devices since it has no radiation loss and thin constraction.
(d)
microstrip conductor
t
W
h
εr
ground plate
εr
It is widely used from low frequency to millimeter wave components, since it is made easily on a substrate as a planar circuit.
(e)
rectangular
circular
It is used for microwave and millimeter wave devices, since it has small loss and it is available to high power.
(f)
L
It is available to lower frequency of microwave and small size components.

Figure 1. Typical waveguides and modes. (a) Parallel wire; (b) coaxial line; (c) strip line; (d) microstrip line; (e) waveguide; (f) ridge waveguide.

the wave is called a TEM wave. The wave radiated from the antenna also becomes a TEM wave at the far distance d ($d/\lambda \gg 1$, λ is a free-space wavelength) from the antenna. The TEM mode of the coaxial line does not take the same values and direction at each point on the same sectional plane (perpendicular to the axis of wave propagation). Such a mode is called a nonuniform TEM plane wave, where the plane wave has the plane equiphase surface at a given instant of time. On the other hand, the far field from the antenna takes the uniform TEM mode because the electromagnetic field far from antenna has the same values and is in the same direction of the field at the plane perpendicular to the wave propagation.

The electric fields of Figs. 1c and 1d only exists in the sectional plane, and the magnetic field exists not only in the sectional plane but also in the direction of wave propagation. Such a field distribution is called as a TE (transverse electric) mode or H mode.

On the other hand, we have the propagation mode where the magnetic field only exists in the sectional plane and the electric field exists not only in the sectional plane but also in the direction of wave propagation. Such a field distribution is called as a TM (transverse magnetic) or E mode.

Finally, we have the mode where the TE and TM mode coexist. Such a mode is called as a hybrid mode, which propagates along the dielectric waveguide.

2 WAVE PROPAGATION ON A DISTRIBUTED TRANSMISSION LINE

2.1 Incident Wave and Reflecting Wave

As an example, when a power source is connected at the left-hand side of a coaxial line of Fig. 2, and a some load impedance R_L is connected at the right-hand side, the electric field $\boldsymbol{E}_0$ and magnetic field $\boldsymbol{H}_0$ at a given instant of time ($t = 0$) is shown by solid arrows and dashed arrows, respectively. The voltage of the center conductor versus the outer conductor, v, and the current flowing in a center conductor toward the load impedance from a source, i, at the same given instant of time ($t = 0$) are also shown in the area below the coaxial line. We consider the case that the distribution along the z-axis (the direction of wave propagation) is sinusoidal and the ratio of v and i takes the same values Z_c as Eq. (1) at any values of Z:

$$\frac{v}{i} = Z_c \tag{1}$$

The voltage, current, electrical field, and magnetic field distribution move

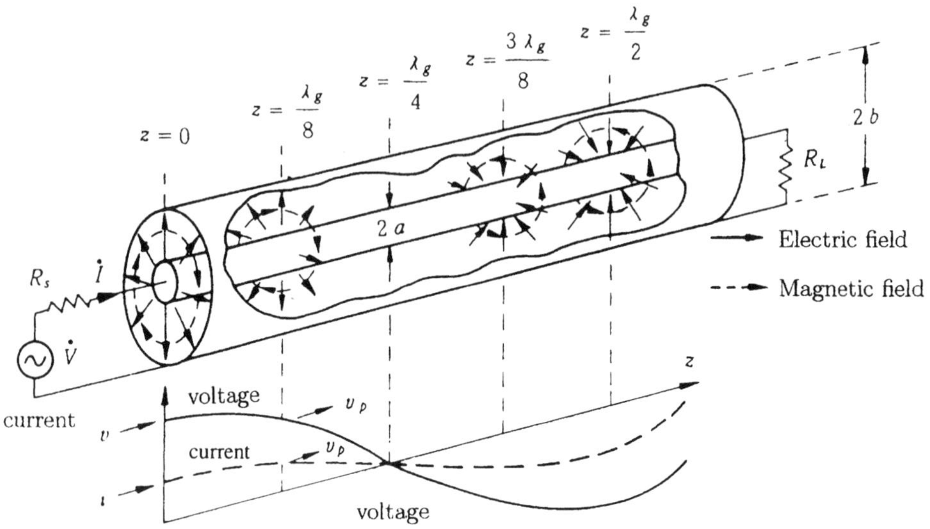

Figure 2. Electrical field distribution in a coaxial line.

toward the load side (right-hand side), keeping the same distribution as time progresses.

The speed of movement is called the phase velocity and it is expressed by v_p. The wave traveling toward the load is called the incident wave. The situation mentioned above is the case when the incident wave only exists in a coaxial line under some special load impedance Z_c. The Z_c is called the characteristic impedance of the coaxial line.

As shown in Fig. 2, v, i, $\boldsymbol{E}$, and $\boldsymbol{H}$ at $Z = 0$ take the same values at $Z = \lambda_g$. λ_g is the guided wavelength. Because the distributions of v, i, $\boldsymbol{E}$, and $\boldsymbol{H}$ move toward the load, it is understood that v, i, $\boldsymbol{E}$, and $\boldsymbol{H}$ at the arbitral point Z takes the same values every $1/f$ sec, where f takes the values

$$f = \frac{v_p}{\lambda_g} \tag{2}$$

The description mentioned above is always available for general two parallel conductors with the arbitral section view as shown in Fig. 3.

Denoting the capacity and conductance per meter between two parallel conductors by C (F/m) and G (S/m), respectively, and the total inductance and series resistance per meter of two parallel conductors by L (H/m) and R (Ω/m), respectively, the equivalent network of the ΔZ region of two parallel conductors becomes that of Figs. 3b and 3c, where

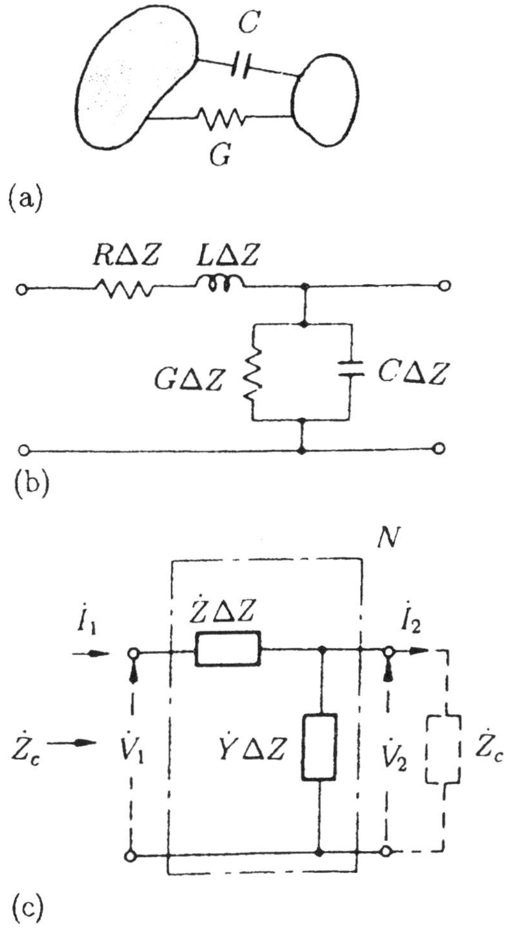

Figure 3. Construction and equivalent network of two parallel conductors. (a) Sectional view; (b) equivalent network of two parallel conductors; (c) equivalent 2-port network of two parallel conductors.

$$
\begin{aligned}
\dot{Z} &= j\omega L + R \\
\dot{Y} &= j\omega C + G
\end{aligned}
\tag{3a}
$$

If an incident wave exists in the two parallel conductors with a termination of $\dot{Z}_c$, the impedance between two conductors should always take the same values of $\dot{Z}_c$, as shown in Fig. 3c. Because the *F* matrix of *N* of Fig.

3c takes the values

$$F = \begin{bmatrix} 1 + \dot{Z}\dot{Y}\Delta z^2 & \dot{Z}\Delta z \\ \dot{Y}\Delta z & 1 \end{bmatrix} \simeq \begin{bmatrix} 1 & \dot{Z}\Delta z \\ \dot{Y}\Delta z & 1 \end{bmatrix} \quad Z_c = \frac{\dot{V}_1}{\dot{I}_1} = \frac{\dot{Z}_c + \dot{Z}\Delta z}{1 + \dot{Y}\dot{Z}_c\Delta z} \tag{3b}$$

Therefore, we obtain

$$\dot{Z}_c = \sqrt{\frac{\dot{Z}}{\dot{Y}}}$$

Using Eq. (3b) together with Eq. (3c), we get

$$\dot{V}_1 = \left(1 + \frac{\dot{Z}}{Z_c}\Delta z\right)\dot{V}_2 = (1 + \sqrt{\dot{Z}\dot{Y}}\Delta z)\dot{V}_2$$

Therefore, considering $\Delta z \ll 1$, we get

$$\dot{V}_2 = \dot{V}_1 e^{-\gamma\Delta z}$$

where γ is the propagation. This means that N is the line with propagation constant γ. In the lossless transmission line, setting $R = 0$ and $G = 0$, we obtain

$$Z_c = \sqrt{\frac{L}{C}} \tag{4a}$$

$$\beta = \omega\sqrt{LC}, \qquad \beta = \frac{\gamma}{j} \tag{4b}$$

where β is the phase constant. Because the guided wavelength is the definition of the distance of repetition of the same field along the line at given instant of the time, the relation of $\beta\lambda_g = 2\pi$ should be satisfied. Therefore,

$$\beta = \frac{2\pi}{\lambda_g} \tag{4c}$$

Considering Eq. (2),

$$\beta = \frac{\omega}{v_p} \tag{4d}$$

From Eqs. (4a), (4b), and (4d),

$$Z_c = \frac{1}{Cv_p} = Lv_p \tag{4e}$$

Equations (3c) and (3d) are generally introduced by Telegrapher's equation as discussed in Chapter 3. We, however, introduced the results by above network concept to make the physical meaning simple.

Because we can prove the propagating wave along two parallel conductors in a homogeneous medium is a TEM wave (see Section 3.2), v_p takes the values

$$v_p = \frac{v_0}{\sqrt{\mu_r \varepsilon_r}} \tag{5a}$$

where $v_0 = 3 \times 10^8$ (m/sec), μ_r is the relative permeability, and ε_r is the relative dielectric constant. Therefore,

$$\lambda_g = \frac{\lambda_0}{\sqrt{\varepsilon_r \mu_r}} \tag{5b}$$

where λ_0 is the free-space wavelength ($=v_0/f$).

Denoting the voltage and the current of a traveling wave by v^+ and i^+, respectively, we can show them by the complex expression

$$\begin{aligned} v^+ &= \text{Re}(\dot{V}^+ e^{j\omega t}) \\ i^+ &= \text{Re}(\dot{I}^+ e^{j\omega t}) \end{aligned} \tag{5c}$$

where Re is the real part of a complex number, and $\dot{V}^+$ and $\dot{I}^+$ can be expressed as

$$\begin{aligned} \dot{V}^+ &= V_0^+ e^{-\gamma z} \\ \dot{I}^+ &= \frac{V_0^+}{Z_c} e^{-\gamma z} \end{aligned} \tag{5d}$$

Next, considering the load Z_c and the source are connected at the left- and right-hand side of a coaxial line, respectively, the wave travels from right to left. In this case, the current corresponding to the same voltage of that of the incident wave takes the sign opposite to that of the incident wave. This matter is easily understood in that a transmission line is excited from the opposite side. In this case, we denote the voltage and current of the wave toward $Z < 0$ as v^- and i^-, respectively; they are expressed by Eqs. (5e) and (5d):

$$\begin{aligned} v^- &= \text{Re}(\dot{V}^- e^{j\omega t}) \\ i^- &= \text{Re}(\dot{I}^- e^{j\omega t}) \end{aligned} \tag{5e}$$

$$\begin{aligned} \dot{V}^- &= \dot{V}_0^{\gamma z} \\ \dot{I}^- &= -\frac{V_0^-}{Z_c} e^{\gamma z} \end{aligned} \tag{5f}$$

The total voltage v and total current i, therefore, are

$$v = \mathrm{Re}(\dot{V}e^{j\omega t})$$
$$i = \mathrm{Re}(\dot{I}e^{j\omega t}) \tag{6a}$$
$$\dot{V} = \dot{V}_0^+ e^{-\gamma z} + \dot{V}_0^- e^{\gamma z}$$
$$\dot{I} = \frac{\dot{V}_0^+}{Z_c} e^{-\gamma z} - \frac{\dot{V}_0^-}{Z_c} e^{\gamma z} \tag{6b}$$
$$\gamma = j\beta \quad \text{in the lossless case} \tag{6c}$$

2.2 Reflection Coefficient and Impedance

When a transmission line is terminated with load impedance Z_L at $z = 0$ as shown in Fig. 4, not only the incident wave but also the wave propagating toward the generator (i.e., the left-hand side) exists. The wave toward the generator is called the reflecting wave. In Eq. (6b), $\dot{V}_0^+$ and $\dot{V}_0^-$ show the voltages of the incident and the reflected waves at $z = 0$ (i.e., the loading point); $\dot{V}_0^- / \dot{V}_0^+$ is called the reflection coefficient, which is denoted by $\dot{\Gamma}_L$. From (6b), we get

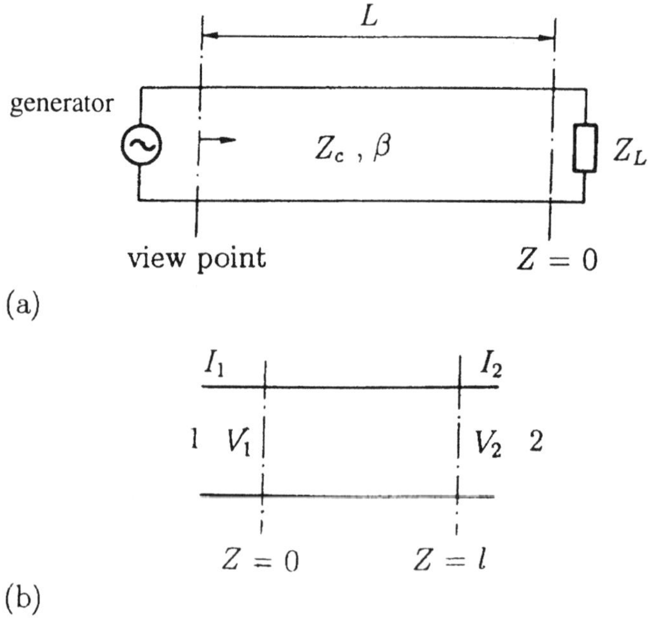

Figure 4. Transmission line. (a) Terminated with load; (b) voltages and currents in transmission line.

$$\dot{Z}_L = \left.\frac{\dot{V}}{\dot{I}}\right|_{Z=0} = \frac{1 + \dot{\Gamma}_L}{1 - \dot{\Gamma}_L} Z_c$$

Then, the load impedance normalized by Z_c, $\dot{Z}_L'$, can be calculated as

$$\dot{Z}_L' = \frac{1 + \dot{\Gamma}_L}{1 - \dot{\Gamma}_L} \tag{7a}$$

Therefore,

$$\dot{\Gamma}_L = \frac{\dot{Z}_L' - 1}{\dot{Z}_L' + 1} \tag{7b}$$

From Eq. (6b), the normalized impedance $\dot{Z}'$ can be expressed by

$$\dot{Z}' = \frac{1 + \dot{\Gamma}}{1 - \dot{\Gamma}} \qquad \dot{\Gamma} = \Gamma_L e^{-2\gamma L} \tag{7c}$$

where L is the distance from load to the viewpoint.

2.3 Standing Wave Ratio, Maximum Impedance, and Minimum Impedance

In the case of a lossless transmission line, Eq. (6b) becomes

$$\dot{V} = \dot{V}_0^+ \exp\left(-j\frac{2\pi}{\lambda_g} z\right) + \dot{V}_0^- \exp\left(j\frac{2\pi}{\lambda_g} z\right) \tag{8a}$$

$$\dot{I} = \frac{\dot{V}_0^+}{Z_c} \exp\left(-j\frac{2\pi}{\lambda_g} z\right) - \frac{\dot{V}_0^-}{Z_c} \exp\left(j\frac{2\pi}{\lambda_g} z\right) \tag{8b}$$

where Z_c and λ_g take the values of Eqs. (4a) and (5b). When we show the vectors $\dot{V}$ and Z, we get $\dot{V}$ as shown in Fig. 5a.

As shown in Fig. 5a, the vector of the incident wave rotates clockwise and that of the reflected wave rotates anticlockwise as z increases.

As for the absolute values of $\dot{V}$, we can consider the vectors as shown in Fig. 5b, where the vector of the incident wave is fixed and the vector of the reflecting wave rotates at the tip of the fixed incident wave vector. As understood from Fig. 5b, the maximum values of $\dot{V}$, V_{max}, and the minimum values of $\dot{V}$, V_{min}, are calculated from

$$V_{max} = |\dot{V}_+| + |\dot{V}_-| \tag{9a}$$

$$V_{min} = |\dot{V}_+| - |\dot{V}_-| \tag{9b}$$

The ratio of V_{max} to V_{min} is called the voltage standing wave ratio (VSWR). Representing VSWR by S, S takes the values from Eq. (10).

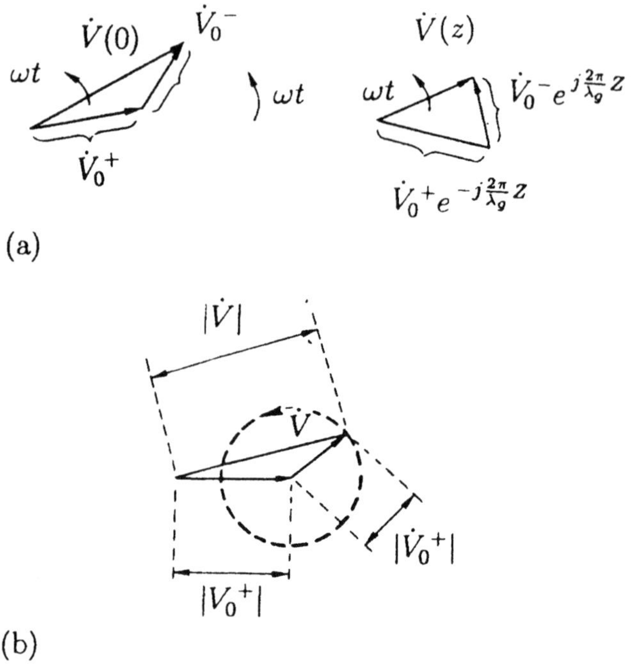

Figure 5. Sum of the vectors of traveling and reflecting waves at Z. (a) Vector **V** at $Z = 0$ and Z; (b) sum of traveling and reflecting wave.

$$S = \frac{V_{max}}{V_{min}} = \frac{|\dot{V}_+| + |\dot{V}_-|}{|\dot{V}_+| - |\dot{V}_-|} = \frac{1 + |\dot{\Gamma}|}{1 - |\dot{\Gamma}|} \tag{10}$$

Next, the vector of $\dot{I}$ can be obtained by changing the sign of the reflected wave in Fig. 5b. This is understood by comparing the sign of reflected waves of the voltage and current as shown in Eqs. (8a) and (8b).

Therefore, the maximum values and minimum values of $\dot{I}$, I_{max} and I_{min}, occur at the points of minimum and maximum values of $\dot{V}$, respectively. I_{max} and I_{min} take the values

$$I_{max} = |I_+| + |I_-| \tag{11a}$$

$$I_{min} = |I_+| - |I_-| \tag{11b}$$

Impedance at $Z = \dot{V}/\dot{I}$ takes different values for z. Denoting the maximum values and minimum values of Z by Z_{max} and Z_{min}, respectively, we get

$$Z_{\max} = \frac{V_{\max}}{I_{\min}} = \frac{|\dot{V}_+| + |\dot{V}_-|}{|\dot{I}_+| - |\dot{I}_-|} = Z_c \frac{1 + |\dot{\Gamma}|}{1 - |\dot{\Gamma}|} = SZ_c$$

$$Z_{\min} = \frac{V_{\min}}{V_{\max}} = Z_c \frac{1 - |\dot{\Gamma}|}{1 + |\dot{\Gamma}|} = Z_c \frac{1 + S}{1 - S}$$

Therefore, the normalized maximum and minimum values of impedance, $Z'_{\max}$ and $Z'_{\min}$, is calculated by

$$Z'_{\max} = \frac{1 + |\dot{\Gamma}|}{1 - |\dot{\Gamma}|} = S \tag{12a}$$

$$Z'_{\min} = \frac{1 - |\dot{\Gamma}|}{1 + |\dot{\Gamma}|} = \frac{1}{S} \tag{12b}$$

2.4 Smith Chart

The normalized impedance Z' can be related to the reflection coefficient $\dot{\Gamma}$ as provided by Eq. (7c). Therefore, $\dot{\Gamma}$ can be obtained also from Z' by

$$\dot{\Gamma} = \frac{\dot{Z}' - 1}{\dot{Z}' + 1} \tag{13}$$

Because $\dot{Z}'$ is a complex number, we can obtain $\dot{Z}'$ by

$$\dot{Z}' = r + jx \quad (r \text{ and } x \text{ are real numbers}) \tag{14}$$

$\dot{Z}'$ can be shown on a complex plane by Fig. 6a. In Eq. (13), the straight line corresponding to r = constant in $\dot{Z}'$ is the locus of a circle, the center which is situated on the real axis of the complex plane of $\dot{\Gamma}$. The straight line corresponding to x = constant in $\dot{Z}'$ is the locus of a circle, the center of which situated on the straight line of $\dot{\Gamma} = 1.0$ on the $\dot{\Gamma}$ plane.

If we represent the matter mentioned above by an equation, the real and imaginary parts of $\dot{\Gamma}$, U, and V can be related by r and x by Eqs. (15a) and (15b), respectively. The results can be easily obtained from Eqs. (13) and (14).

$$\left(U - \frac{r}{r+1}\right)^2 + V^2 = \left(\frac{1}{r+1}\right)^2 \tag{15a}$$

$$(U - 1)^2 + \left(V - \frac{1}{x}\right)^2 = \left(\frac{1}{x}\right)^2 \tag{15b}$$

Equation (15a) is the equation of circles the center of which is $r/(r + 1)$ and radius is $1/(r + 1)$. In Eq. (15a), $U = 1$ when $V = 0$, which means the circle always pass the point (1, 0) on the $\dot{\Gamma}$ plane. Equation (15b) is the equation of circles the center of which is situated at $(1, 1/x)$ of the $\dot{\Gamma}$ plane,

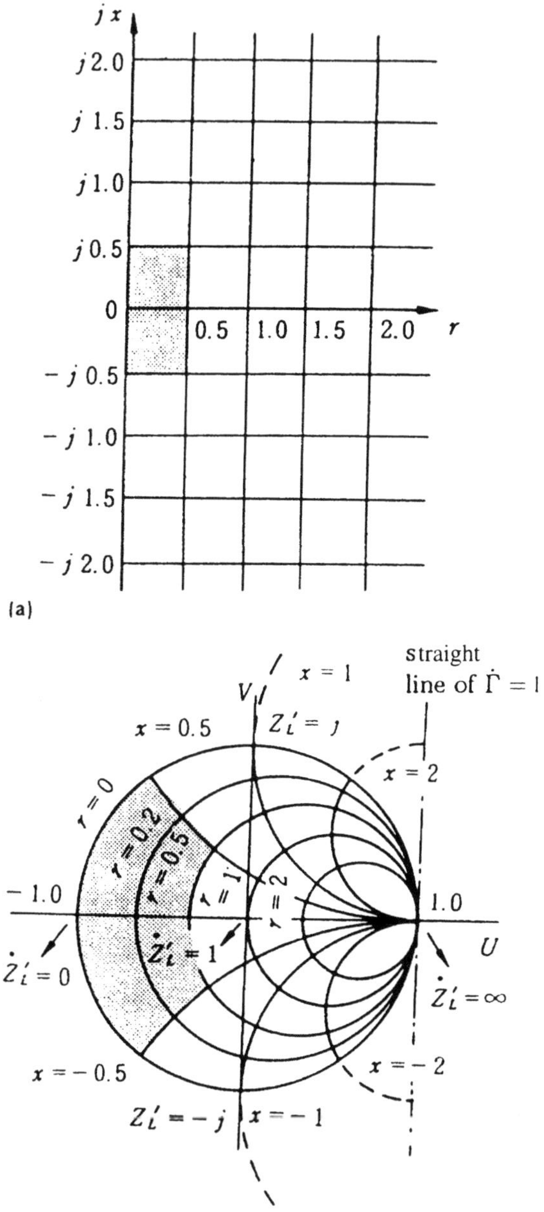

Figure 6. Z' and the corresponding Γ. (a) $\dot{Z}'_L(=r + jx)$. (b) $\dot{\Gamma} = \dot{Z}'_L - 1/(\dot{Z}'_L + 1)(=U + jV)$.

the radius of which is $1/|x|$. Therefore, the circles always passes (1, 0) of the $\dot{\Gamma}$ plane. The results are shown in Fig. 6b.

Because $|\dot{\Gamma}| \leq 1$, any values of impedance must situated inside the circle of $|\dot{\Gamma}| \leq 1$ in passive circuits. Figure 6b is called a Smith Chart (Impedance Smith Chart). As understood from the above description, the vector bound between the specified point and the origin of the $\dot{\Gamma}$ plane shows the reflection coefficient $\dot{\Gamma}$, and the number shown at the specified point is the normalized impedance corresponding to $\dot{\Gamma}$. As a reference, the regions of $0 < r < 0.5$ and $-0.5 < x < 0.5$, and $1.0 < r < 2.0$ and $-0.5 < x < 0.5$ in the $\dot{Z}'$ plane are shown in the $\dot{\Gamma}$ plane.

In a similar manner, the values of normalized admittance $\dot{Y}' = Z_c\dot{Y}$ can be related to $\dot{\Gamma}$ by

$$\dot{\Gamma} = \frac{1 - \dot{Y}'}{1 + \dot{Y}'} \tag{16}$$

Using

$$\dot{Y}' = g + jb \tag{17}$$

the $\dot{Y}$ plane can be transformed to the $\dot{\Gamma}$ − plane by Eq. (16) in the same way; the results are shown in Figs. 7a and 7b. Figure 7b is called the Admittance Smith Chart.

2.5 *F* Matrix of a Distributed Line

In Fig. 4b, we showed the transmission line of length of l with the characteristic impedance Z_c and the propagation constant γ, where V_1 and I_1 are the voltage and the current of port 1 at $z = 0$ and V_2 and I_2 are those of port 2 at $Z = l$. From Eq. (6b), we get

$$\begin{bmatrix} V_1 \\ I_1 \end{bmatrix} = \begin{bmatrix} 1 & 1 \\ \dfrac{1}{Z_c} & \dfrac{-1}{Z_c} \end{bmatrix} \begin{bmatrix} \dot{V}_+ \\ \dot{V}_- \end{bmatrix}, \qquad \begin{bmatrix} V_2 \\ I_2 \end{bmatrix} = \begin{bmatrix} e^{-\gamma l} & e^{\gamma l} \\ \dfrac{e^{-\gamma l}}{Z_c} & \dfrac{-e^{\gamma l}}{Z_c} \end{bmatrix} \begin{bmatrix} \dot{V}_+ \\ \dot{V}_- \end{bmatrix}.$$

From these equations, we get

$$\begin{bmatrix} V_1 \\ I_1 \end{bmatrix} = \begin{bmatrix} 1 & 1 \\ \dfrac{1}{Z_c} & \dfrac{-1}{Z_c} \end{bmatrix} \begin{bmatrix} e^{-\gamma l} & e^{\gamma l} \\ \dfrac{e^{-\gamma l}}{Z_c} & \dfrac{-e^{\gamma l}}{Z_c} \end{bmatrix}^{-1} \begin{bmatrix} V_2 \\ I_2 \end{bmatrix} = [F] \begin{bmatrix} V_2 \\ I_2 \end{bmatrix}$$

Therefore, we get

$$[F] = \begin{bmatrix} \cosh \gamma l & Z_c \sinh \gamma l \\ \dfrac{1}{Z_c} \sinh \gamma l & \cosh \gamma l \end{bmatrix} \tag{18a}$$

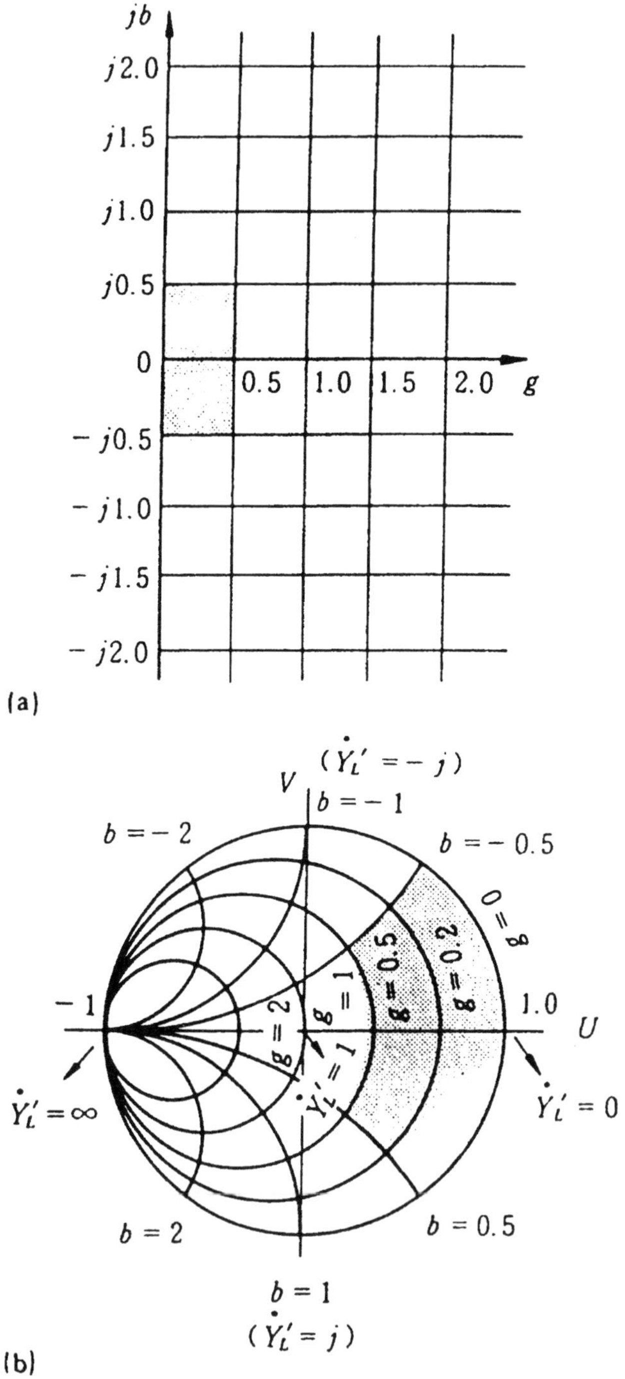

Figure 7. $\dot{Y}'$ and the corresponding $\dot{\Gamma}$. (a) $\dot{Y}'_L(=g + jb)$; (b) $\dot{\Gamma} = 1 - \dot{Y}'_L/(1 + \dot{Y}'_L)(=U + jV)$.

In the case of the lossless transmission line, we have

$$[F] = \begin{bmatrix} \cos \beta l & jZ_c \sin \beta l \\ j\dfrac{1}{Z_c} \sin \beta l & \cos \beta l \end{bmatrix} \tag{18b}$$

In the case of a quarter wavelength,

$$[F] = \begin{bmatrix} 0 & jZ_c \\ j\dfrac{1}{Z_c} & 0 \end{bmatrix} \tag{18c}$$

3 CLASSIFICATION OF WAVES AND PERFORMANCES

3.1 Existence of TM, TE, and TEM Modes from the Viewpoint of the Helmholtz Equation in a Vector Field

In Section 1.2, we discussed the distributed transmission line of a TEM wave. As mentioned in Section 1.1, TM and TE waves can be transmitted in a hollow waveguide as shown in Fig. 8. The reason why two kinds of wave exist is explained as follows.

The electric field and magnetic field propagating to the Z axis as shown in Fig. 8 can be expressed as $\boldsymbol{E}(x, y)e^{-\gamma z}$ and $\boldsymbol{H}(x, y)e^{-\gamma z}$, where γ is a propagating constant and $\boldsymbol{E}$ and $\boldsymbol{H}$ are the electric and magnetic fields at

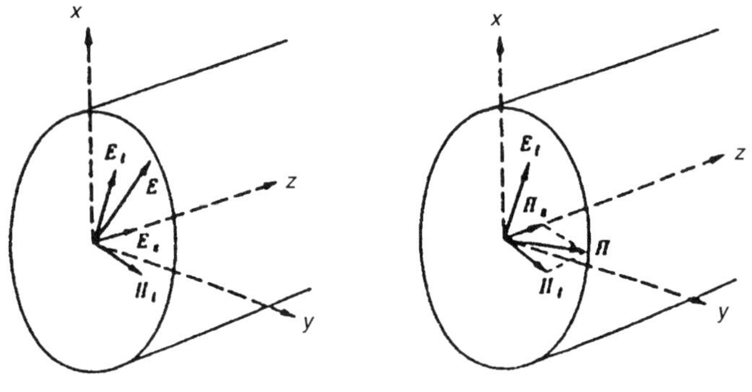

Figure 8. TM and TE waves propagating in a hollow waveguide.

$z = 0$, respectively. Because $\boldsymbol{E}$ and $\boldsymbol{H}$ have their tangential components $\boldsymbol{E}_t$ and $\boldsymbol{H}_t$ and z components $i_z E_z$ and $i_z H_z$, the electrical field $\boldsymbol{E}e^{-\gamma z}$ and magnetic field $\boldsymbol{H}e^{-\gamma z}$ can be expressed as

$$\boldsymbol{E}e^{-\gamma z} = (\boldsymbol{E}_t + i_z E_z)e^{-\gamma z} \tag{19a}$$

$$\boldsymbol{H}e^{-\gamma z} = (\boldsymbol{H}_t + i_z H_z)e^{-\gamma z} \tag{19b}$$

where $\boldsymbol{E}_t$ and $\boldsymbol{H}_t$ are two-dimensional vector fields.

A general vector field $\boldsymbol{A}$, however, can be expressed by the sum of an irrotational field (or lamella field) $\boldsymbol{F}^i$ and a rotational field (or solenoidal field) $\boldsymbol{F}^r$, where $\boldsymbol{F}^i$ and $\boldsymbol{F}^r$ are defined as follows:

$$\nabla \times \boldsymbol{F}^i = 0 \tag{20a}$$

$$\nabla \cdot \boldsymbol{F}^r = 0 \tag{20b}$$

This is called the Helmholtz equation in the vector field.

$\boldsymbol{E}_t$ of Eq. (19a), therefore, can be decomposed into the irrotational field $\boldsymbol{E}_t^i$ and the rotational field $\boldsymbol{E}_t^r$ as follows:

$$\boldsymbol{E}_t = \boldsymbol{E}_t^i + \boldsymbol{E}_t^r \tag{21}$$

We will consider the two cases (i) $\boldsymbol{E}_t^r = 0$ and (ii) $\boldsymbol{E}_t^i = 0$. The electromagnetic fields for the corresponding case mentioned above can easily obtained through Maxwell's equation for a guidewave. Maxwell's equation for instant values of an electromagnetic field, however, have the following relation:

$$\nabla \times \mathcal{H} = \sigma\mathcal{E} + \varepsilon \frac{\partial \mathcal{E}}{\partial t} + \boldsymbol{i} \tag{22a}$$

$$\nabla \times \mathcal{E} = -\mu \frac{\partial \mathcal{H}}{\partial t} \tag{22b}$$

$$\nabla \cdot \mathcal{H} = 0 \tag{22c}$$

$$\nabla \cdot \mathcal{E} = 0 \tag{22d}$$

where ε, $\mathcal{H}$, and $\boldsymbol{i}$ are the constant vectors of electrical field, magnetic field, and current density. σ, ε, and μ are the conductivity, dielectric constant, and permeability of the medium.

When $\mathcal{E}$ and $\mathcal{H}$ have the sinusoidal changing corresponding to time, they can expressed by complexed vectors $\dot{\boldsymbol{E}}$ and $\dot{\boldsymbol{H}}$ as follows:

$$\begin{aligned} \mathcal{E} &= \sqrt{2}\,\mathrm{Re}(\boldsymbol{E}e^{j\omega t}) \\ \mathcal{H} &= \sqrt{2}\,\mathrm{Re}(\boldsymbol{H}e^{j\omega t}) \\ \boldsymbol{i} &= \sqrt{2}\,\mathrm{Re}(\boldsymbol{J}e^{j\omega t}) \end{aligned} \tag{23}$$

where Re( ) is the real part of a complex number and ω is the angular frequency.

Substituting Eqs. (23) into Eqs. (22), we obtain

$$\nabla \times \boldsymbol{H} = (\sigma + j\omega\varepsilon)\boldsymbol{E} + \boldsymbol{J} \tag{24a}$$

$$\nabla \times \boldsymbol{E} = -j\omega\mu\boldsymbol{H} \tag{24b}$$

$$\nabla \cdot \boldsymbol{H} = 0 \tag{24c}$$

$$\nabla \cdot \boldsymbol{E} = 0 \tag{24d}$$

In the case of the incident wave in a waveguide, $\boldsymbol{E}$ and $\boldsymbol{H}$ of Eqs. (24a)–(24d) take Eqs. (19a) and (19b), and ∇ can be expressed by

$$\nabla = \nabla_t + \boldsymbol{i}_z \frac{\partial}{\partial z} = \nabla_t - \gamma \boldsymbol{i}_z$$
$$\left(\nabla_t = \boldsymbol{i}_x \frac{\partial}{\partial x} + \boldsymbol{i}_z \frac{\partial}{\partial y}\right) \tag{24e}$$

Using the relation (24e) and the values of $\boldsymbol{E}$ and $\boldsymbol{H}$ of Eqs. (19a) and (19b) in Eqs. (24a)–(24d), we can obtain

$$\nabla \times \boldsymbol{E} = (\underbrace{\nabla_t \times \boldsymbol{E}_t} - \underbrace{\gamma \boldsymbol{i}_z \times \boldsymbol{E}_t + \nabla_t \times \boldsymbol{i}_z E_z})e^{-\gamma z}$$
$$= -j\omega\mu(\boldsymbol{i}_z H_z + \boldsymbol{H}_t)e^{-\gamma z} \tag{25a}$$

$$\nabla \times \boldsymbol{H} = (\underbrace{\nabla_t \times \boldsymbol{H}_t} - \underbrace{\gamma \boldsymbol{i}_z \times \boldsymbol{H}_t + \nabla_t \times \boldsymbol{i}_z H_z})e^{-\gamma z}$$
$$= (j\omega\varepsilon + \sigma)(\boldsymbol{i}_z E_z + \boldsymbol{E}_t)e^{-\gamma z} \tag{25b}$$

$$\nabla \cdot \boldsymbol{E} = (\nabla_t \cdot \boldsymbol{E}_t - \gamma E_z)e^{-\gamma z} = 0 \tag{25c}$$

$$\nabla \cdot \boldsymbol{H} = (\nabla_t \cdot \boldsymbol{H}_t - \gamma H_z)e^{-\gamma z} = 0 \tag{25d}$$

Using Maxwell's equation for a guided wave [Eq. (25)], we will discuss the cases when $\boldsymbol{E}_t$ is irrotational and rotational.

(i) $\boldsymbol{E}_t$ is irrotational: $\boldsymbol{E}_t = \boldsymbol{E}_t^i$ ($\boldsymbol{E}_t^r = 0$)
Because $\nabla_t \times \boldsymbol{E}_t = \nabla_t \times \boldsymbol{E}_t^i = 0$, we get

$$H_z = 0 \tag{26a}$$

This means the wave should a TM wave. Substituting Eq. (26a) into Eq. (25d), we get

$$\nabla_t \cdot \boldsymbol{H}_t = 0 \tag{26b}$$

Then $\boldsymbol{H}_t$ is rotational. Therefore, we can have fields such as

$$\nabla_t \times \boldsymbol{H}_t \neq 0, \quad E_z \neq 0 \tag{26c}$$

From this, we see that TM wave is also called an E wave. Substituting Eq. (26a) into Eq. (25b), we get

$$\boldsymbol{E}_t = Z_e \boldsymbol{H}_t \times \boldsymbol{i}_z$$

$$Z_e = \frac{\gamma}{j\omega\varepsilon + \sigma} \tag{26d}$$

Z_e is the so-called wave impedance of the TM wave. In the case of a lossless guided wave, setting $\gamma = jk_z$ and $\sigma = 0$,

$$Z_e = \frac{k_z}{\omega\varepsilon} \tag{26e}$$

(ii) $\boldsymbol{E}_t$ is rotational: $\boldsymbol{E}_t = \boldsymbol{E}_t^{\gamma}$ ($\boldsymbol{E}_t = 0$)
Because $\boldsymbol{\nabla}_t \cdot \boldsymbol{E}_t = 0$, from Eq. (25c) we get

$$\boldsymbol{E}_z = 0 \tag{27a}$$

This means that the wave should a TE wave. Substituting Eq. (27a) into Eq. (25b), we get

$$\boldsymbol{\nabla}_t \times \boldsymbol{H}_t = 0 \tag{27b}$$

Then, $\boldsymbol{H}_t$ is irrotational. Therefore, we can have a field such as

$$\boldsymbol{\nabla}_t \cdot \boldsymbol{H}_t \neq 0$$

Substituting this relation into Eq. (25d),

$$\boldsymbol{H}_z \neq 0 \tag{27c}$$

For this reason, the TE wave is also called an H wave. Substituting Eq. (27a) into Eq. (25a), we obtain

$$\boldsymbol{E}_t = Z_m \boldsymbol{H}_t \times \boldsymbol{i}_z$$

$$Z_m = \frac{j\omega\mu}{\gamma} \tag{27d}$$

Z_m is the so-called wave impedance of the TE wave. In the case of a lossless guided wave, setting $\gamma = jk_z$,

$$Z_m = \frac{\omega\mu}{k_z} \tag{27e}$$

(iii) The case of a TEM wave ($E_z = H_z = 0$)
Substituting this condition into Eqs. (25a) and (25b), we get

$$\boldsymbol{\nabla}_t \times \boldsymbol{E}_t = 0$$

$$\boldsymbol{\nabla}_t \times \boldsymbol{H}_t = 0 \tag{28a}$$

and

$$\gamma \boldsymbol{i}_z \times \boldsymbol{E}_t = j\omega\mu \boldsymbol{H}_t$$
$$\gamma \boldsymbol{i}_z \times \boldsymbol{H}_t = -(j\omega\varepsilon + \sigma)\boldsymbol{E}_t \tag{28b}$$

Equation (28a) shows that $\boldsymbol{E}_t$ and $\boldsymbol{H}_t$ are irrotational. From Eq. (28b), we get

$$\boldsymbol{E}_t = \dot{Z}\boldsymbol{H}_t \times \boldsymbol{i}_z, \qquad \dot{Z} = \sqrt{\frac{j\omega\mu}{\sigma + j\omega\varepsilon}}$$
$$\gamma = \sqrt{j\omega\mu\sigma - \omega^2\mu\varepsilon} \tag{28c}$$

In the lossless case, setting $\sigma = 0$ and $\gamma = jk_z$, we get

$$Z = \sqrt{\frac{\mu}{\varepsilon}} = 120\pi\sqrt{\frac{\mu_r}{\varepsilon_r}}$$
$$\gamma = jk_z, \qquad k_z = \omega\sqrt{\mu\varepsilon} \tag{28d}$$

The above results are summarized in Table 1.

3.2 TEM Wave and the Characteristics

Uniform Plane Wave Is a TEM Wave

From the conditions of a uniform plane wave, $\partial/\partial x = \partial/\partial y = 0$ which results in $\boldsymbol{\nabla}_t = 0$. Substituting this relation into Eqs. (25a) and (25b), we get $H_z = 0$ and $E_z = 0$. Therefore, the wave is a TEM wave.

Nonuniform Plane Wave Existing in Multiple Parallel Lines in a Homogeneous Medium

As a simple case, we consider two parallel conductors, where the electric charges $+Q$ (C/m) and $-Q$ (C/m) are supplied as shown in Fig. 9a. In this case, there are electrical fields as shown by solid lines and an equipotential line shown by dotted lines, where the both lines crossed at 90°.

Table 1. $\boldsymbol{E}_t$, $\boldsymbol{H}_t$, E_z, and H_z TM, TE, and TEM Modes

	$\boldsymbol{E}_t$	$\boldsymbol{H}_t$	E_z	H_z
TM wave (E wave)	$\nabla_t \times \boldsymbol{E}_t = 0$ (irrotational)	$\nabla_t \cdot \boldsymbol{H}_t = 0$ (rotational)	$\neq 0$	0
TE wave (H wave)	$\nabla_t \cdot \boldsymbol{E}_t = 0$ (rotational)	$\nabla_t \times \boldsymbol{H}_t = 0$ (irrotational)	0	$\neq 0$
TEM wave	$\nabla_t \times \boldsymbol{E}_t = 0$ (irrotational)	$\nabla_t \times \boldsymbol{H}_t = 0$ (irrotational)	0	0

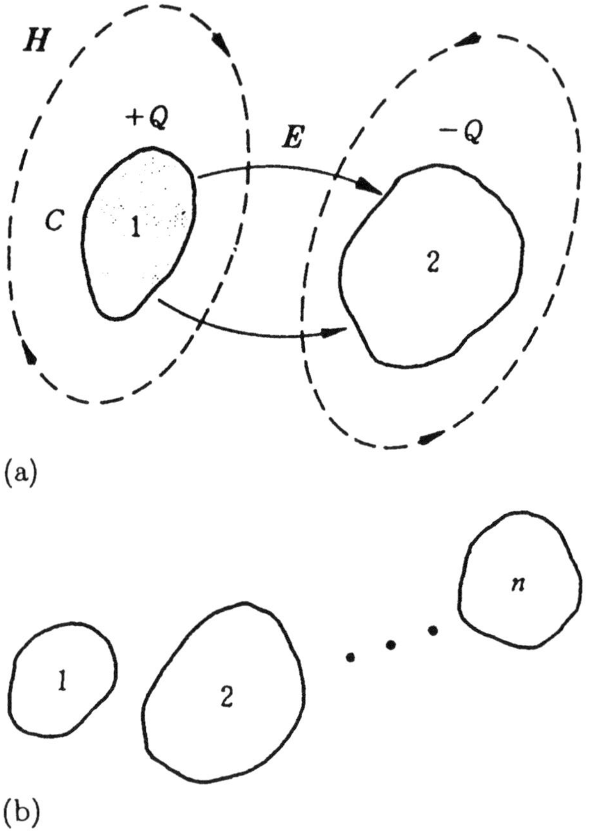

Figure 9. Sectional view of parallel conductor. (a) Two parallel conductors; (b) *N* parallel conductor.

If the dotted lines are assumed to be the magnetic field in the case of a TEM wave, $|\boldsymbol{H}| = \sqrt{\varepsilon/\mu}|\boldsymbol{E}|$ from Eq. (28c) and values on the conductor should take

$$|\boldsymbol{H}| = \sqrt{\frac{\varepsilon}{\mu}}\frac{\rho_s}{\varepsilon} = \frac{\rho_s}{\sqrt{\mu\varepsilon}}$$

by the Gauss law. Integrating $|\boldsymbol{H}|$ along the circumference of the conductor, we get

$$\oint_c |\boldsymbol{H}|\, dl = \frac{1}{\sqrt{\mu\varepsilon}}\int_c \rho_s\, dl = \frac{Q}{\sqrt{\mu\varepsilon}}$$

On the other hand, the above values should be the current flowing on a conductor and it must be Qv_p, where v_p is the phase velocity. Therefore, v_p should take the values $1/\sqrt{\mu\varepsilon}$, which coincides with the values of Eq. (28c) because $v_p = \omega/\beta = 1/\sqrt{\mu\varepsilon}$.

Next, in the case of N parallel conductors as shown in Fig. 9b, we first consider that all conductors except the ith conductor are connected to ground and the voltage V_i is supplied between the ith conductor and the ground. We will call this state the ith state. In this case the TEM wave propagates between the ith conductor and the ground. When arbitrary voltages are added to each conductor, such a situation can be obtained by summation of the solutions of the ith state ($i = 1, 2, \ldots n$).

In all states, the propagating wave is a TEM wave with the same velocity, v_p. Therefore, there exists a TEM wave for any kind of excitation of N conductors.

When we show the capacity between ith conductor and the ground by C_{ii} and the capacity between the ith and jth conductor by C_{ij}, as shown in Fig. 10, the charge on the kth conductor q_k ($k = 1, \ldots, n$) can be expressed by the voltage between the lth conductor and the ground, V_l ($l = 1, \ldots, n$) as Eqs. (29a)–(29e):

$$\boldsymbol{q} = |C'|\boldsymbol{v} \tag{29a}$$

$$\boldsymbol{q} = \begin{bmatrix} q_1 \\ \vdots \\ q_n \end{bmatrix}, \quad |C'| = \begin{bmatrix} C'_{11} & \cdots & C'_{1i} & \cdots & C'_{1n} \\ \vdots & & \vdots & & \vdots \\ C'_{j1} & \cdots & C'_{ji} & \cdots & C'_{jn} \\ \cdots & & \vdots & & \vdots \\ C'_{n1} & \cdots & C'_{ni} & \cdots & C'_{nn} \end{bmatrix} \quad \boldsymbol{v} = \begin{bmatrix} V_1 \\ \vdots \\ V_n \end{bmatrix} \tag{29b}$$

$$C'_{ii} = \sum_{k=1}^{n} C_{ik} \qquad C'_{ji} = -C_{ji} \tag{29c}$$

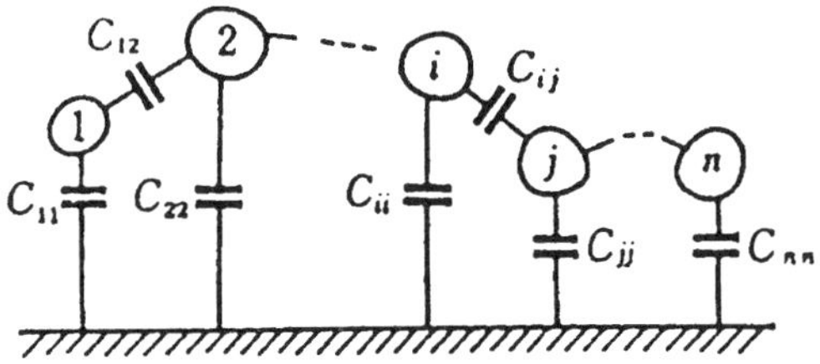

Figure 10. Capacities in N parallel conductors.

Because

$$q v_p = \frac{q}{\sqrt{\mu\varepsilon}} = i = \begin{bmatrix} i_1 \\ \vdots \\ i_n \end{bmatrix} \tag{29d}$$

we have the relation

$$i = \frac{|C'|}{\sqrt{\mu\varepsilon}} v \tag{29e}$$

In the case $N = 2$, the characteristic impedance Z_c can be obtained by the capacity C (F/m) for a length of 1 meter by Eq. (4e). The results of a typical TEM line are shown in Table 2.

Why a TEM Does Not Exist in a Hollow Waveguide

In a hollow waveguide made by one conductor as shown in Figs. 11a and 11c–11e, a TEM wave does not exist. The reason can be easily explained by Ampere's law and Faraday's law in an electromagnetic theory.

In the case of a TM wave as shown in Fig. 11a, magnetic fields exist only in the sectional plane perpendicular to the z direction of wave propagation. By Ampere's law, we must have currents penetrating the magnetic loop shown by dotted lines. Because the hollow waveguide does not include any conductor, the currents should be displacement currents, which require electrical fields along the z direction. Therefore, the wave is an E wave and not a TEM wave. In the case of a coaxial line, currents required from Ampere's law flow into the center conductor as a conductive current as in Fig. 11b.

As another example, when magnetic fields exist in the plane including the z axis as shown in Fig. 11c, the wave is already not a TEM wave. Electric fields naturally exist to satisfy Faraday's law as shown in solid lines of Fig. 11c. As a practical example of a TM wave, the $TM_{11}^{\square}$ mode is shown in Fig. 11d, where electrical fields penetrate the closed loop of the magnetic field at the center. As another example of a TE wave, the $TE_{20}^{\square}$ mode is shown in Fig. 11e, where magnetic fields exist in the z direction. This magnetic field produces the electrical fields on both sides of the center of magnetic field according to Faraday's law.

3.3 Characteristics of TE and TM Modes

Cutoff Frequency and Cutoff Wavelength

First, we describe so-called wave equations. In the source free ($\boldsymbol{J} = 0$) and lossless region, multiplying by $\nabla \times$ Eqs. (24a) and (24b) and considering

Table 2. Characteristic Impedance of Several TEM Lines

Construction	Characteristic impedance
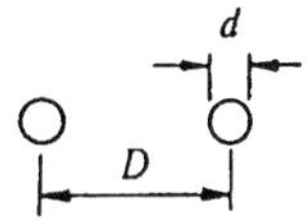	$Z = \frac{1}{\pi\eta} \ln\left\{\frac{D}{d} + \sqrt{\left(\frac{D}{d}\right)^2 - 1}\right\} = \frac{1}{\pi\eta} \cosh^{-1}\left(\frac{D}{d}\right)$ $\simeq \frac{1}{\pi\eta} \ln \frac{2D}{d}$ [in the case of $d \ll D$]
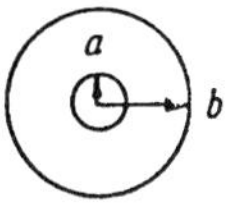	$Z = \frac{1}{2\pi\eta} \ln \frac{b}{a}$
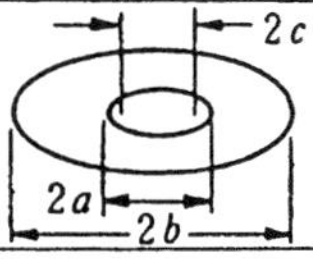	$Z = \frac{1}{2\pi\eta} \ln \frac{b + \sqrt{b^2 - c^2}}{a + \sqrt{a^2 - c^2}}$
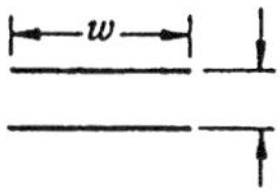	$Z \simeq \frac{1}{\eta} \frac{d}{w}$ [in the case of $b \ll W$]
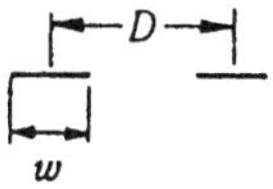	$Z \simeq \frac{1}{\pi\eta} \ln \frac{4D}{w}$, [in the case of $W \ll D$]
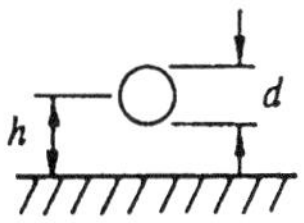	$Z = \frac{1}{2\pi\eta} \ln\left\{\frac{2h}{d} + \sqrt{\left(\frac{2h}{d}\right)^2 - 1}\right\} = \frac{1}{2\pi\eta} \cosh^{-1}\left(\frac{2h}{d}\right)$ $\simeq \frac{1}{2\pi\eta} \ln \frac{4h}{d}$ [in the case of $d \ll h$]
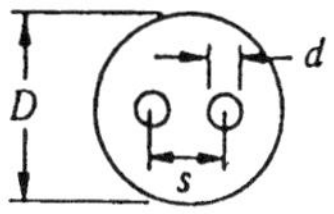	$Z \simeq \frac{1}{\pi\eta} \ln\left(\frac{2s}{d} \frac{D^2 - s^2}{D^2 + s^2}\right)$, $D \gg d$, $s \gg d$

$$\frac{1}{\eta} = 120\pi \sqrt{\frac{\mu_r}{\varepsilon_r}} \qquad \frac{1}{\pi\eta} = 120 \sqrt{\frac{\mu_r}{\varepsilon_r}}$$

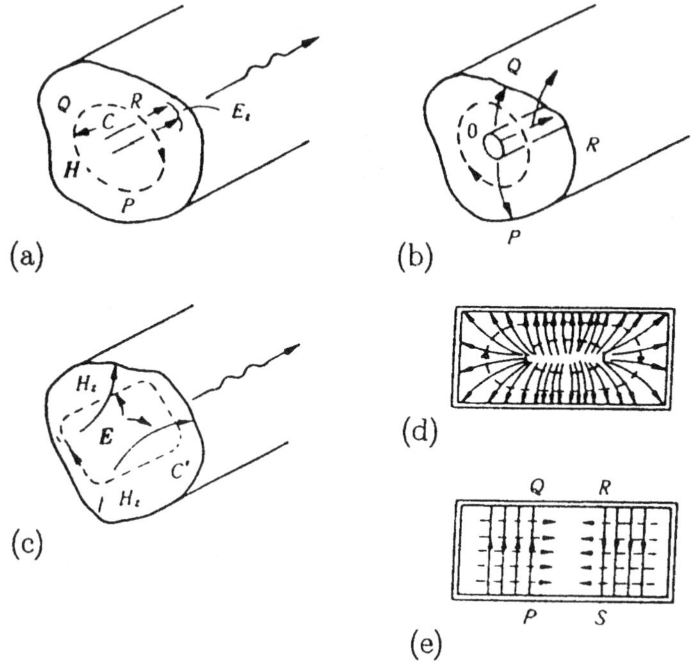

Figure 11. Explanation of the reason why a TEM wave does not exist in a hollow waveguide. (a) TM wave; (b) TEM wave; (c) TE wave; (d) TM_{11} mode; (e) TE_{20} mode.

the conditions of Eqs. (24c) and (24d), we get

$$\nabla \times \nabla \times \boldsymbol{E} = \nabla(\nabla \cdot \boldsymbol{E}) - \nabla^2 \boldsymbol{E} = -\nabla^2 \boldsymbol{E} = -j\omega\mu\nabla \times \boldsymbol{H}$$
$$\nabla \times \nabla \times \boldsymbol{H} = \nabla(\nabla \cdot \boldsymbol{H}) - \nabla^2 \boldsymbol{H} = -\nabla^2 \boldsymbol{H} = j\omega\varepsilon\nabla \times \boldsymbol{E}$$

Substituting the relations in Eqs. (24a) and (24b) into the above equations, we get

$$\nabla^2 \boldsymbol{E} + k^2 \boldsymbol{E} = 0 \quad (30a)$$
$$\nabla^2 \boldsymbol{H} + k^2 \boldsymbol{H} = 0 \quad (30b)$$
$$k^2 = \omega^2 \mu\varepsilon \quad (30c)$$

Equations (30a) and (30b) are called vector wave equations.

When we consider the z components, we get the relation of follows.

$$\nabla^2 E_z + k^2 E_z = 0 \quad (31a)$$
$$\nabla^2 H_z + k^2 H_z = 0 \quad (31b)$$

Considering Eq. (25), we get the following equations:

$$\nabla_t^2 E_z + k_t^2 E_z = 0 \tag{31c}$$

$$\nabla_t^2 H_z + k_t^2 H_z = 0 \tag{31d}$$

$$k_t^2 = k^2 + \gamma^2 \tag{31e}$$

TM Mode (E Mode) Because $\nabla_t \times \boldsymbol{E}_t = 0$, as shown in Table 1, we can express $\boldsymbol{E}_t$ by the scalar function φ^e as

$$\boldsymbol{E}_t = -\nabla_t \varphi^e \tag{32a}$$

Substituting Eq. (32a) into the equation derived from Eq. (24d), that is,

$$\nabla \cdot \boldsymbol{E} = \nabla_t \cdot \boldsymbol{E}_t - \gamma E_z = 0,$$

we get

$$\nabla_t^2 \varphi^e + \gamma \boldsymbol{E}_z = 0 \tag{32b}$$

Comparing Eqs. (31c) and (32b), we get

$$\nabla_t^2 \varphi^e + k_t^2 \varphi^e = 0 \tag{33a}$$

$$E_z = \frac{k_t^2}{\gamma} \varphi^e \tag{33b}$$

Equation (33a) is called the Helmholtz equation.

The differential equation (33a), however, should satisfy the boundary condition

$$\varphi^e = 0 \quad \text{on } S \tag{33c}$$

because $E_z = 0$ on S (metal wall of waveguide); therefore, $\varphi^e = 0$ from the relation Eq. (33b).

In the Helmholtz equation (33a) with the boundary condition Eq. (33c), k_t should take special values called eigenvalues or cutoff wave numbers associated with the guide cross section, which is always a positive number. (See Appendix to this chapter.)

There exists φ^e corresponding to k_t; φ^e is called an eigenfunction. Because we have an infinite number of k_t, we can denote the ith eigenvalues by $k_{t,i}$. In a lossless waveguide, the phase constant k_z $(=\gamma/j)$ can be obtained from Eq. (31e) and it takes the values

$$k_z^2 = \omega^2 \mu\varepsilon - k_{t,i}^2 \tag{33d}$$

It is understood that k_z is a function of ω as shown in Fig. 12, and we have the angular frequency ω_c for $k_z = 0$. Using ω_c, k_z can be expressed by

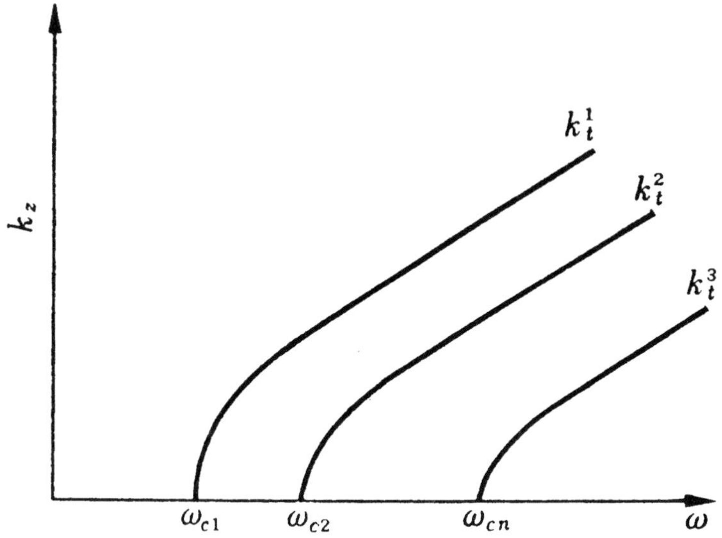

Figure 12. ω–k diagram.

$$k_z = \omega\sqrt{\mu\varepsilon}\sqrt{1 - \left(\frac{\omega_c}{\omega}\right)^2}, \qquad \omega_c = 2\pi f_c \tag{33e}$$

In the frequency region $\omega < \omega_c$, the guided wave decreases as an evanescent mode. f_c of Eq. (33e) is called a cutoff frequency and

$$\lambda_c = \frac{v_0}{f_c} \quad (v_0 = \text{light velocity}) \tag{33f}$$

is called the cutoff wavelength.

TE Mode (H Mode) Because $\nabla_t \times \boldsymbol{H}_t = 0$, as shown in Table 1, $\boldsymbol{H}_t$ can be expressed by the scalar function φ^m as

$$\boldsymbol{H}_t = \nabla_t \varphi^m \tag{34a}$$

Substituting Eq. (34a) into the equation derived from Eq. (24c), that is,

$$\nabla \cdot \boldsymbol{H} = \nabla_t \cdot \boldsymbol{H}_t - \gamma H_z = 0,$$

we get

$$\nabla_t^2 \varphi^m + \gamma H_z = 0 \tag{34b}$$

Comparing Eqs. (31d) and (34b), we get

$$\nabla_t^2 \varphi^m + k_t^2 \varphi^m = 0 \tag{34c}$$

$$H_z = \frac{k_t^2}{\gamma} \varphi^m \tag{34d}$$

As the magnetic field perpendicular to the metal wall of a waveguide periphery must be zero, we get the boundary condition

$$\frac{\partial \varphi^m}{\partial n} = 0 \tag{34e}$$

from Eq. (34a), where $\partial/\partial n$ shows the deviation for the normal direction to a periphery.

Equation (34c) with the boundary condition (34e) should have the eigenvalues of k_t and the corresponding eigenfunctions φ^m. Therefore, there exists a cutoff frequency f_c and cutoff wavelength λ_c.

In the frequency region $\omega < \omega_c$, the wave does not propagate and decays as an evanescent mode.

In the results mentioned above, the waveguides of the TM and TE mode are high-pass filters with the cutoff frequency f_c.

Wave Impedance of TM and TE Mode

Substituting the values of k_z and of Eq. (33e) into Eqs. (26e) and (27e), we get

$$Z_e = Z \sqrt{1 - \left(\frac{\omega_c}{\omega}\right)^2} \tag{35a}$$

$$Z_m = \frac{Z}{\sqrt{1 - (\omega_c/\omega)^2}} \tag{35b}$$

It is understood that the wave impedance of TM mode takes lower values for lower frequency and that of TE modes takes higher values for lower frequency.

Because in the frequency range $\omega < \omega_c$ the propagation constant takes real values γ, Z_e becomes capacitive and Z_m becomes inductive as understood from Eqs. (26d) and (27d). The impedances corresponding to ω are shown in Fig. 13.

Because TM mode and TE mode have zero and infinite values of wave impedance, respectively, cutoff frequencies of TM and TE modes are coincident with those of two-dimensional resonators of Figs. 14a and 14b, respectively. Figure 14a shows a resonator which consists of two sectional planes S_1 and S_2 and periphery metal wall, where S_1 and S_2 are electrical

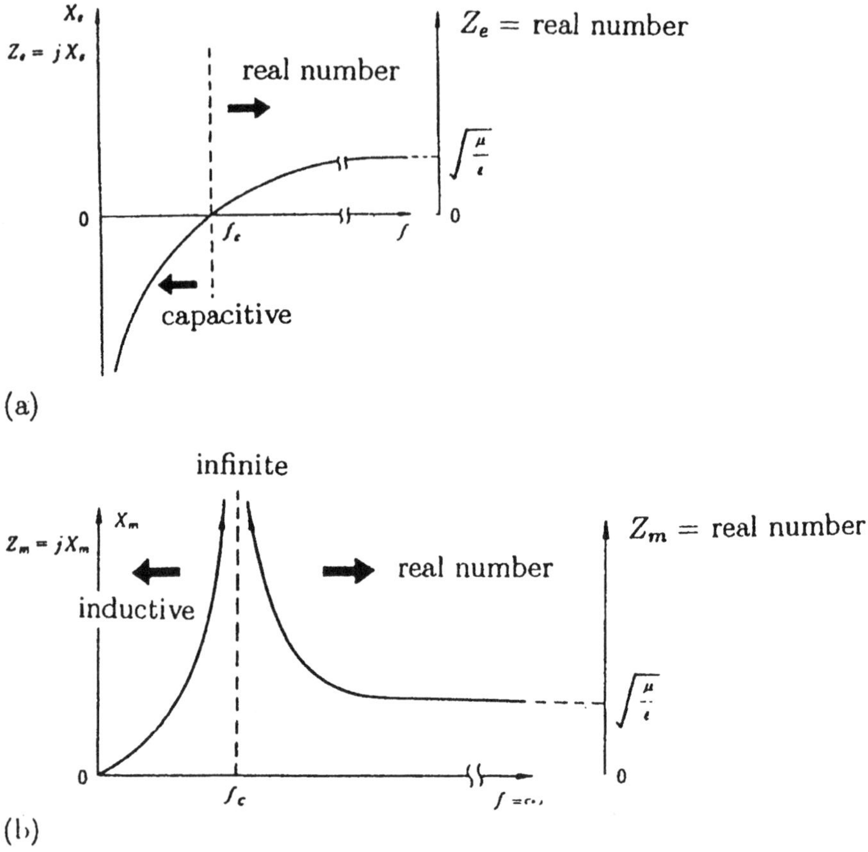

Figure 13. Frequency performance of wave impedance of TM and TE modes.

walls. The resonator of Fig. 14b is the same construction as that of Fig. 14a; the only difference is that S_1 and S_2 are magnetic walls.

Phase Velocity and Guided Wavelength

The incident wave in a lossless waveguide is expressed by

$$\mathscr{E} = \mathrm{Re}(\boldsymbol{E}_0 \exp[-j(k_z z - \omega t)]), \qquad \mathscr{H} = \mathrm{Re}(\boldsymbol{H}_0 \exp[-j(k_z z - \omega t)]).$$

Considering the position of the same phase, the equation

$$kz - \omega t = \text{constant} \tag{36}$$

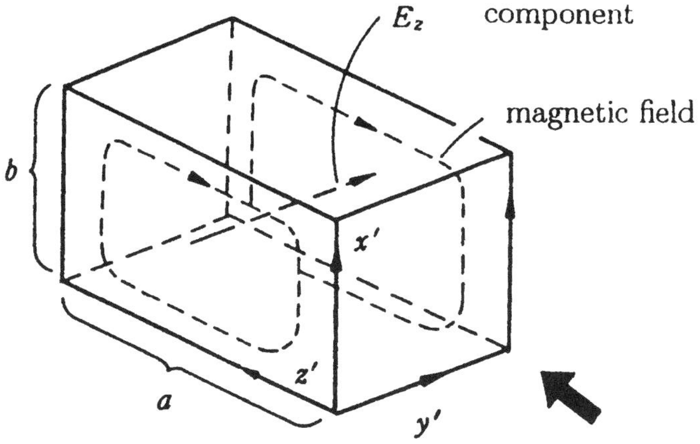

(a)

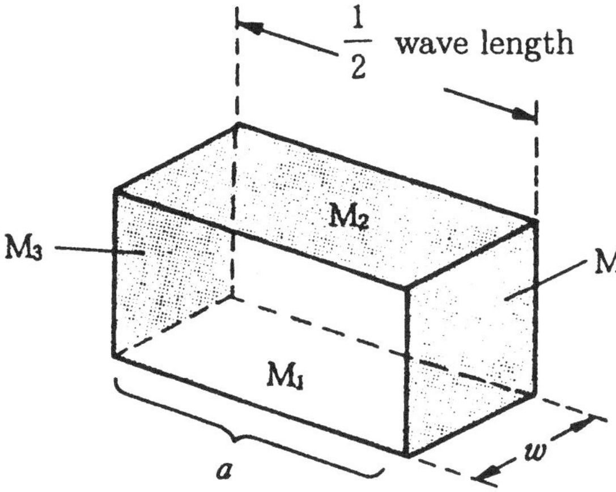

(b)

Figure 14. Two-dimensional resonators with resonant frequency of f_c. (a) The cavity with a two-dimensional mode made by cutting a TM waveguide to make the two parallel sections, which are replaced by metal walls, that is, the electric walls; (b) The cavity with a two-dimensional mode by cutting a TE waveguide to make the two parallel sections, which are replaced by magnetic walls.

is required. Differentiating both sides, we get the relations

$$k_z \Delta z = \omega \Delta t$$

$$v_p = \frac{\Delta z}{\Delta t} = \frac{\omega}{k_z} \tag{37a}$$

where v_p is called the phase velocity. Substituting Eq. (33e) into Eq. (37a), we get

$$v_p = \frac{v}{\sqrt{1 - (\omega_c/\omega)^2}}, \qquad v = \frac{1}{\sqrt{\mu\varepsilon}} = \frac{v}{\sqrt{\mu_r \varepsilon_r}} \tag{37b}$$

where μ_r is the relative permeability, v is the light velocity, and ε_r is the relative dielectric constant. It is understood that v_p is faster than the TEM wave in the same medium, and it takes infinite values at the cutoff frequency. This is true in for both TM and TE modes.

The guided wavelength λ_g is the wavelength in a waveguide, which means the length between the same phase repetition. Therefore, from Eq. (37b) we can obtain

$$\lambda_g = \frac{\lambda}{\sqrt{1 - (\omega_c/\omega)^2}}, \qquad \lambda = \frac{v}{f} \tag{37c}$$

Poynting Vector, Energy Flow, and Group Velocity

Poynting Vector and Energy Flow We consider a vector $\boldsymbol{S}$ as

$$\boldsymbol{S} = \mathscr{E} \times \mathscr{H} \tag{38}$$

$\boldsymbol{S}$ is called the Poynting vector. The vector $\boldsymbol{S}$ shows the flow of electromag-

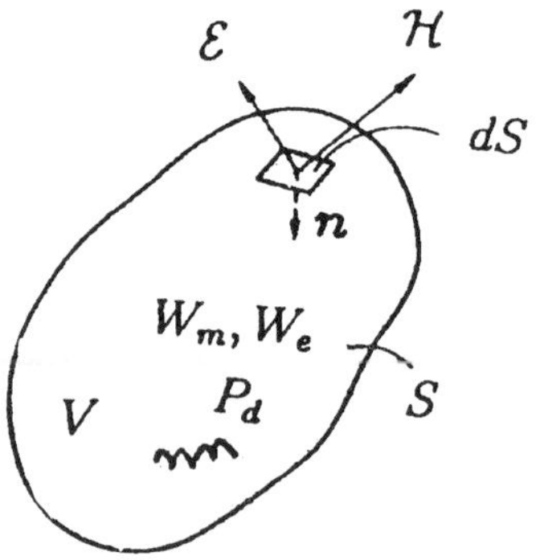

Figure 15. Reactive energy and dissipation power in a closed surface S and $\boldsymbol{E}$ and $\boldsymbol{H}$ on S. $\boldsymbol{n}$ points toward the inside of S.

netic energy per unit time and in the direction of flow. It is described as follows. By the vector formula,

$$\nabla\cdot(\mathscr{E}\times\mathscr{H})=\mathscr{H}\cdot\nabla\times\mathscr{E}-\mathscr{E}\cdot\nabla\times\mathscr{H} \tag{39a}$$

Substituting Maxwell's equation of (22a) and (22b) into Eq. (39a), we get

$$\nabla\cdot\mathscr{E}\times\mathscr{H}=-\mu\mathscr{H}\cdot\frac{\partial\mathscr{H}}{\partial t}-\sigma\mathscr{E}^2-\varepsilon\mathscr{E}\cdot\frac{\partial\mathscr{E}}{\partial t} \tag{39b}$$

Integrating Eq. (39b) over the region V surrounded by S,

$$\begin{aligned}-\int_V\nabla\cdot\mathscr{E}\times\mathscr{H}\,dV&=\int_S\mathscr{E}\times\mathscr{H}\cdot\boldsymbol{n}\,dS\\&=\int_v\sigma\mathscr{E}^2\,dV+\frac{\partial}{\partial t}\int_V(\mathscr{W}_e+\mathscr{W}_m)\,dV\end{aligned} \tag{39c}$$

$$\mathscr{W}_e=\frac{\varepsilon\mathscr{E}^2}{\epsilon},\qquad \mathscr{W}_m=\frac{\mu\mathscr{H}^2}{\epsilon} \tag{39d}$$

The first and second terms of the right-hand side of Eq. (39c) are joules loss and the increment of reactive energy per second, respectively, where the reactive energy is the sum of electrical energy $\mathscr{W}_e$ and magnetic energy $\mathscr{W}_m$. When $\mathscr{E}$ and $\mathscr{H}$ are sinusoidal charging, we can consider the vector $\boldsymbol{S}$ as

$$\boldsymbol{S}=\boldsymbol{E}\times\boldsymbol{H}^* \tag{40}$$

$\boldsymbol{S}$ is called the complex Poynting vector. In the same way as for Eq. (39a),

$$\nabla\cdot(\boldsymbol{E}\times\boldsymbol{H}^*)=\boldsymbol{H}^*\cdot\nabla\times\boldsymbol{E}-\boldsymbol{E}\cdot\nabla\times\boldsymbol{H}^* \tag{41a}$$

Substituting Eqs. (24a) and (24b) into Eq. (41a), we get

$$\int_S\boldsymbol{E}\times\boldsymbol{H}^*\cdot ds=\sigma|\boldsymbol{E}|^2+2j\omega(\tilde{W}_m-\tilde{W}_e) \tag{41b}$$

$$\tilde{W}_e=\frac{\varepsilon|\boldsymbol{E}|^2}{2},\qquad \tilde{W}_m=\frac{\mu|\boldsymbol{H}|^2}{2} \tag{41c}$$

where $\tilde{W}_e$ and $\tilde{W}_m$ are the time averages of the electric and magnetic energy, respectively.

When the electric field and the conjugate of the magnetic field of an incident wave at $Z = 0$ are $\boldsymbol{E}$ and $\boldsymbol{H}^*$, respectively, they become $\boldsymbol{E}e^{-jk_zL}$ and $\boldsymbol{H}^*e^{jk_zL}$ at $Z = L$ in a lossless waveguide, as shown in Fig. 16.

If we consider the closed surface S surrounded by S_1, S_2, and S_3 of Fig. 16, the left-hand side of Eq. (41b) becomes zero because $\int_{S_1} = -\int_{S_3}$. There-

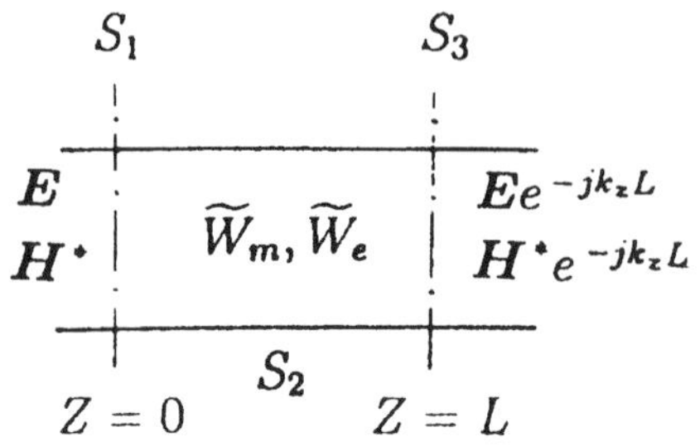

Figure 16. Electric field and magnetic field of traveling wave at $Z = 0$ and $Z = L$.

fore, under the consideration of a lossless waveguide, that is, $\sigma = 0$, we get

$$\tilde{W}_m = \tilde{W}_e \tag{42}$$

Denoting the time average of the reactive total energy of an incident wave by $\tilde{W}_t$, we get $\tilde{W}_t = 2\tilde{W}_m = 2\tilde{W}_e$. Therefore, in the case of a TM mode, we get

$$\tilde{W}_t = 2\tilde{W}_m = \int\int_{S_1} \mu |H_t|^2 \, dS = \frac{\mu}{Z_e^2} \int\int_{S_1} |\nabla_t \varphi^e|^2 \, dS \tag{43a}$$

[from Eqs. (26d) and (32a)], and in the case of a TE mode, we get

$$\tilde{W}_t = 2\tilde{W}_e = \int\int_{S_1} \varepsilon |E_t|^2 \, dS = Z_m^2 \varepsilon \int\int_{S_1} |\nabla_t \varphi^m|^2 \, dS \tag{43b}$$

[from Eqs. (27d) and (34a)].

When we consider the surface S surrounded by S_1, S_2, and S_3 of Fig. 17, $\sigma|E^2|$ of Eq. (41b) is the absorbed power P in an absorber at S_3, and P is

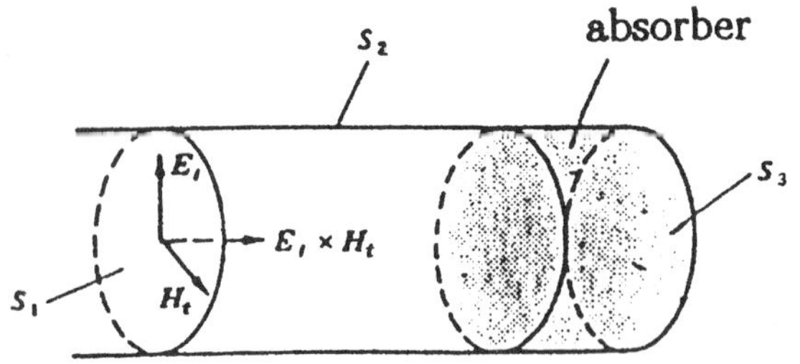

Figure 17. Waveguide-terminated absorber.

the Poynting power flowing through S_1 because $\tilde{W}_m = \tilde{W}_e$ from Eq. (41b). Therefore, P can be obtained as follows:

$$P = \int\int_{S_1} \boldsymbol{E}_t \times \boldsymbol{H}_t^* \cdot \boldsymbol{n}\, dS = \frac{1}{Z_e} \int\int_{S_1} |\boldsymbol{\nabla}_t \varphi^e|^2\, dS \quad \text{(TM)} \tag{43c}$$

$$= Z_m \int\int_{S_1} |\boldsymbol{\nabla}_t \varphi^m|^2\, dS \quad \text{(TE)} \tag{43d}$$

Representing the velocity of energy flow by v_e, we have the relation of

$$P = \tilde{W} v_e \tag{43e}$$

Substituting Eqs. (43a) and (43b) into Eq. (43e), and comparing them with Eqs. (43c) and (43d), we get

$$v_e = \frac{Z_e}{\mu} = \frac{k_z}{\omega\mu\varepsilon} = \frac{v_0^2}{v_p} \quad \text{(TM)}$$

$$v_e = \frac{1}{Z_m \varepsilon} = \frac{k_z}{\omega\mu\varepsilon} = \frac{v_0^2}{v_p} \quad \text{(TE)}$$

Therefore,

$$v_e v_p = v_0^2$$

where

$$v_0 = \frac{1}{\sqrt{\mu\varepsilon}} \tag{44}$$

Group Velocity When a carrier is modulated by a signal, the signal is transmitted to the output terminal with a velocity v_g. In this case, v_g is called a group velocity and it takes the values (Appendix 1)

$$v_g = \left(\frac{dk_z}{d\omega}\right)^{-1} \tag{45}$$

Differentiating Eq. (33d),

$$2k_z\, dk_z = 2\omega\mu\varepsilon\, d\omega$$

and we get

$$v_g v_p = v_0^2$$

$$v_g = v_e \tag{46}$$

Evanescent Mode and the Related Reactive Energy of the TM and TE Modes

In the frequency region below the cutoff frequency, the wave applied to the waveguide does not propagate and decreases exponentially in a waveguide, where reactive energies exist.

In reactive energies, the electrical energy is dominated in the TM cutoff waveguide and the magnetic energy is dominated in the TE cutoff waveguide; the waveguide operated with a frequency below the cutoff frequency is called the cutoff waveguide.

The reason, however, can be explained as follows. In the closed surface S surrounded by S_1, S_1', and the cutoff waveguide, we consider that the wave supplied at S_1 sufficiently decreases at S_1'. In this case, we get

$$\begin{aligned}\iint_{S_1} \boldsymbol{E} \times \boldsymbol{H}^* \cdot \boldsymbol{n}\, dS &= \iint_{S_1} \boldsymbol{E}_t \times \boldsymbol{H}_t^* \cdot \boldsymbol{n}\, dS \\ &= Z_m \iint_{S_1} |\boldsymbol{H}_t|^2\, dS = \frac{j\omega\mu}{\gamma} \iint_{S_1} |\boldsymbol{H}_t|^2\, dS \\ &= P_d + 2j\omega(\tilde{W}_m - \tilde{W}_e) \quad \text{for the TE mode}\end{aligned}$$

Considering $\gamma > 0$ in an evanescent mode and $P_d = 0$ for the lossless waveguide, we get

$$\tilde{W}_m > \tilde{W}_e \tag{47a}$$

Therefore, the total reactive energy becomes magnetic. In the same manner, in a TM cutoff waveguide, we get

$$\tilde{W}_e > \tilde{W}_m \tag{47b}$$

Therefore, the total reactive energy becomes capacitive.

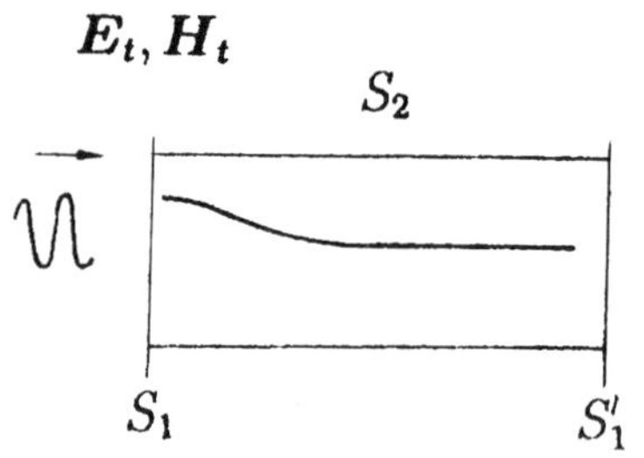

Figure 18. Evanescent mode.

4 EQUIVALENT DISTRIBUTED LINES OF A HOLLOW WAVEGUIDE

4.1 Definitions of Voltage and Current of a Hollow Waveguide

In the case of TEM lines as a coaxial line, we have the voltage between two conductors and the current flowing in the center conduct. In a hollow waveguide, however, we can define the voltage and the current in many ways. The main definitions will be introduced as follows.

(a) Definition by Mode Voltage and Mode Current. This is the most generally used definition and the typical one will be introduced. When we define the mode functions for the sectional component of an electrical field vector and magnetic field vector as $\boldsymbol{e}_t$ and $\boldsymbol{h}_t$, we can express the actual existing electrical field and magnetic field by

$$\boldsymbol{E}_t = V\boldsymbol{e}_t \tag{48a}$$

$$\boldsymbol{E}_t = I\boldsymbol{h}_t \tag{48b}$$

In the equations, V and I are called the mode voltage and mode current, respectively, which are defined corresponding to the mode vectors $\boldsymbol{e}_t$ and $\boldsymbol{h}_t$.

Let us consider the case of the normalization of mode vectors by

$$\begin{aligned} &\int\int_S |\boldsymbol{e}_t|^2\, dS = \int\int_S |\boldsymbol{h}_t|^2\, dS = 1 \\ &\int\int_S \boldsymbol{e}_t \times \boldsymbol{h}_t^* \cdot \boldsymbol{i}_z\, dS \end{aligned} \tag{48c}$$

In the case of an incident wave, we have the relation

$$\frac{|\boldsymbol{E}_t^+|}{|\boldsymbol{H}_t^+|} = Z_{m,e} = \frac{V^+|\boldsymbol{e}_t|}{I^+|\boldsymbol{h}_t|}$$

Because $|\boldsymbol{e}_t| = |\boldsymbol{h}_t|$ from Eq. (48c), we get

$$\frac{V^+}{I^+} = Z_{m,e} \tag{48d}$$

In the case of a reflecting wave, using the same method, we get

$$\frac{V^-}{I^-} = -Z_{m,e} \tag{48e}$$

We, therefore, get the equivalent distributing lines, of which the characteristic impedance is the wave impedance and the phase constant is that of the waveguide, as shown in Fig. 19.

As an example, we will obtain $\boldsymbol{e}_t$ and $\boldsymbol{h}_t$ of the $TE_{10}^{\square}$ mode shown in Fig. 20. Defining $\boldsymbol{e}_t$ and $\boldsymbol{h}_t$ by

$$\boldsymbol{e}_t(x, y) = \boldsymbol{i}_y A \sin \pi \frac{x}{a}$$

$$\boldsymbol{h}_t(x, y) = -\boldsymbol{i}_x B \sin \pi \frac{x}{a}$$

and integrating them over the sectional area of the waveguide, we get

$$\int_0^b \int_0^a A^2 \sin^2 \left(\pi \frac{x}{a}\right) dx\, dy = \frac{abA^2}{2} = 1$$

$$\int_0^b \int_0^a B^2 \sin^2 \left(\pi \frac{y}{a}\right) dx\, dy = \frac{abB^2}{2} = 1$$

Therefore,

$$\boldsymbol{e}_t(x, y) = \boldsymbol{i}_y \sqrt{\frac{2}{ab}} \sin \pi \frac{x}{a}$$

$$\boldsymbol{h}_t(x, y) = \boldsymbol{i}_x \sqrt{\frac{2}{ab}} \sin \pi \frac{x}{a} \qquad (48f)$$

It is also understood that $|\boldsymbol{e}_t| = |\boldsymbol{h}_t|$.

(b) Determination of V and I by the Traveling Power P and the Voltage V Between Two Specified Points and the Defined Pass. In Fig. 21a, we choose points A and B at the center of the H plane of the $TE_{10}^{\square}$ waveguide. The voltage is determined by the voltage difference measured from B to A along the pass perpendicular the H plane. In Fig. 21b, we choose points A and B at the centers of the ridge parts. In Fig. 21c, we choose point A at the center of the ridge part, and point B at the center of the H plane. The pass to measure voltage is the straight line between A and B.

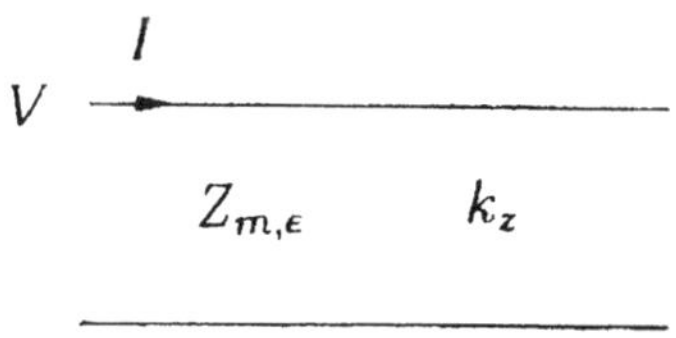

Figure 19. Equivalent distributing lines.

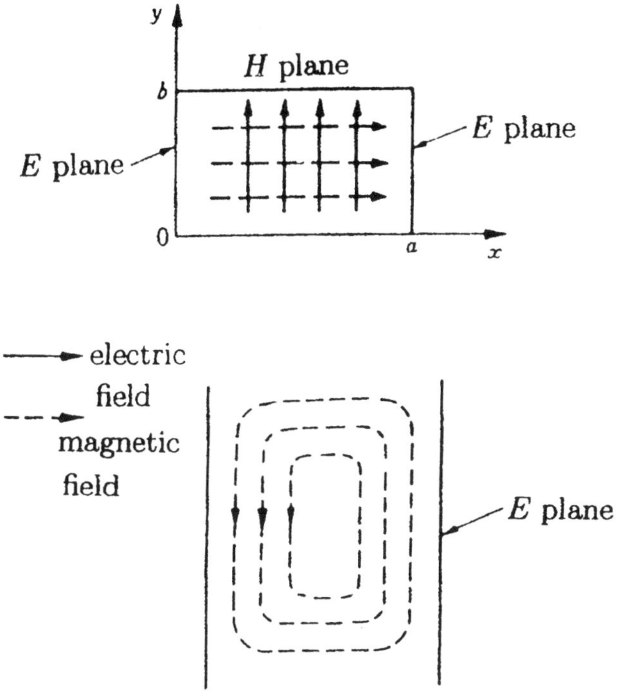

Figure 20. Sectional distribution of $\boldsymbol{e}_t$ and $\boldsymbol{h}_t$.

Using the incident power P, I and the characteristic impedance of the equivalent distributing line, Z_{VP}, can be obtained by

$$I^+ = \frac{P}{V^+}$$

and

$$Z_{VP} = \frac{V^+}{I^+} = \frac{|V^+|^2}{P} \tag{49a}$$

where V^+ is the effective values. Z_{VP} of Fig. 21a, can be calculated as

$$Z_{VP} = 2\frac{b}{a}Z_m \tag{49b}$$

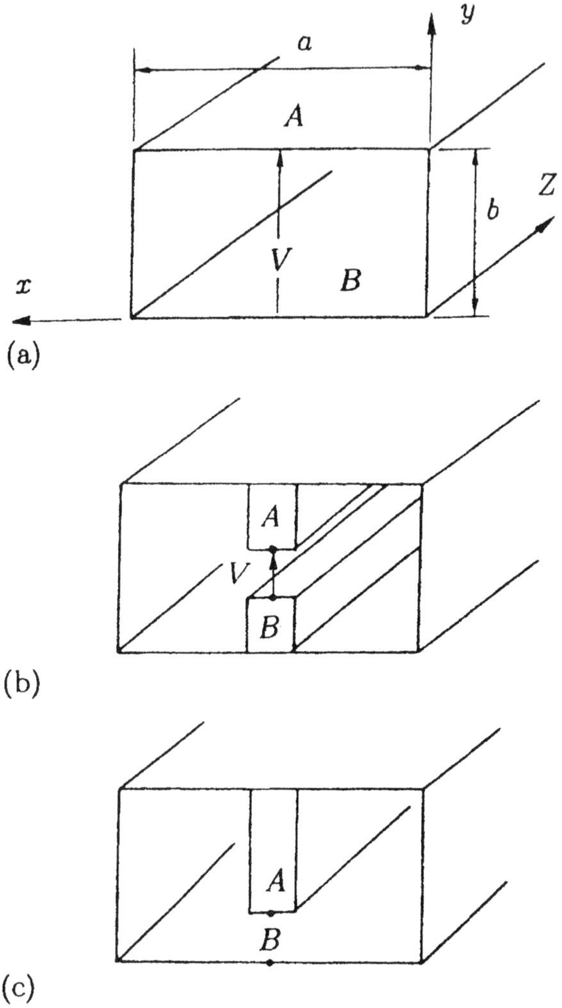

Figure 21. Example of the two specified points to determine the voltage *V*. (a) $TE_{10}^{\square}$ rectangular waveguide; (b) double ridge waveguide; (c) single ridge waveguide.

By this definition, we can get the physical image as the impedance of the waveguide becomes lower in the case of a smaller height (smaller *b*) of the waveguide. In the case of the ridge guides, the definition in Eq. (49a) is frequently used.

4.2 The *L, C* Equivalent Circuit of the TEM Line, and the TE and TM Waveguides

In Fig. 19, we obtained the equivalent distributing line of the TM and TE waveguides.

When the distributed line is expressed by the series impedance with the reactance of jX and the parallel admittance with the susceptance jB in a lossless case, we get the equivalent network of Fig. 22b.

In this case, we must have the relations (50a) and (50b) as understood by Eqs. (3c) and (3d):

$$Z_c = \sqrt{\frac{X}{B}}, \qquad \beta = \sqrt{XB} \tag{50a}$$

$$X = Z_c\beta, \qquad B = \frac{\beta}{Z_c} \tag{50b}$$

We, however, know the values of Z_c and β for TEM, TE, and TM modes as follows:

$$\begin{aligned} Z_c &= \sqrt{\frac{\mu}{\varepsilon}} \quad \text{TEM mode} \\ &= \frac{\sqrt{\mu/\varepsilon}}{\sqrt{1-(\omega_c/\omega)^2}} \quad \text{TE mode} \\ &= \sqrt{\frac{\mu}{\varepsilon}}\sqrt{1-\left(\frac{\omega_c}{\omega}\right)^2} \quad \text{TM mode} \\ \beta &= \omega\sqrt{\mu\varepsilon} \quad \text{TEM mode} \\ &= \omega\sqrt{\mu\varepsilon}\sqrt{1-\left(\frac{\omega_c}{\omega}\right)^2} \quad \text{TE and TM modes} \end{aligned} \tag{50c}$$

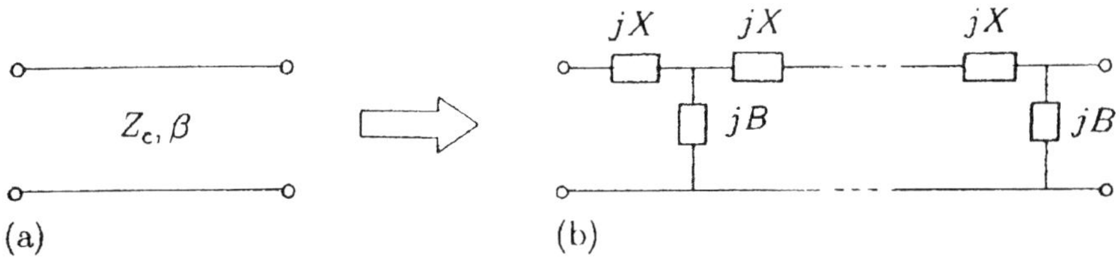

Figure 22. Distributing line with two parallel conductors and the equivalent network. (a) Distributing line with two parallel conductors; (b) equivalent network constructed by series reactance jX and parallel jB.

Substituting the values of Eq. (50c) into Eq. (50b), we get the results of Table 3.

The equivalent circuits of several transmission lines, therefore, are as shown in Fig. 23.

From Fig. 23, we can understand the following. In the cutoff TE waveguide, it becomes the induction reactive attenuator because the parallel tuned network becomes inductance, which is shown in Fig. 23b. In the cutoff TM waveguide, it becomes the capacitive reactive attenuator because the series tuned network becomes capacitance, which is shown in Fig. 23c. Next, when we consider the equivalent circuit corresponding to the definition in Section 4.1, we get the following results.

In the case of the TEM mode,

$$Z_c = \sqrt{\frac{L}{C}} \quad \text{from Eq. (4a),} \qquad \beta = \omega\sqrt{LC} \quad \text{from Eq. (4b)}$$

where L is the inductance per meter (H/m) and C is the capacitance per meter (F/m).

In the case of TE waveguide,

$$\mu \to 2\frac{b}{a}\mu, \qquad \varepsilon \to \frac{a}{2a}\varepsilon \quad \text{from Eq. (49b)}$$

5 LOSS OF TRANSMISSION LINES

5.1 Loss Mechanism and the Relation of Transmission Loss and Q Values

The loss of a transmission line is caused by a conducting loss on a conductor, a dielectric loss of a substrate such as a microstrip line, and a radiation loss in the case of open guides.

The transmission power at Z, $P(Z)$, decreases with attenuation constant α when Z increases, as shown by

$$P(Z) = P_0 e^{-2\alpha z} \tag{51}$$

From Eq. (51), we get

$$\alpha = \frac{dP/dZ}{2P(Z)} = \frac{P_c + P_d + P_r}{2P(Z)} \left(\frac{NP}{m}\right) \tag{52a}$$

where P_c, P_d, and P_r are lost power for 1 m by a conducting, dielectric, and radiation loss, respectively.

When we, therefore, denote the attenuation constants corresponding to P_c, P_d, and P_r by α_c, α_d, and α_r, respectively, we can express them as

Table 3. Lumped Element (L, C) Equivalent Circuit of Several Transmission Lines (in the Case of the Mode Voltage and Mode Current used for V and I)

Transmission Lines	X	Equivalent circuit of X	B	Equivalent circuit of B
TEM-line	$\omega\mu$	 μ	$\omega\mu$	 ϵ
TE waveguide	$\omega\mu$	 μ	$\omega\epsilon\left\{1-\left(\frac{\omega_c}{\omega}\right)^2\right\}$	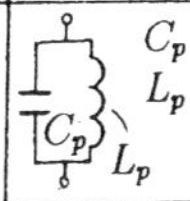 C_p, L_p; $C_p=\epsilon$, $L_p=\frac{1}{\omega_c^2\epsilon}$
TM waveguide	$\omega\mu\left\{1-\left(\frac{\omega_c}{\omega}\right)^2\right\}$	 L_s C_s; $L_s=\mu$, $C_s=\frac{1}{\omega_c^2\mu}$	$\omega\epsilon$	 ϵ

$$\alpha_c = \frac{P_c}{2p(Z)}, \qquad \alpha_d = \frac{P_d}{2P(Z)}, \qquad \alpha_r = \frac{P_r}{2P(Z)} \tag{52b}$$

The values of attenuation constants can be related to corresponding Q values of Q_c, Q_d, and Q_r by

$$\alpha_{c,d,r} = \frac{\pi\lambda_g}{\lambda^2 Q_{c,d,r}} \tag{53}$$

where λ and λ_g are the wavelength of the plane wave in an infinite medium and the guided wavelength, respectively.

Q_c, Q_d, and Q_r in Eq. (53) are also defined as

$$Q_{c,d,r} = \frac{\omega\tilde{W}_t}{P_{c,d,r}} \tag{54}$$

In Eq. (54), $\tilde{W}_t$ is the average time of the total energy included in unit length (1 m) of waveguide to propagate an incident wave. However, relation (54) can be, easily obtained by

$$P(Z) = v_e\tilde{W}_t = \frac{v^2}{v_p}\tilde{W}_t = \frac{\lambda^2}{2\pi\lambda_g}\omega\tilde{W}_t$$

By using

$$\frac{1}{Q} = \frac{1}{Q_c} + \frac{1}{Q_d} + \frac{1}{Q_r} \tag{55a}$$

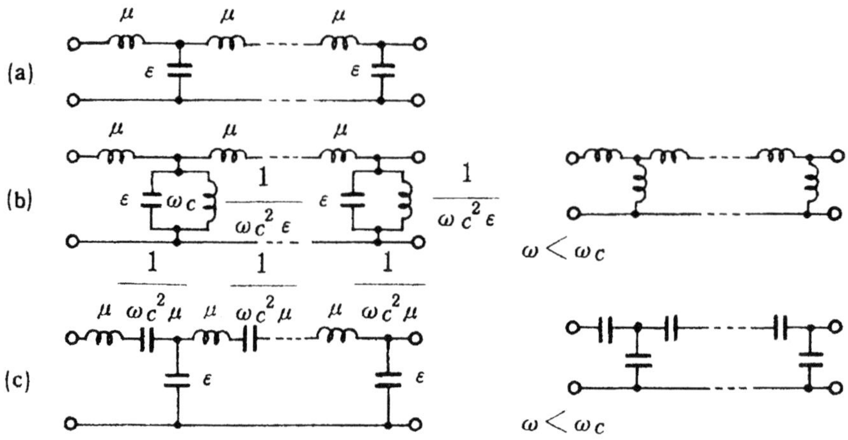

Figure 23. Equivalent lumped element networks of several waveguides. (a) TEM; (b) TE; (c) TM.

we get

$$\alpha = \alpha_c + \alpha_d + \alpha_r \tag{55b}$$

$$\alpha = \frac{\pi \lambda_g}{\lambda^2 Q} \tag{55c}$$

When we construct the ring resonator with the length of $n(\lambda/2)$ of the waveguide, the resonator takes the same values of Q.

5.2 Conductive Loss

The high-frequency currents flow on the thin surface region with thickness δ_s by the so-called skin effect, where δ_s is called the skin depth. Because at a higher frequency the thickness decreases, the ohmic resistance of a waveguide increases, which results in an increasing α_c and a decreasing Q_c.

Skin Depth

In the case of metal, the values of σ and ε are obtained from

$$\sigma \simeq (1\text{–}6) \times 10^7 \quad \text{(S/m)}$$
$$\varepsilon \simeq 10^{-11} \quad \text{(F/m)}$$

This means that $\omega\varepsilon = 6 \times 10^{-3}$ (at 100 MHz) $\gg \sigma$.

In the case of a plane wave, γ and $\dot{Z}$ of the wave can be obtained by substituting the above relation into Eq. (28c):

$$\gamma = \alpha + j\beta, \qquad \alpha = \beta = \sqrt{\frac{\omega\mu\sigma}{2}} \tag{56a}$$

$$\dot{Z} = (1 + j)\sqrt{\frac{\omega\mu}{2\sigma}} \tag{56b}$$

Because the electromagnetic field decays as $e^{-\alpha z}$, the current I_c also decays as $e^{-\alpha z}$.

Setting the values of Z where I_c decays as $e^{-\alpha z}$, the current I_c also decays as $e^{-\alpha z}$. Setting the values of Z where I_c decays to the values of $I_c e^{-1}$ and δ_s, we get

$$\delta_s = \frac{1}{\sqrt{\pi f \mu \sigma}} \tag{57}$$

In the case of metal, for example, δ_s takes the value of about 10 μm at 100 MHz, although λ is 3 m.

When we consider the wave that exists on the surface of a metal exposed to air, the electromagnetic field rapidly charges inside the metal on the order of $1/\delta_s$, whereas it charges on the surface on the order of $1/\lambda$. For this reason, the values of δ_s obtained by the assumption of a plane wave perpendicular to the metal surface is available for the depth of the current flow on the metal surface for any kind of guided wave. The results mentioned above can be also directly obtained from Maxwell's equation by using the relation

$$\nabla^2 \simeq \frac{\partial^2}{\partial x^2} \quad \left(\text{neglecting of term of } \frac{\partial}{\partial x} \text{ and } \frac{\partial}{\partial y}\right)$$

Denoting the surface resistance by R_s, we get

$$R_s = \frac{1}{\sigma\delta_s} = \sqrt{\frac{\pi f \mu}{\sigma}} \quad (\Omega) \tag{58}$$

We show the values of δ_s and R_s of several metals at 1 GHz in Table 4.

Because R_s is proportional to $\sqrt{f}$ as in Eq. (58),

$$R_{s,\mathrm{Cu}} = 8.29 \times 10^{-3}\sqrt{f} \quad (f \text{ in GHz}) \tag{59a}$$

$$R_s = 8.29 \times 10^{-3}\, R_{s,r}\sqrt{f} \quad (f \text{ in GHz}) \tag{59b}$$

The value of R_s increases with increasing surface roughness.

Attenuation of Some Practical Waveguides

Coaxial TEM Mode The attenuation constant is obtained as

$$\alpha = \frac{R_s}{4\pi Z_c}\left(\frac{1}{a} + \frac{1}{b}\right) \quad \left(\frac{\mathrm{N}}{\mathrm{m}}\right) \tag{60a}$$

Z_c is the characteristic impedance.

Table 4. σ, δ_s, R_s, and $R_{s,r}$ of Several Metals ($R_{s,r} = R_{s,\mathrm{metal}})/E_{s,\mathrm{Cu}}$

Metal	σ (Ω/m)	δ_s(1 GHz) (m)	R_s(1 GHz) (Ω)	$R_{s,\gamma}$
Ag	6.17×10^{7}	2.03×10^{-6}	7.98×10^{-3}	0.97
Cu	5.80×10^{7}	2.08×10^{-6}	8.29×10^{-3}	1.00
Au	4.10×10^{7}	2.84×10^{-6}	9.84×10^{-3}	1.19
Cr	3.84×10^{7}	2.565×10^{-6}	10.15×10^{-3}	1.23
Al	3.72×10^{7}	2.612×10^{-6}	10.29×10^{-3}	1.25
Brass	1.57×10^{7}	4.016×10^{-6}	15.86×10^{-3}	1.92
Solder	0.17×10^{7}	5.85×10^{-6}	24.08×10^{-3}	2.86

In the case of Cu, we obtain the performance shown in Fig. 24. The attenuation constant is obtained by

$$\alpha = \frac{R_s}{120\pi}\,\frac{1}{b}\,\frac{1 + 2(b/a)(\lambda/2a)^2}{\sqrt{1 - (\lambda/2a)^2}} \quad \left(\frac{\mathrm{N}}{\mathrm{m}}\right) \tag{60b}$$

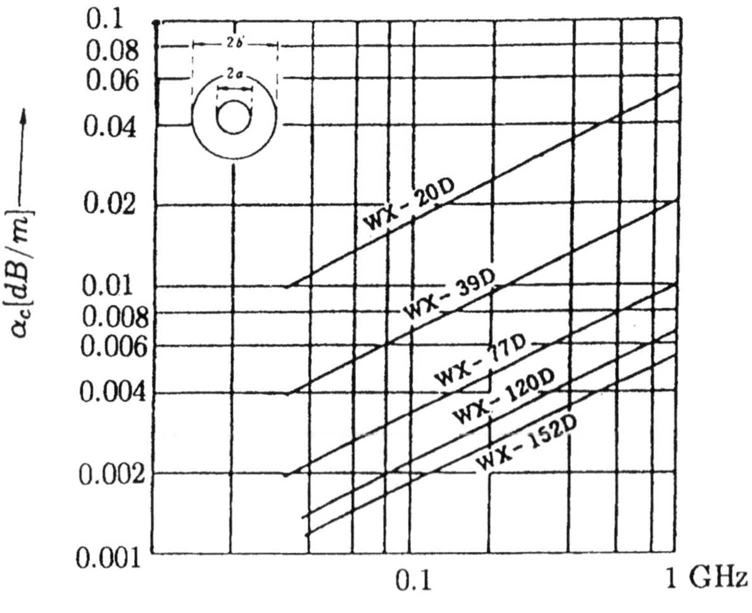

Figure 24. Attenuation of a coaxial line by conductive loss.

In the case of Cu, we obtain the performance shown in Fig. 25. Figure 26 shows the size of the $TE_{10}^{\square}$ waveguide cross section.

Microstrip Line The attenuation of a microstrip line caused by conductive loss was obtained in the case of the characteristic impedance of 50 Ω; this is shown in Fig. 27. From the vertical scale showing the values of α_c (dB/m)/h (mm), the attenuation of h = 2 mm, for example, is $\alpha_c/2$ (dB/m).

In Figs. 24, 25, and 26, the attenuation was obtained for the case of Cu. However, because R_s is inversely proportional to $\sqrt{\sigma}$ (v/m) from Eq. (58), the attenuation in the case of another material, α_c', must be obtained by

$$\alpha_c' = \sqrt{\frac{\sigma_{Cu}}{\sigma_M}}\,\alpha_{c,Cu} \quad \left(\frac{dB}{m}\right) \tag{61}$$

5.3 Dielectric Loss

When a dielectric material used for a substrate has a dielectric loss, the dielectric constance can be expressed by a complex number $\dot{\varepsilon}$ as shown in

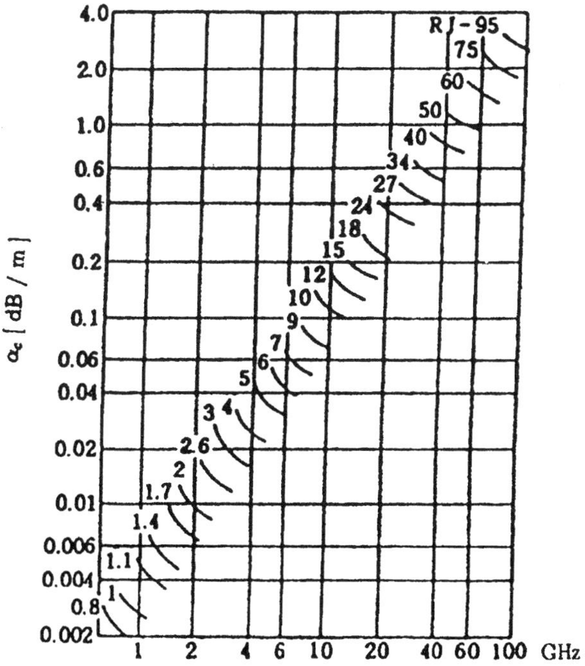

Figure 25. Attenuation of $TE_{10}^{\square}$ waveguide by conductive loss.

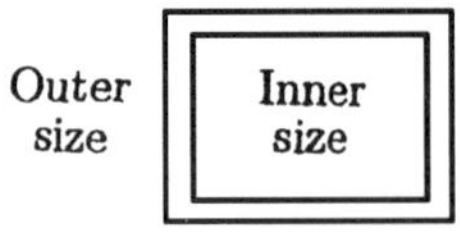

	(Gc)	Inner size (mm)	Outer size (mm)
WRJ- 4	3.3 ~ 4.9	58.1 × 29.1	63.1 × 32.3
WRJ- 6	4.9 ~ 7.05	40.0 × 20.0	43.2 × 23.2
WRJ- 7	5.85 ~ 8.20	34.85 × 15.85	38.05 × 19.05
WRJ- 9	7.05 ~ 10.0	28.5 × 12.6	31.7 × 15.8
WRJ-10	8.20 ~ 12.4	22.9 × 10.2	25.4 × 12.7
WRJ-12	10 ~ 15	19.00 × 9.50	21.60 × 12.10
WRJ-15	12 ~ 18	15.80 × 7.90	17.80 × 9.90
WRJ-18	15 ~ 22	13.00 × 6.50	15.00 × 8.50
WRJ-24	18 ~ 27	10.70 × 4.30	12.70 × 6.30
WRJ-27	22 ~ 33	8.60 × 4.30	10.60 × 6.30
WRJ-34	26 ~ 40	7.10 × 3.55	9.10 × 5.55
WRJ-40	33 ~ 50	5.70 × 2.85	7.70 × 4.85
WRJ-50	40 ~ 60	4.78 × 2.39	6.78 × 4.39
WRJ-60	50 ~ 75	3.76 × 1.88	5.76 × 3.88
WRJ-75	60 ~ 90	3.10 × 1.55	5.10 × 3.55
WRJ-95	75 ~ 110	2.54 × 1.27	4.54 × 3.27

Figure 26. The size of the $TE^{\square}_{10}$ waveguide cross section.

$$\dot{\varepsilon} = \varepsilon' - j\varepsilon'' \tag{62a}$$

where ε' and ε'' are the lossless and loss terms, respectively. The ratio $\varepsilon''/\varepsilon'$ is called the loss angle or $\tan \delta_\varepsilon$, that is,

$$\tan \delta_\varepsilon = \frac{\varepsilon''}{\varepsilon'} = \frac{\varepsilon_r''}{\varepsilon_r'} \tag{62b}$$

where ε_r' and ε_r'' are the lossless and loss terms, respectively, of the relative dielectric constant. $1/\tan \delta_\varepsilon$ is called the quality factor of the dielectric material. Denoted by Q_ε,

$$Q_\varepsilon = \frac{1}{\tan \delta_\varepsilon} \tag{62c}$$

Generally, Q_ε decreases as the frequency, f, increases. In a paraelectric material, we have the relation between Q_ε and f as

$$Q_\varepsilon = K\frac{1}{f} \tag{63}$$

where K is a constant.

In the case of glass epoxy and fluorocarbon polymers (PTFE) usually used for print circuits, Q_ε does not generally follow Eq. (63). For example, we have the measured values of Q_ε of 1100 at 1 MHz and 384 at 1 GHz

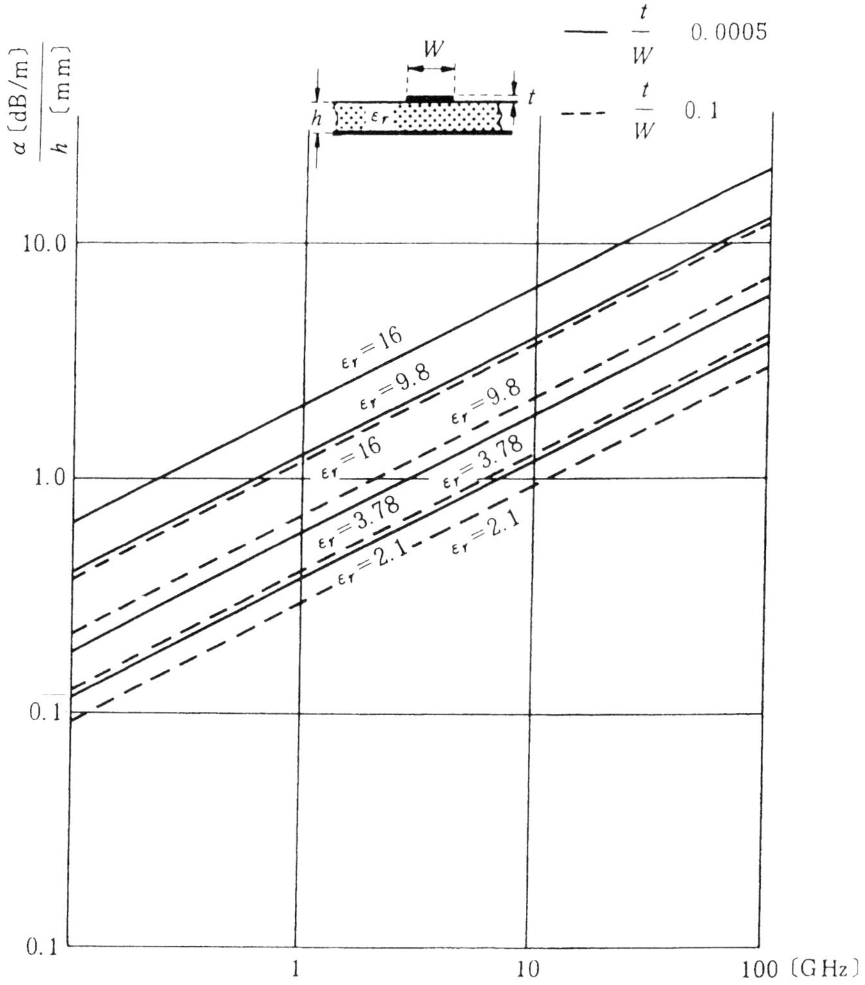

Figure 27. Attenuation of a microstrip line of 50 Ω caused by conductive loss.

with $\varepsilon_r = 2.6$ in the case of glass containing fluorocarbon. As another example, in the case of fluorocarbon polymers, we have the measured values of Q_ε of 1600–1400 in the frequency range below 2 GHz and 1400 in the range below 20 GHz with $\varepsilon_r = 2.0$.

Next, in the case of a microstrip line, the electric fields exist not only in the dielectric material but also in air. The effective relative dielectric constant, therefore, has the complex values (see Section 4.2).

$$\dot{\varepsilon}_{r,\text{eff}} = \varepsilon'_{r,\text{eff}} - j\varepsilon''_{r,\text{eff}} \tag{64}$$

$$\varepsilon'_{r,\text{eff}} = \frac{\varepsilon'_r + 1}{2} + \frac{\varepsilon'_r - 1}{2}\left(1 + \frac{10h}{W}\right)^{-1/2} \tag{65a}$$

$$\varepsilon''_{r,\text{eff}} = \frac{\varepsilon''_r}{2}\left[1 + \left(1 + \frac{10h}{W}\right)^{-1/2}\right] \tag{65b}$$

Using the definition

$$\frac{\varepsilon''_r}{\varepsilon'_r} = \tan\delta_\varepsilon = \frac{1}{Q_\varepsilon}$$

and Eq. (53), we get

$$\alpha_\varepsilon = \frac{\pi}{\lambda_g}\frac{\varepsilon'_r \tan\delta_\varepsilon[1 + (1 + 10h/W)^{-1/2}]}{\varepsilon'_r + 1 + (\varepsilon'_r - 1)(1 + 10h/W)^{-1/2}} \tag{66}$$

where we used the relation of $\lambda_g = \lambda$ because we consider the wave TEM mode. is filled in a guide such as coaxial line and waveguide, the Q values caused by dielectric loss, $Q_{\varepsilon,G}$, is calculated by

$$Q_{\varepsilon,G} = Q_\varepsilon \tag{67}$$

5.4 Radiation Loss

In the case when the guided line is not completely surrounded by metal, a part of the transmitted wave is radiated from the line, especially from the part where there exists a discontinuity, such as the bending part of a line, an open end, and a short end. For the radiation power P_{rad} and the incident power P_{in}, the ratio $P_{\text{rad}}/P_{\text{in}}$ of a microstrip line is obtained by

$$\frac{P_{\text{rad}}}{P_{\text{in}}} = 240\pi^2\left(\frac{h}{\lambda_0}\right)^2\frac{F}{Z_c} \tag{68}$$

where F takes the constant values corresponding to. In the case of a right-angle bend, F is calculated from

$$F \simeq \frac{4}{3\varepsilon_{r,\text{eff}}} \tag{69}$$

As for the radiation from microstrip line, it is described in Section 4.2.

APPENDIX: THE EIGENVALUES OF THE HELMHOLTZ EQUATION IN A WAVEGUIDE

Green's Theorem:

$$\iiint_v (\psi \nabla^2 \varphi + \nabla\psi \cdot \nabla\varphi)\, dv = \iint \psi \frac{d\varphi}{dn}\, ds \tag{A1}$$

In the case of the two-dimensional Green's theorem,

$$\iint_s (\psi \nabla_t^2 \varphi + \nabla_t\psi \cdot \nabla_t\varphi)\, ds = \oint \psi \frac{d\varphi}{dn}\, ds \tag{A2}$$

For

$$\psi = \varphi \tag{A3}$$

$$\iint_s (\nabla_t\varphi)^2\, ds = -\iint_s \varphi \nabla_t^2 \varphi \cdot ds + \oint \varphi \frac{\delta\varphi}{\delta n} \tag{A4}$$

The Helmholtz equation for the TM and TE modes are

$$\nabla_t^2 \varphi^e = -k_t^2 \varphi^e \tag{A5}$$

$$\nabla_t^2 \varphi^m = -k_t^2 \varphi^m \tag{A6}$$

Substituting Eqs. (5) and (6) into Eq. (4), we get

$$k_t^2 = \left(\frac{2\pi}{\lambda_c}\right)^2 = \frac{\displaystyle\iint_s \nabla_t^2 \varphi^{e,m}\, ds}{\displaystyle\iint_s (\varphi^{e,m})^2\, ds}$$

$$= \frac{\displaystyle\iint_s [(\delta\varphi^{e,m}/\delta x)^2 + (\delta\varphi^{e,m}/\delta y^2)]\, ds}{\displaystyle\iint_s (\varphi^{e,m})^2\, ds} \tag{A7}$$

because $\varphi^e = 0$ for the TM mode and $\delta\varphi^m/\delta n = 0$ for the TE mode.

If the quantity of Eq. (7) is always positive, then the transmission mode is always positive.

2

Basic Knowledge of Microwave Network Theory

1 CIRCUIT MATRIX AND THE CHARACTERISTICS

1.1 Impedance Matrix and Admittance Matrix

Putting the voltage and the current at the kth port of an n-port network as shown in Fig. 1, we can express the relationship between the voltage and the current of the network by the following equations:

$$\boldsymbol{V} = [Z]\boldsymbol{I} \tag{1a}$$

$$\boldsymbol{I} = [Y]\boldsymbol{V} \tag{1b}$$

$$\boldsymbol{V} = \begin{bmatrix} V_1 \\ V_2 \\ \vdots \\ V_n \end{bmatrix}, \qquad \boldsymbol{I} = \begin{bmatrix} I_1 \\ I_2 \\ \vdots \\ I_n \end{bmatrix} \tag{1c}$$

$$[Z] = \begin{bmatrix} Z_{11} & Z_{12} & \cdots & Z_{1n} \\ Z_{21} & Z_{22} & \cdots & Z_{2n} \\ \vdots & \vdots & & \vdots \\ Z_{n1} & Z_{n2} & \cdots & Z_{nn} \end{bmatrix} \tag{1d}$$

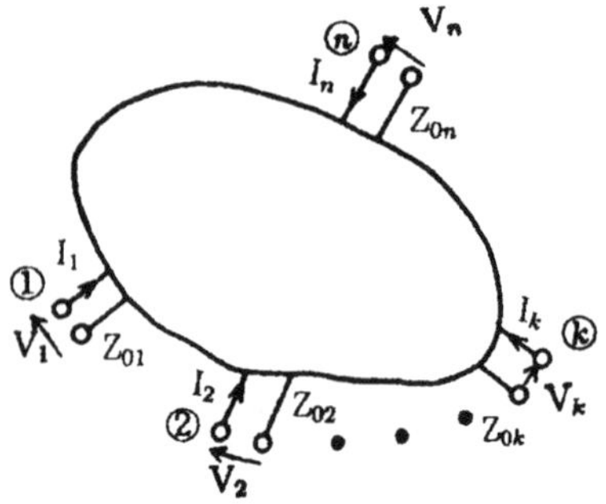

Figure 1. An n-port network and the voltage V_k and current I_k ($k = 1, 2, \ldots n$).

$$[Y] = \begin{bmatrix} Y_{11} & Y_{12} & \cdots & Y_{1n} \\ Y_{21} & Y_{22} & \cdots & Y_{2n} \\ \vdots & \vdots & & \vdots \\ Y_{n1} & Y_{n2} & \cdots & Y_{nn} \end{bmatrix} \tag{1e}$$

$[Z]$ in Eqs. (1a) and (1d) is called an impedance matrix, and $[Y]$ in Eqs. (1b) and (1e) is called an admittance matrix of the n-port network. $\boldsymbol{V}$ in Eqs. (1a) and (1e) expresses the vector of voltages, and $\boldsymbol{I}$ in Eqs. (1b) and (1c) is the vector of currents at each port.

$[Z]$ and $[Y]$ of a passive network have several important characteristics as shown in the following:

1. $[Z]$ and $[Y]$ can be shown by the Hermitian matrix (see Note 1 in the Appendix to this chapter)

$$[Z] = [Z_H] + j[Z'_H] \tag{2a}$$

$$[Y] = [Y_H] + j[Y'_H] \tag{2b}$$

where $[Z_H]$ and $[Y_H]$ represent the non-negative Hermitian matrix (see Note 2) or the positive Hermitian matrix for the lossy network (see Note 3) and $[Z_{H'}]$ and $[Y_{H'}]$ present the Hermitian matrix.

2. $[Z]$ and $[Y]$ of a reciprocal network are symmetric matrices:

$$Z_{ij} = Z_{ji} \tag{2c}$$

$$Y_{ij} = Y_{ji} \tag{2d}$$

where Z_{ij} and Y_{ij} are the ith row and jth column of $[Z]$ and $[Y]$, respectively.

3. $[Z]$ and $[Y]$ of a lossless network take the forms

$$[Z] = j[Z'_H] \tag{2e}$$

$$[Y] = j[Y'_H] \tag{2f}$$

and

$$[Z_{H'}]' = \frac{\partial}{\partial\omega}[Z_{H'}] \quad \text{and} \quad [Y_{H'}]' = \frac{\partial}{\partial\omega}[Y_{H'}]$$

are the Positive Hermitian Matrices (see Note 3).

4. $[Z]$ and $[Y]$ of a lossless reciprocal network take the form

$$[Z] = j[X] \tag{2g}$$

$$[Y] = j[B] \tag{2h}$$

where X and B are real symmetric matrices, and also from the condition of characteristic 3,

$$[X]' = \frac{\partial}{\partial\omega}[X] \quad \text{and} \quad [Y]' = \frac{\partial}{\partial\omega}[Y]$$

are the non-negative Hermitian Matrices. In the case of an ideal transformer network, the principal minor becomes zero.

1.2 Scattering Matrix

The voltage V_k and the current I_k at the kth port of an n-port network of Fig. 1 is defined by Eqs. (48a) and (48b) of Chapter 1. The values of V_k and I_k, however, can be expressed by the incident wave V_k^+ and I_k^+ and the reflected wave V_k^- and I_k^- as in Eqs. (3a) and (3b), similar to that in Eqs. (6b) and (6c) of Chapter 1.O

$$V_k = V_k^+ e^{-\gamma z} + V_k^- e^{\gamma z} \tag{3a}$$

$$I_k = \frac{V_k^+}{Z_{ok}} e^{-\gamma z} - \frac{V_k^-}{Z_{ok}} e^{\gamma z} \tag{3b}$$

where Z_{ok} is the wave impedance of the waveguide and also the characteristic impedance of the equivalent distributed lines of the waveguide.

Therefore, we can show the above relationship of voltage and current vectors as follows:

$$\boldsymbol{V} = \boldsymbol{V}^+ + \boldsymbol{V}^- \tag{4a}$$

$$\boldsymbol{I} = [Z_0]^{-1}(\boldsymbol{V}^+ - \boldsymbol{V}^-) = [Y_0](\boldsymbol{V}^+ - \boldsymbol{V}^-) \tag{4b}$$

$$[Z_0] = \begin{bmatrix} Z_{01} & & 0 \\ & \ddots & \\ 0 & & Z_{0n} \end{bmatrix}, \qquad [Y_0] = [Z_0]^{-1} \tag{4c}$$

$$\boldsymbol{V}^{\pm} = \begin{bmatrix} V_1^{\pm} \\ \vdots \\ V_n^{\pm} \end{bmatrix}, \qquad \boldsymbol{I}^{\pm} = \begin{bmatrix} I_1^{\pm} \\ \vdots \\ I_n^{\pm} \end{bmatrix} \tag{4d}$$

Because we have

$$\boldsymbol{V} = [Z]\boldsymbol{I}, \qquad \boldsymbol{I} = [Y]\boldsymbol{V} \tag{4e}$$

we obtain Eqs. (4f) and (4g) from Eqs. (4a), (4b), and (4d):

$$\boldsymbol{V}^{-} = [S_V]\boldsymbol{V}^{+} \tag{4f}$$

$$\begin{aligned} [S_V] &= ([Z] - [Z_0])([Z] + [Z_0])^{-1} \\ &= ([Y] + [Y_0])^{-1}([Y_0] - [Y]) \end{aligned} \tag{4g}$$

$[S_V]$ show the relation between the incident wave vector to a network and the reflected wave vector from the network, called the voltage scattering matrix.

When we consider the incident wave and reflected wave with amplitudes a_k and b_k, respectively, taking the values

$$a_k = \frac{1}{\sqrt{Z_{ok}}} V_k^{+}$$

$$b_k = \frac{1}{\sqrt{Z_{ok}}} V_k^{-} \tag{5a}$$

by using the vector relation

$$[\sqrt{Z_0}]\boldsymbol{a} = \boldsymbol{V}^{+}, \qquad [\sqrt{Z_0}]\boldsymbol{b} = \boldsymbol{V}^{-} \tag{5b}$$

$$\boldsymbol{a} = \begin{bmatrix} a_1 \\ \vdots \\ a_n \end{bmatrix}, \qquad \boldsymbol{b} = \begin{bmatrix} b_1 \\ \vdots \\ b_n \end{bmatrix} \tag{5c}$$

we get Eq. (5d) by substituting Eq. (5b) into Eq. (4f):

$$\boldsymbol{b} = [S]\boldsymbol{a} \tag{5d}$$

$$[S] = [\sqrt{Z_0}]^{-1}[S_V][\sqrt{Z_0}] \tag{5e}$$

For the case in which the characteristic impedance of the equivalent distributing lines of the waveguide connected to a network takes the same values, Eq. (5e) is replaced by (5f):

$$[S] = [S_V] \tag{5f}$$

In this case, Eqs. (4g) become Eqs. (5g):

$$[S] = ([Z'] - [I])([Z'] + [I])^{-1}$$
$$= ([Y'] + [I])^{-1}([I] - [Y'])$$
$$[Z'] = [Z_0]^{-1}[Z], \qquad [Y'] = [Z_0][Y]$$
$$\sqrt{Z_0} = \begin{bmatrix} \sqrt{Z_{01}} & & 0 \\ & \ddots & \\ 0 & & \sqrt{Z_{0n}} \end{bmatrix}$$
$$[I] = \begin{bmatrix} 1 & & 0 \\ & \ddots & \\ 0 & & 1 \end{bmatrix} \quad \text{(unit matrix)} \tag{5g}$$

The scattering matrix of a passive network has the following main characteristics:

1. $[I] - [\tilde{S}]^*[S]$ is a non-negative Hermitian matrix; it results in

   $$|[I] - [\tilde{S}]^*[S]| \geq 0 \tag{6a}$$

   This result is caused by the fact that the consumed power P_d in a passive network takes the values zero or positive, that is,

   $$P_d \geq 0 \tag{6b}$$

2. In the lossless network, we have the relation

   $$[\tilde{S}]^*[S] = [I] \tag{6c}$$

   which is called the unitary matrix. Equation (6c) is the case $P_d = 0$ in Eq. (6b).
3. In the case of an ideal transformer network,

   $$[S][S] = [I] \tag{6d}$$

4. In the case of the reciprocal network with the same characteristic impedance,

   $$[\tilde{S}] = [S], \qquad S_{ij} = S_{ji} \tag{6e}$$

   where S_{ij} is the ith row and jth column element of $[S]$.

1.3 Relationships Between Circuit Parameters of Typical Matrix of a Two-Port Network

Typical Matrix of a Two-Port Network

When we denote the values of voltages, currents, incident, wave and reflected wave as shown in Fig. 2, we have several matrices showing the

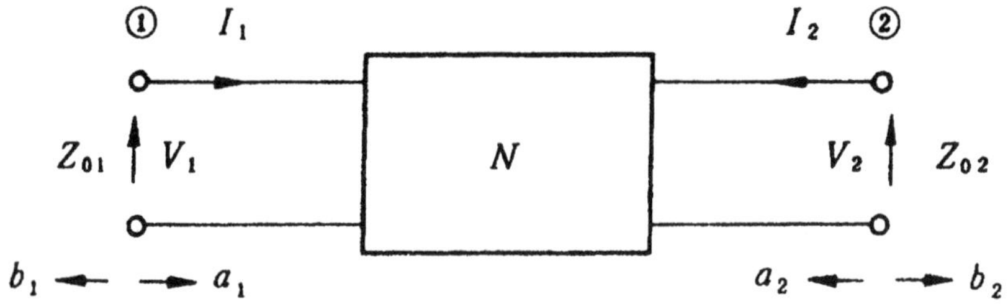

Figure 2. Voltages, currents, incident waves, and reflecting waves of a two-port network N.

relation between the values mentioned earlier. The matrix expressions are summarized in Table 1.

The parameters of several matrices shown in Table 1 can be related as follows:

$$F = \begin{bmatrix} A & B \\ C & D \end{bmatrix} = \frac{1}{Z_{21}} \begin{bmatrix} Z_{11} & |Z| \\ 1 & Z_{22} \end{bmatrix} \tag{7a}$$

$$Y = \begin{bmatrix} Y_{11} & Y_{12} \\ Y_{21} & Y_{22} \end{bmatrix} = \frac{1}{B} \begin{bmatrix} D & -|F| \\ -1 & A \end{bmatrix} \tag{7b}$$

Table 1. Several Matrices of a Two-Port Network.

F matrix	Impedance matrix	Admittance matrix
$\begin{bmatrix} V_1 \\ I_1 \end{bmatrix} = [F] \begin{bmatrix} V_2 \\ -I_2 \end{bmatrix}$	$\begin{bmatrix} V_1 \\ V_2 \end{bmatrix} = [Z] \begin{bmatrix} I_1 \\ I_2 \end{bmatrix}$	$\begin{bmatrix} I_1 \\ I_2 \end{bmatrix} = [Y] \begin{bmatrix} V_1 \\ V_2 \end{bmatrix}$
$[F] = \begin{bmatrix} A & B \\ C & D \end{bmatrix}$	$[Z] = \begin{bmatrix} Z_{11} & Z_{12} \\ Z_{21} & Z_{22} \end{bmatrix}$	$[Y] = \begin{bmatrix} Y_{11} & Y_{12} \\ Y_{21} & Y_{22} \end{bmatrix}$

H matrix	*G* matrix	Scattering matrix	Wave
$\begin{bmatrix} V_1 \\ I_2 \end{bmatrix} = [H] \begin{bmatrix} I_1 \\ V_2 \end{bmatrix}$	$\begin{bmatrix} I_1 \\ V_2 \end{bmatrix} = [G] \begin{bmatrix} V_1 \\ I_2 \end{bmatrix}$	$\begin{bmatrix} b_1 \\ b_2 \end{bmatrix} = [S] \begin{bmatrix} a_1 \\ a_2 \end{bmatrix}$	$\begin{bmatrix} b_1 \\ a_1 \end{bmatrix} = [W] \begin{bmatrix} a_2 \\ b_2 \end{bmatrix}$
$H = \begin{bmatrix} H_{11} & H_{12} \\ H_{21} & H_{22} \end{bmatrix}$	$G = \begin{bmatrix} G_{11} & G_{12} \\ G_{21} & G_{22} \end{bmatrix}$	$[S] = \begin{bmatrix} S_{11} & S_{12} \\ S_{21} & S_{22} \end{bmatrix}$	$[W] = \begin{bmatrix} r_{11} & r_{12} \\ r_{21} & r_{22} \end{bmatrix}$

$$Z = Y^{-1} = \frac{1}{C}\begin{bmatrix} A & |F| \\ 1 & D \end{bmatrix} \tag{7c}$$

$$H = \frac{1}{Y_{11}}\begin{bmatrix} 1 & -Y_{12} \\ Y_{21} & |Y| \end{bmatrix} \tag{7d}$$

$$G = H^{-1} = \frac{1}{Y_{22}}\begin{bmatrix} |Y| & Y_{12} \\ -Y_{21} & 1 \end{bmatrix} \tag{7e}$$

$$S = \begin{bmatrix} S_{11} & S_{12} \\ S_{21} & S_{22} \end{bmatrix} = \frac{1}{r_{22}}\begin{bmatrix} r_{12} & (r_{11}r_{22} - r_{12}r_{21}) \\ 1 & -r_{21} \end{bmatrix} \tag{7f}$$

$$W = \begin{bmatrix} r_{11} & r_{12} \\ r_{21} & r_{22} \end{bmatrix} = \frac{1}{S_{21}}\begin{bmatrix} -(S_{11}S_{22} - S_{12}S_{21}) & S_{11} \\ -S_{22} & 1 \end{bmatrix} \tag{7g}$$

$$\begin{bmatrix} S_{11} & S_{12} \\ S_{21} & S_{22} \end{bmatrix} = \left[\left(\frac{Z_{11}}{Z_{01}} + 1\right)\left(\frac{Z_{22}}{Z_{02}} + 1\right) - \frac{Z_{12}Z_{21}}{Z_{01}Z_{02}}\right]^{-1}$$
$$\cdot\begin{bmatrix} \left(\frac{Z_{11}}{Z_{01}} - 1\right)\left(\frac{Z_{22}}{Z_{02}} + 1\right) - \frac{Z_{12}Z_{21}}{Z_{01}Z_{02}} & 2\frac{Z_{12}}{Z_{02}} \\ 2\frac{Z_{21}}{Z_{01}} & \left(\frac{Z_{11}}{Z_{01}} + 1\right)\left(\frac{Z_{22}}{Z_{02}} - 1\right) - \frac{Z_{12}Z_{21}}{Z_{01}Z_{02}} \end{bmatrix} \tag{7h}$$

$$\begin{bmatrix} S_{11} & S_{12} \\ S_{21} & S_{22} \end{bmatrix} = \left[\left(1 + \frac{Y_{11}}{Y_{01}}\right)\left(1 + \frac{Y_{22}}{Y_{02}}\right) - \frac{Y_{12}Y_{21}}{Y_{01}Y_{02}}\right]^{-1}$$
$$\cdot\begin{bmatrix} \left(1 - \frac{Y_{11}}{Y_{01}}\right)\left(1 + \frac{Y_{22}}{Y_{02}}\right) + \frac{Y_{12}Y_{21}}{Y_{01}Y_{02}} & -2\frac{Y_{12}}{Y_{01}} \\ -2\frac{Y_{21}}{Y_{02}} & \left(1 + \frac{Y_{11}}{Y_{01}}\right)\left(1 - \frac{Y_{22}}{Y_{02}}\right) + \frac{Y_{12}Y_{21}}{Y_{01}Y_{02}} \end{bmatrix} \tag{7i}$$

$$\begin{bmatrix} S_{11} & S_{12} \\ S_{21} & S_{22} \end{bmatrix} = [(B + CZ_{01}Z_{02})(AZ_{02} + DZ_{01})]^{-1}$$
$$\cdot\begin{bmatrix} (B - CZ_{01}Z_{02}) + (AZ_{02} - DZ_{01}) & 2Z_{01}(AD - BC) \\ 2Z_{02} & (B - CZ_{01}Z_{02}) - (AZ_{02} - DZ_{01}) \end{bmatrix} \tag{7j}$$

$$\begin{bmatrix} Z_{11} & Z_{12} \\ Z_{21} & Z_{22} \end{bmatrix} = [(1 - S_{11})(1 - S_{22}) - S_{12}S_{21}]^{-1}$$
$$\cdot\begin{bmatrix} Z_{01}[(1 + S_{11})(1 - S_{22}) + S_{12}S_{21}] & 2Z_{01}Z_{21} \\ 2Z_{02}S_{12} & Z_{02}[(1 - S_{11})(1 + S_{22}) + S_{12}S_{21}] \end{bmatrix} \tag{7k}$$

$$\begin{bmatrix} Y_{11} & Y_{12} \\ Y_{21} & Y_{22} \end{bmatrix} = [(1 + S_{11})(1 + S_{22}) - S_{12}S_{21}]^{-1}$$

$$\begin{bmatrix} Y_{01}[(1 - S_{11})(1 + S_{22}) + S_{12}S_{21}] & -2Y_{01}S_{12} \\ -2Y_{02}S_{21} & Y_{02}[(1 + S_{11})(1 - S_{22}) + S_{12}S_{21}] \end{bmatrix} \tag{7l}$$

$$\begin{bmatrix} A & B \\ C & D \end{bmatrix} = \frac{1}{2S_{21}}$$

$$\begin{bmatrix} [(1 + S_{11})(1 - S_{22}) + S_{12}S_{21}] & Z_{02}[(1 + S_{11})(1 + S_{22}) - S_{12}S_{21}] \\ \dfrac{1}{Z_{01}}[(1 - S_{11})(1 - S_{22}) - S_{12}S_{21}] & \dfrac{Z_{01}}{Z_{02}}[(1 - S_{11})(1 + S_{22}) + S_{12}S_{21}] \end{bmatrix} \tag{7m}$$

$$\begin{bmatrix} A & B \\ C & D \end{bmatrix} = \frac{1}{2}\begin{bmatrix} [(r_{22} + r_{11}) + (r_{21} + r_{12})] & Z_{02}[(r_{22} - r_{11})(r_{21} - r_{12})] \\ \dfrac{1}{Z_{01}}[(r_{22} - r_{11}) + (r_{21} - r_{12})] & \dfrac{Z_{01}}{Z_{02}}[(r_{22} + r_{11})(r_{21} - r_{12})] \end{bmatrix} \tag{7n}$$

Connection of a Two-Port Network

Parallel Connection The admittance matrix Y of a network surrounded by a broken line (see Fig. 3) is

$$Y = Y' + Y'' \tag{8a}$$

Series Connection The impedance matrix Z of a network surrounded by a broken line (see Fig. 4) is

$$Z = Z' + Z'' \tag{8b}$$

Series and Parallel Connection of a Two-Port Network The H matrix of a network surrounded by a broken line (see Fig. 5) is

$$H = H' + H'' \tag{8c}$$

Parallel and Series Connection of a Two-Port Network The G matrix of a network surrounded by a broken line (see Fig. 6) is

$$G = G' + G'' \tag{8d}$$

Cascade Connection of a Two-Port Network The F matrix and W matrix of a network surrounded by a broken line (see Fig. 7) are

$$F = F'F'' \tag{8e}$$

$$W = W'W'' \tag{8f}$$

Admittance in Parallel and Impedance in Series Connected to a Two-Port Network

Matrices of a network surrounded by a broken line are shown in Eqs. (9a)–(9d) (see Figs. 8–11).

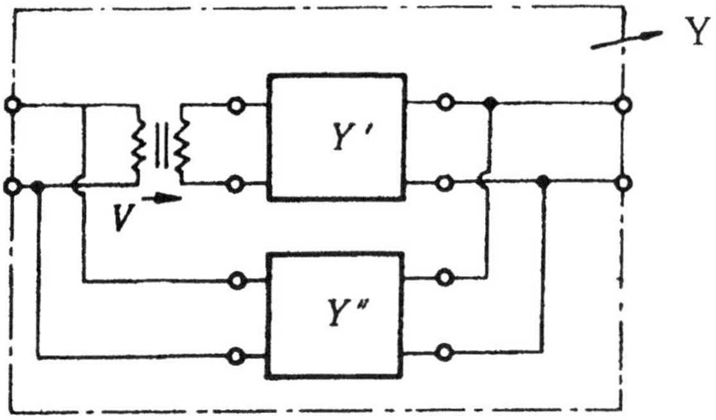

(a)

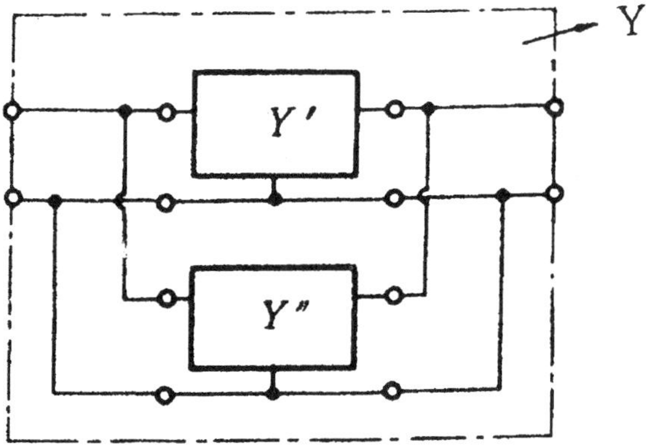

(b)

Figure 3. Parallel connection of a two-port network with admittance matrices Y' and Y''. (a) Parallel connection with ideal transformer; (b) parallel connection without ideal transformer.

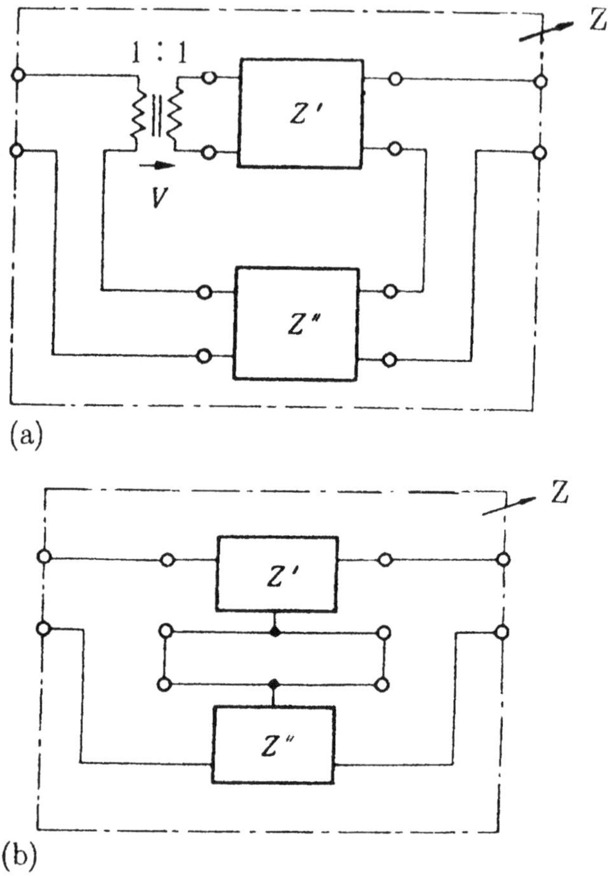

Figure 4. Series connection of a two-port network with impedance matrices Z' and Z''. (a) Series connection with ideal transformer; (b) series connection without ideal transformer.

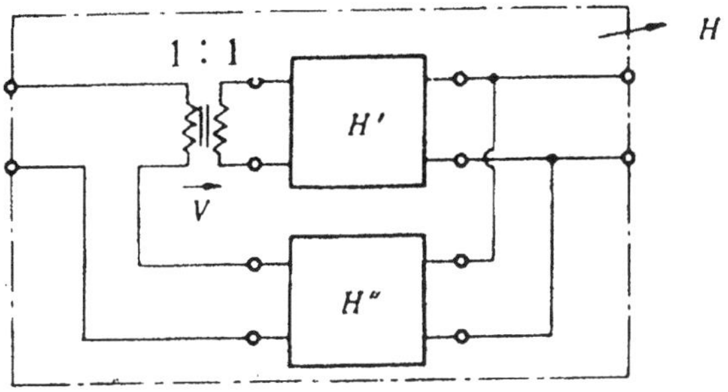

$$H = H' + H'' \qquad (8\text{-c})$$

Figure 5. Series and parallel connection.

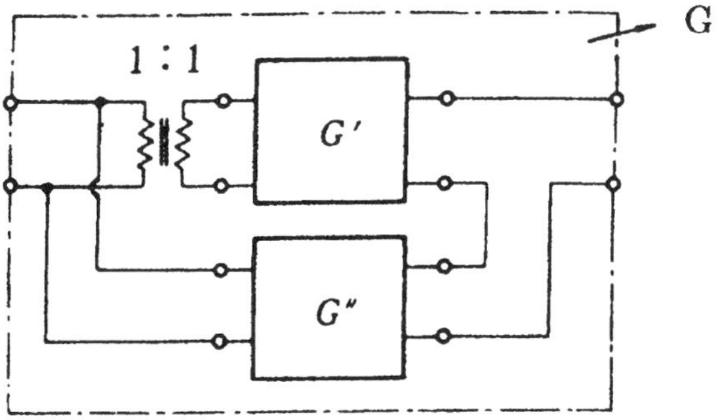

$$G = G' + G'' \qquad (8\text{-d})$$

Figure 6. Parallel and series connection.

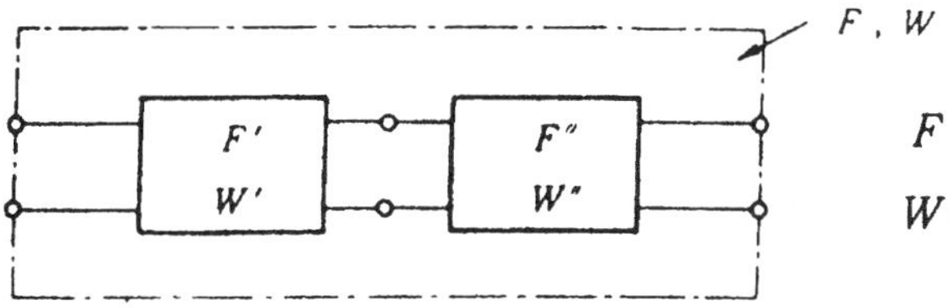

$$F = F'F'' \qquad (8\text{-e})$$

$$W = W'W'' \qquad (8\text{-f})$$

Figure 7. Cascade connection.

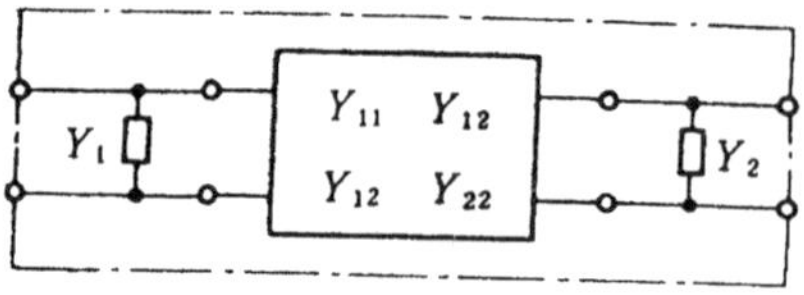

$$Y = \begin{bmatrix} Y_1 + Y_{11} & Y_{12} \\ Y_{12} & Y_2 + Y_{22} \end{bmatrix} \qquad (9\text{-a})$$

Figure 8. Admittances Y_1 and Y_2 are connected at input and output.

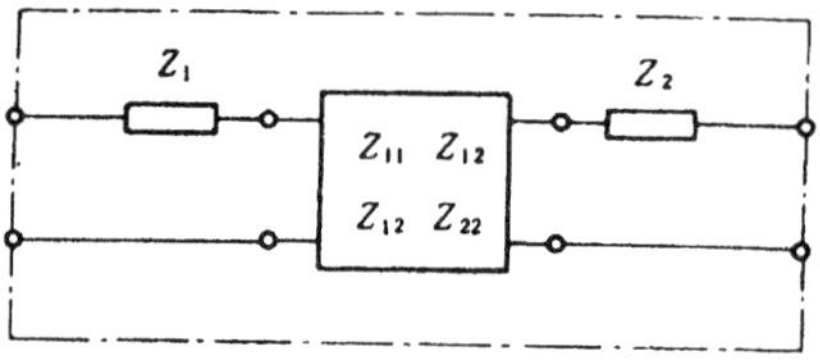

$$Z = \begin{bmatrix} Z_1 + Z_{11} & Z_{12} \\ Z_{12} & Z_2 + Z_{22} \end{bmatrix} \qquad (9\text{-b})$$

Figure 9. Impedances Z_1 and Z_2 are connected at input and output.

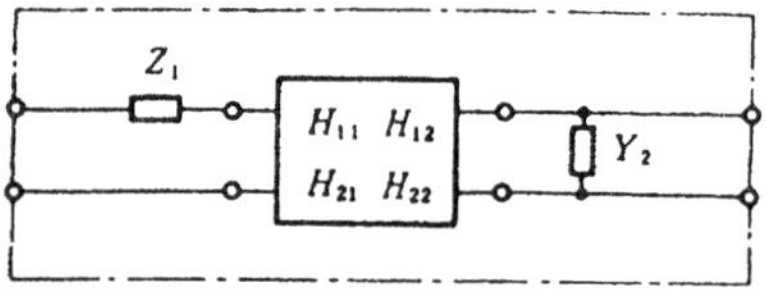

$$H = \begin{bmatrix} Z_1 + H_{11} & H_{12} \\ H_{21} & Y_2 + H_{22} \end{bmatrix} \qquad (9\text{-c})$$

Figure 10. Impedance Z_1 and admittance Y_2 are connected at input and output.

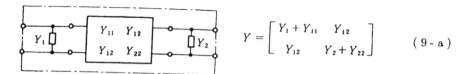

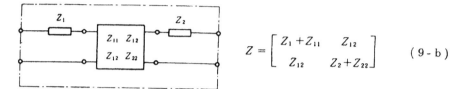

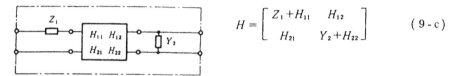

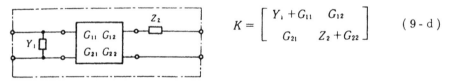

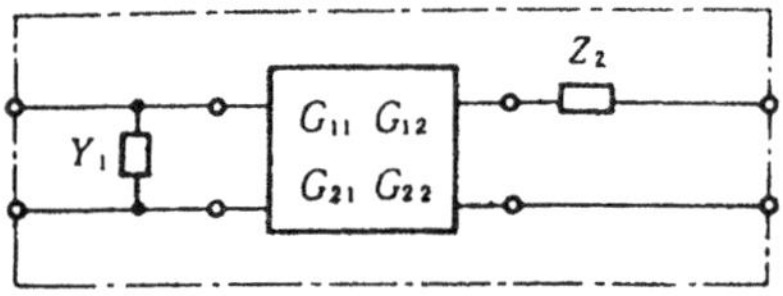

$$K = \begin{bmatrix} Y_1 + G_{11} & G_{12} \\ G_{21} & Z_2 + G_{22} \end{bmatrix} \qquad (9\text{-d})$$

Figure 11. Admittance Y_1 and impedance Z_2 are connected at input and output.

$$Y = \begin{bmatrix} Y_1 + Y_{11} & Y_{12} \\ Y_{12} & Y_2 + Y_{22} \end{bmatrix} \qquad (9a)$$

$$Z = \begin{bmatrix} Z_1 + Z_{11} & Z_{12} \\ Z_{12} & Z_2 + Z_{22} \end{bmatrix} \qquad (9b)$$

$$H = \begin{bmatrix} Z_1 + H_{11} & H_{12} \\ H_{21} & Y_2 + H_{22} \end{bmatrix} \qquad (9c)$$

$$K = \begin{bmatrix} Y_1 + G_{11} & G_{12} \\ G_{21} & Z_2 + G_{22} \end{bmatrix} \qquad (9d)$$

2 EIGENVECTORS AND EIGENVALUES OF A NETWORK MATRIX

2.1 Eigenvectors and Eigenvalues of *Z, Y, S* Matrix and the Physical Meaning

When the voltages $V_1^{(s)}, V_2^{(s)}, \ldots, V_n^{(s)}$ are supplied to ports 1, 2, . . . , n of the n-port network of Fig. 1, we assume the impedance at each port takes the same values of $z^{(s)}$. In this case, the currents $I_1^{(s)}, I_2^{(s)}, \ldots, I_n^{(s)}$ at ports 1, 2, . . . , n should take values satisfying Eq. (10a):

$$\frac{V_1^{(s)}}{I_1^{(s)}} = \frac{V_2^{(s)}}{I_2^{(s)}} = \cdots = \frac{V_n^{(s)}}{I_n^{(s)}} = z^{(s)} \tag{10a}$$

When we use the vector showing the voltage and the current at each terminal, we get

$$\boldsymbol{V}^{(s)} = z^{(s)}\boldsymbol{I}^{(s)} \quad (s = 1, 2, \ldots, n) \tag{10b}$$

$$\boldsymbol{V}^{(s)} = \begin{bmatrix} V_1^{(s)} \\ V_2^{(s)} \\ \vdots \\ V_n^{(s)} \end{bmatrix}, \qquad \boldsymbol{I}^{(s)} = \begin{bmatrix} I_1^{(s)} \\ I_2^{(s)} \\ \vdots \\ I_n^{(s)} \end{bmatrix} \tag{10c}$$

Using Eq. (1a), we get Eq. (11a):

$$[Z - z^{(s)}\boldsymbol{I}]\boldsymbol{I}^{(s)} = 0 \tag{11a}$$

where $\boldsymbol{I}$ is the unit matrix. Because $|\boldsymbol{I}| \neq 0$, we get the condition of

$$\begin{bmatrix} Z_{11} - z^{(s)} & Z_{12} & \cdots & Z_{1n} \\ Z_{21} & Z_{22} - z^{(s)} & & \vdots \\ \vdots & & \ddots & \vdots \\ Z_{n1} & \cdots & \cdots & Z_{nn} - z^{(s)} \end{bmatrix} = 0 \tag{11b}$$

Solving Eq. (11b), we get $z^1, z^2, \ldots, z^n$, which are called the eigenvalues of the Z matrix. Next, substituting the values of $z^{(s)}$ ($k = 1, 2, \ldots, n$) into Eq. (11a) again, we get Eq. (11c):

$$\begin{aligned} (Z_{11} - z^{(s)})I_1^{(s)} + Z_{12}I_2^{(s)} + \cdots + Z_{1n}I_n^{(s)} &= 0 \\ Z_{21}I_1^{(s)} + (Z_{22} - z^{(s)}) + \cdots + Z_{2n}I_n^{(s)} &= 0 \\ Z_{n1}I_1 + \cdots + (Z_{nn} - z^{(s)}I_n^{(s)}) &= 0 \end{aligned} \tag{11c}$$

From Eq. (11c), we can obtain the $n - 1$ sets of the relative values of the currents. For example, assuming the values of $I_1^{(s)}$, we get $I_2^{(s)}/I_1^{(s)}$, . . . , $I_n^{(s)}/I_1^{(s)}$. In other words, we get $\boldsymbol{I}^{(s)}$ of Eq. (10c), where only the relative values are defined. $\boldsymbol{I}^{(s)}$ is called the eigenvector of the Z matrix corresponding to the eigenvalue of $z^{(s)}$. When $z^{(s)}$ takes different values for the different s, we have n different eigenvectors.

When we have the i pieces of eigenvalues with the same values, that is,

$$z^{(1)} = z^{(2)} = \cdots = z^{(i)} = z_0 \tag{11d}$$

such an eigenvalues is called degenerate. In this case, denoting i pieces of eigenvector corresponding to the same eigenvalues, $\boldsymbol{U}^{(1)}, \boldsymbol{U}^{(2)}, \ldots, \boldsymbol{U}^{(i)}$, we must have the relation

$$\begin{aligned} &[Z - z_0 I]\boldsymbol{U}^{(1)} = 0 \\ &\qquad \vdots \\ &[Z - z_0 I]\boldsymbol{U}^{(i)} = 0 \end{aligned} \tag{11e}$$

Therefore, a new eigenvector $\boldsymbol{U}$ of

$$m_1\boldsymbol{U}^{(1)} + m_2\boldsymbol{U}^{(2)} + \cdots + m_i\boldsymbol{U}^{(i)} = \boldsymbol{U}$$

can also be the eigenvector corresponding to z_0. For example, the three-port reciprocal network with rotational symmetry, as shown in Fig. 12, has a Z matrix as in Eq. (12a) and eigenvectors as in Eq. (12b), which will be described in Eq. (17k).

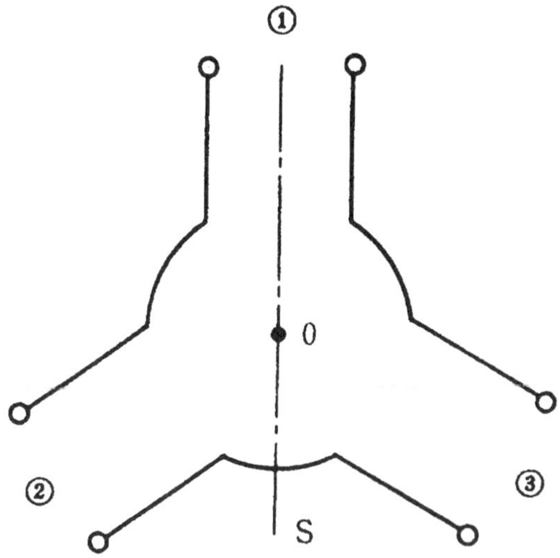

Figure 12. Three-port network with rotationally symmetry.

$$Z = \begin{bmatrix} Z_{11} & Z_{12} & Z_{12} \\ Z_{12} & Z_{11} & Z_{12} \\ Z_{12} & Z_{12} & Z_{11} \end{bmatrix} \tag{12a}$$

$$\boldsymbol{U}^{(1)} = \frac{1}{\sqrt{3}} = \begin{bmatrix} 1 \\ 1 \\ 1 \end{bmatrix}, \qquad \boldsymbol{U}^{(2)} = \frac{1}{\sqrt{3}} = \begin{bmatrix} 1 \\ \alpha^2 \\ \alpha \end{bmatrix} \tag{12b}$$

$$\boldsymbol{U}^{(3)} = \frac{1}{\sqrt{3}} = \begin{bmatrix} 1 \\ \alpha \\ \alpha^2 \end{bmatrix}$$

$$\alpha = \exp\left(j\,\frac{2\pi}{3}\right)$$

where the eigenvector is normalized to $|\boldsymbol{u}^{(s)}| = 1$.

Using the matrix U of Eq. (12c),

$$U = [\boldsymbol{U}^{(1)} \quad \boldsymbol{U}^{(2)} \quad \boldsymbol{U}^{(3)}] \tag{12c}$$

we get Eq. (12d) (see Note 5):

$$Z = U\Lambda_z U^{-1}, \qquad \Lambda = \begin{bmatrix} z^{(1)} & & 0 \\ & z^{(2)} & \\ 0 & & z^{(3)} \end{bmatrix} \tag{12d}$$

From Eqs. (12b) and (12d), we get

$$\begin{aligned} Z_{11} &= \tfrac{1}{3}(z^{(1)} + z^{(2)} + z^{(3)}) \\ Z_{12} &= \tfrac{1}{3}(z^{(1)} + \alpha z^{(2)} + \alpha^2 z^{(3)}) \\ Z_{13} &= \tfrac{1}{3}(z^{(1)} + \alpha^2 z^{(2)} + \alpha z^{(3)}) \end{aligned} \tag{12e}$$

Because $Z_{12} = Z_{13}$ from Eq. (12a), we get

$$z^{(2)} = z^{(3)} \tag{12f}$$

Therefore, we get the new eigenvector of the following equations:

$$\boldsymbol{U}^{(2)\prime} = \frac{1}{\sqrt{2}}\,\boldsymbol{U}^{(2)} + \frac{1}{\sqrt{2}}\,\boldsymbol{U}^{(3)} = \frac{1}{\sqrt{2}} \begin{bmatrix} 2 \\ -1 \\ -1 \end{bmatrix} \tag{12g}$$

$$\boldsymbol{U}^{(3)\prime} = \frac{j}{\sqrt{2}}\,\boldsymbol{U}^{(2)} - \frac{j}{\sqrt{2}}\,\boldsymbol{U}^{(3)} = \frac{1}{\sqrt{2}} \begin{bmatrix} 0 \\ 1 \\ -1 \end{bmatrix} \tag{12h}$$

From $\boldsymbol{U}^{(1)}$ of Eq. (12b), $\boldsymbol{U}^{(2)\prime}$ of Eq. (12g), and $\boldsymbol{U}^{(3)\prime}$ of Eq. (12h), we get a new matrix consisting of the eigenvectors

$$U' = [\boldsymbol{U}^{(1)} \quad \boldsymbol{U}^{(2)} \quad \boldsymbol{U}^{(3)}]$$

$$= \begin{bmatrix} 1 & 2 & 0 \\ 1 & -1 & 1 \\ 1 & -1 & -1 \end{bmatrix} \tag{12i}$$

For the excitation corresponding to $\boldsymbol{U}^{(3)'}$, we have the electrical wall at the symmetrical plane S, as shown in Fig. 13b. It is understood that $z^{(2)}$ $(= z^{(3)})$ can be obtained from the impedance at port 2 of Fig. 13b.

Next, we will discuss the Y matrix and S matrix. We, however, need the mathematical theorem as follows (see Note 6). In a rational function of square matrix M,

$$f(M) = C_0(M - C_1I)(M - C_2I) \cdots (M - C_{-1}I)^{-1}(M - C_{-2}I)^{-1} \tag{13a}$$

where I is the unit matrix, the eigenvectors of $f(M)$ are the same to those of M and the eigenvalues $\lambda_F^{(s)}$ are obtained as

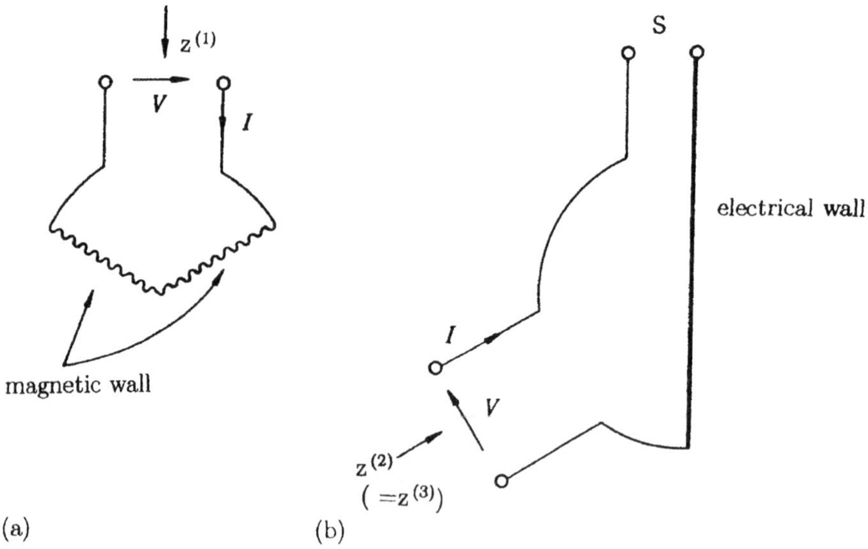

Figure 13. Equivalent impedance of eigenvalues $z^{(1)}$ and $z^{(2)} = z^{(3)}$. (a) $z^{(1)}$; (b) $z^{(2)} = z^{(3)}$.

$$\lambda_F^{(s)} = C_0(\lambda^{(s)} - C_2)(\lambda^{(s)} - C_2) \cdots (\lambda^{(s)} - C_{-1})^{-1}(\lambda^{(s)} - C_{-2})^{-1} \tag{13b}$$

However, from Eqs. (4c) and (5g), we have the relation

$$Y = Z^{-1}$$

$$S = (Z' + I)^{-1}(Z' - I), \qquad Z' = \frac{Z}{Z_0} \tag{14a}$$

where Z_0 is the characteristic impedance.

Therefore, the eigenvectors of the Y and S matrices, are the same as that of the Z matrix. Eigenvalues of the Y and S matrices are also obtained as

$$y^{(s)} = \frac{1}{z^{(s)}}, \qquad s^{(s)} = \frac{z'^{(s)} - 1}{z'^{(s)} + 1} = \frac{1 - y'^{(s)}}{1 + y'^{(s)}}$$

$$z'^{(s)} = \frac{z^{(s)}}{Z_0}, \qquad y'^{(s)} = \frac{y^{(s)}}{Y_0} \tag{14b}$$

where Y_0 is the characteristic admittance (= $1/Z_0$).

Generally, we have to solve Eq. (11c) to get the eigenvalues. Next, we can obtain eigenvectors from Eq. (11c). In microwave circuits, however, we frequently encounter the symmetrical construction. In this case, we can obtain the eigenvector simply. By using the obtained eigenvectors, we can get the eigenvalues from Eq. (11c).

We will introduce the method to obtain eigenvectors for the symmetrical structure. We have the theorem as follows (see Note 7).

When square matrices P and Q are satisfied, the equation

$$PQ = QP \tag{15}$$

the eigenvector of P, $\boldsymbol{u}^{(s)}$ is also the eigenvector of Q, where $\boldsymbol{u}^{(s)}$ is the eigenvector corresponding to the nondegenerated eigenvalues.

Example 1. Eigenvectors of a Symmetrical Two-Port Network

The Z matrix of a symmetrical two-port network is not changed by interchanging port 1 and port 2 in Fig. 14.

When we represent voltage and current vectors in the case of Figs. 14a and 14b by $\boldsymbol{v}$, $\boldsymbol{i}$ and $\boldsymbol{v}'$, $\boldsymbol{i}'$, respectively, we get

$$\boldsymbol{v} = [Z]\boldsymbol{i}, \qquad \boldsymbol{v} = \begin{bmatrix} v_1 \\ v_2 \end{bmatrix}, \qquad \boldsymbol{i} = \begin{bmatrix} i_1 \\ i_2 \end{bmatrix} \tag{16a}$$

$$\boldsymbol{v}' = [Z]\boldsymbol{i}', \qquad \boldsymbol{v}' = \begin{bmatrix} v_2 \\ v_1 \end{bmatrix}, \qquad \boldsymbol{i}' = \begin{bmatrix} i_2 \\ i_1 \end{bmatrix} \tag{16b}$$

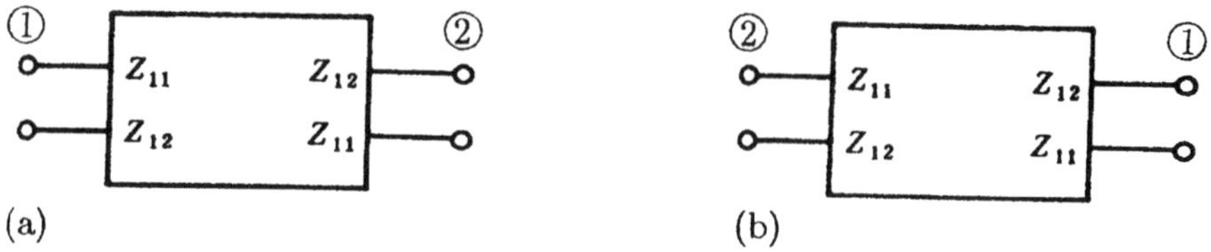

Figure 14. Symmetrical two-port network. (a) Symmetrical network; (b) network with interchanged ports 1 and 2.

$\boldsymbol{v}$, $\boldsymbol{v}'$, $\boldsymbol{i}$, and $\boldsymbol{i}'$ are related by $[F]$:

$$[F]\boldsymbol{v} = \boldsymbol{v}', \qquad [F]\boldsymbol{i} = \boldsymbol{i}', \qquad [F] = \begin{bmatrix} 0 & 1 \\ 1 & 0 \end{bmatrix} \tag{16c}$$

Substituting Eqs. (16c) and (16a) into Eq. (16b), we get

$$[F][Z]\boldsymbol{i} = [Z][F]\boldsymbol{i}$$
$$[F][Z] = [Z][F] \tag{16d}$$

Therefore, the eigenvectors of $[Z]$ are obtained from the eigenvectors of $[F]$. Denoting the eigenvalues of $[F]$ by f^1 and f^2, from $|[F] - f^i[I]| = 0$, we get

$$f^1 = 1, \qquad f^2 = -1 \tag{16e}$$

Substituting the eigenvalues of Eq. (16e) into

$$[F]\boldsymbol{u}^i = f^i[I]\boldsymbol{u}^i = 0$$

or

$$-f^i u_1^i + u_2^i = 0$$
$$u_1^i - f u_2^i = 0 \quad (i = 1, 2)$$

we get

$$\boldsymbol{u}^1 = \begin{bmatrix} \dfrac{1}{\sqrt{2}} \\ \dfrac{1}{\sqrt{2}} \end{bmatrix}, \qquad \boldsymbol{u}^2 = \begin{bmatrix} \dfrac{1}{\sqrt{2}} \\ -\dfrac{1}{\sqrt{2}} \end{bmatrix} \tag{16f}$$

Equation (16f) gives the eigenvectors not only of the matrix Z but also the $[Y]$ and $[S]$ matrices from Eqs. (13a) and (14a).

Example 2. Eigenvectors of a Three-Port Network with Rotational Symmetry

The circuit matrix of the Y junction with the rotational symmetrical construction as shown in Fig. 15 does not change even if the each terminal is named as 1, 2, 3, or 2, 3, 1 or 3, 2, 1.

Using the matrix

$$[R] = \begin{bmatrix} 0 & 0 & 1 \\ 1 & 0 & 0 \\ 0 & 1 & 0 \end{bmatrix} \tag{17a}$$

the new voltage and current vectors $\boldsymbol{v}'$ and $\boldsymbol{i}'$ are expressed as

$$\begin{aligned} [R]\boldsymbol{i} &= \boldsymbol{i}' \\ [R]\boldsymbol{v} &= \boldsymbol{v}' \end{aligned} \tag{17b}$$

Because we have the relation

$$\begin{aligned} \boldsymbol{v}' &= [Z]\boldsymbol{i}' \\ \boldsymbol{v} &= [Z]\boldsymbol{i} \end{aligned} \tag{17c}$$

we obtain Eq. (17d) from Eqs. (17b) and (17c):

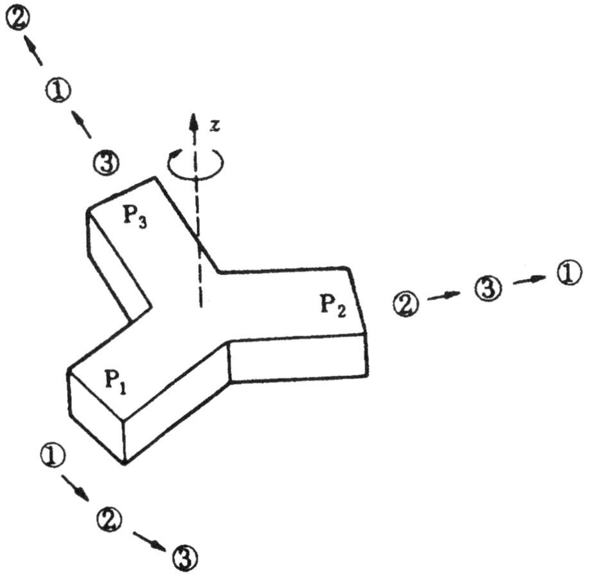

Figure 15. *H*-plane *Y* junction with the rotationally symmetrical construction.

$$[R][Z] = [Z][R] \tag{17d}$$

As Eq. (17d) is the same relation as Eq. (16d), the eigenvectors of $[Z]$ are same as those of $[R]$.

To obtain the eigenvectors of $[R]$, we consider the following relation:

$$[R]^3 = \begin{bmatrix} 0 & 0 & 1 \\ 1 & 0 & 0 \\ 0 & 1 & 0 \end{bmatrix} \begin{bmatrix} 0 & 0 & 1 \\ 1 & 0 & 0 \\ 0 & 1 & 0 \end{bmatrix} \begin{bmatrix} 0 & 0 & 1 \\ 1 & 0 & 0 \\ 0 & 1 & 0 \end{bmatrix}$$

$$= \begin{bmatrix} 0 & 0 & 1 \\ 1 & 0 & 0 \\ 0 & 1 & 0 \end{bmatrix} \begin{bmatrix} 0 & 1 & 0 \\ 0 & 0 & 1 \\ 1 & 0 & 0 \end{bmatrix} = \begin{bmatrix} 1 & 0 & 0 \\ 0 & 1 & 0 \\ 0 & 0 & 1 \end{bmatrix} = [I]$$

We, however, get Eq. (17d) by substituting $[R]$ into $[M]$, $[I]$ into $F[M]$, and 0 for C_k in Eq. (13a). Representing the eigenvalues of $[I]$ and $[R]$ as $\boldsymbol{i}$ and $\boldsymbol{r}^{(s)}$, we get

$$i = (r^{(s)})^3 \tag{17e}$$

Because values of $\boldsymbol{i}$, however, are the solution of

$$|[I] - i[I]| = 0$$

and

$$\begin{bmatrix} 1-i & 0 & 0 \\ 0 & 1-i & 0 \\ 0 & 0 & 1-i \end{bmatrix} = 0$$

we get $i = 1$. Substituting these values into Eq. (17e), we have

$$(r^{(s)})^3 = 1$$

To satisfy the above equation, $r^{(s)}$ can be obtained as

$$r^{(1)} = 1$$

$$r^{(2)} = \alpha$$

$$r^{(3)} = \alpha^2, \qquad \alpha = \exp\left(j\frac{2\pi}{3}\right) \tag{17f}$$

Inserting the eigenvalues of Eq. (17f) into

$$([R] - r[I])\boldsymbol{u} = 0 \tag{17g}$$

we obtain $u^{(s)}$. For example, substituting $r^{(1)} = 1$ into Eq. (17g), we get

$$u_1^{(1)} + 0u_2^{(1)} + u_3^{(1)} = 0, \qquad u_1^{(1)} - u_2^{(1)} + u_3^{(1)} = 0$$

and

$$u_1^{(1)} = u_2^{(1)} = u_3^{(1)} = 1 \tag{17h}$$

Next, substituting $r^{(2)} = \alpha$ into Eq. (17g), we get

$$-\alpha u_1^{(2)} + 0u_2^{(2)} + u_3^{(2)} = 0, \qquad u_1^{(2)} - \alpha u_2^{(2)} + 0u_3^{(2)} = 0$$

then

$$u_1^{(2)} = 1, \qquad u_2^{(2)} = \alpha^2, \qquad u_3^{(2)} = \alpha \tag{17i}$$

Finally, substituting $r^{(3)} = \alpha^2$ into Eq. (17g), we get

$$-\alpha^2 u_1^{(3)} + 0u_2^{(3)} + u_3^{(3)} = 0, \qquad u_1^3 - \alpha^2 u_2^2 + 0u_3^3 = 0$$

then

$$u_1^{(3)} = 1, \qquad u_2^{(3)} = \alpha, \qquad u_3^{(3)} = \alpha^2 \tag{17j}$$

Summarizing the results of Eqs. (17h), (17i), and (17j), we obtain the following eigenvectors:

$$\boldsymbol{u}^{(1)} = \begin{bmatrix} 1 \\ 1 \\ 1 \end{bmatrix}, \qquad \boldsymbol{u}^{(2)} = \begin{bmatrix} 1 \\ \alpha^2 \\ \alpha \end{bmatrix}, \qquad \boldsymbol{u}^{(3)} = \begin{bmatrix} 1 \\ \alpha \\ \alpha^2 \end{bmatrix} \tag{17k}$$

These results were already used in Eq. (12b) to obtain the eigenvalues of a three-port network with rotationally symmetric construction. In Eq. (12b), the eigenvectors are normalized, as its length is unitary where the constant $1/\sqrt{3}$ is multiplied by Eq. (17k).

2.2 Eigenvalues of *Z* and *Y* Matrices Expressed by Energies in a Network

The tangential component of the electrical field and the magnetic field at the *m*th port of a *n*-port network, $\boldsymbol{E}_{tm}$ and $\boldsymbol{H}_{tm}$, respectively, are expressed by using the mode voltage V_m and the mode current I_m together with the mode vectors $\boldsymbol{e}_{tm}$ and $\boldsymbol{h}_{tm}$ as follows:

$$\boldsymbol{E}_{tm} = V_m \boldsymbol{e}_{tm}$$

$$\boldsymbol{H}_{tm} = I_m \boldsymbol{h}_{tm} \tag{18a}$$

Therefore, the complex power $\dot{P}$ flowing toward the network is obtained as Eq. (18b) taking into consideration Eq. (48c) of Chapter 1:

$$\dot{P} = \sum_{m=1}^{n} I_m^* V_m = \tilde{I}^* V \tag{18b}$$

$\dot{P}$, however, can be expressed by the consumed power in the network, P_R, the time average of the magnetic energy $\tilde{W}_m$, and the electrical energy $\tilde{W}_e$ in the network:

$$\dot{P} = P_R + 2j\omega(\tilde{W}'_m - \tilde{W}'_e) \tag{18c}$$

If V and I are the sth eigenvectors $V^{(s)}$ and $I^{(s)}$, respectively, by using the relation $V^{(s)} = [Z]I^{(s)} = z^{(s)}I^{(s)}$, we get

$$\dot{P} = \tilde{I}^* z^{(s)} I^{(s)} = P_R + 2j\omega(\tilde{W}'_m - \tilde{W}'_e)$$

We, therefore, obtain the expression of the sth eigenvalues of the Z matrix as

$$z^{(s)} = \frac{P_R + 2j\omega(\tilde{W}_m - \tilde{W}_e)}{\sum_{m=1}^{n} |I_m^{(s)}|^2} \tag{18d}$$

When we normalize I_m as

$$\sum_{m=1}^{n} |I_m^{(s)}|^2 = 1 \tag{18e}$$

we get

$$z^{(s)} = P_R + 2j\omega(\tilde{W}_m - \tilde{W}_e) \tag{18f}$$

In the lossless network, because $P_R = 0$, $Z^{(s)}$ becomes pure imaginary and takes the values

$$z^{(s)} = jX^{(s)}, \qquad X^{(s)} = 2\omega(\tilde{W}_m - \tilde{W}_e) \tag{18g}$$

Next, the sth eigenvalues of the Y matrix, $y^{(s)}$, can be obtained in the same way:

$$y^{(s)} = \frac{P_R + 2j\omega(\tilde{W}_e - \tilde{W}_m)}{\sum_{m=1}^{n} |V_m|^2} \tag{19a}$$

For the case when $V_m^{(s)}$ is normalized as

$$\sum_{m=1}^{n} |V_m^{(s)}|^2 = 1 \tag{19b}$$

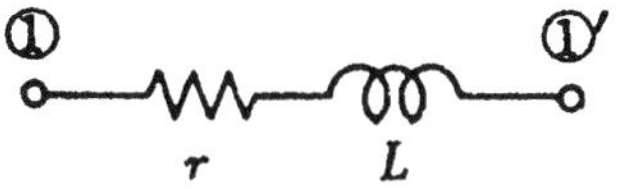

Figure 16. r,L series connection.

$y^{(s)}$ can be expressed by

$$y^{(s)} = P_R + 2j\omega(\tilde{W}_e - \tilde{W}_m) \tag{19c}$$

In the lossless network, we get

$$y^{(s)} = jB^{(s)}, \qquad B^{(s)} = 2\omega(\tilde{W}_e - \tilde{W}_m) \tag{19d}$$

Example

In the case of $n = 1$, $z^{(s)}$ becomes the impedance of a one-port or two-terminal network. For example, considering the impedance of an r,L series connection as shown in Fig. 16, we obtain.

$$P_r = |I|^2 r, \qquad \tilde{W}_m = \frac{L|I|^2}{2}, \qquad \tilde{W}_e = 0$$

Substituting these values into Eq. (18d), we get

$$z = \frac{|I|^2 r + 2j\omega(L|I|^2/2)}{|I|^2} = r + j\omega L \tag{19e}$$

This is the well-known result.

2.3 Relation Between the Frequency Variation of Eigenvalues of the *Z* and *Y* Matrices and the Reactive Energies in a Lossless Circuit

When each port of a lossless n-port network is excited by voltages $V_1^{(s)}, V_2^{(s)}, \ldots, V_n^{(s)}$ corresponding to the sth eigenvector, we denote the time average of the total reactive energies included in the lossless circuit by $\tilde{W}_t^{(s)}$. In this case, we can show that the frequency variation of the sth eigenvalues of the Y matrix is proportional to $\tilde{W}_t^{(s)}$ (see Note 9):

$$\frac{\partial y}{\partial \omega} = j\frac{2(\tilde{W}_e^{(s)} + \tilde{W}_m^{(s)})}{\sum_{m=1}^{n} |V_m^{(s)}|^2} = j\frac{2\tilde{W}_t^{(s)}}{\sum_{m=1}^{n} |V_m^{(s)}|^2} \tag{20a}$$

In the case of

$$\sum_{m=1}^{n} |V_m^{(s)}|^2 = 1 \tag{20b}$$

we get

$$\frac{\partial y^{(s)}}{\partial \omega} = 2j\tilde{W}_t^{(s)} \tag{20c}$$

$$\delta y^{(s)} = 2j\delta\omega\tilde{W}_t^{(s)} \tag{20d}$$

In the same way, when each port is excited by currents $I_1^{(s)}, I_2^{(s)}, \ldots, I_n^{(s)}$ corresponding to the sth eigenvector, we denote the time average of the reactive energies included in a network by $\tilde{W}_t'^{(s)}$.

In this case, the frequency variation of the sth eigenvalues of the Z matrix is proportional to $\tilde{W}_t^{(s)}$:

$$\frac{\partial z^i}{\partial \omega} = j\frac{2(\tilde{W}_e'^i + \tilde{W}_m'^i)}{\sum_{m=1}^{n} |I_m^i|^2} = j\frac{2\tilde{W}_t'}{\sum_{m=1}^{n} |I_m^i|^2} \tag{21a}$$

For the case

$$\sum_{m=1}^{n} |I_m^{(s)}|^2 = 1 \tag{21b}$$

we get

$$\frac{\partial z^{(s)}}{\partial \omega} = 2j\tilde{W}_t'^{(s)} \tag{21c}$$

$$\delta z^{(s)} = 2j\delta\omega\tilde{W}_t'^{(s)} \tag{21d}$$

In the case $n = 1$, the frequency variation of the admittance and impedance of a given two-terminal network are proportional to the time average of the total reactive energies for a 1-V excitation and a 1-A current excitation respectively, that is,

$$\delta y = 2j\delta\omega(\tilde{W}_m + \tilde{W}_e) = 2j\delta\omega\tilde{W}_t \tag{22a}$$

$$\delta z = 2j\delta\omega(\tilde{W}_m' + \tilde{W}_e') = 2j\delta\omega\tilde{W}_t' \tag{22b}$$

The results are important for estimating the slope of the reactance corresponding to the variation of frequency.

3 RECIPROCAL CIRCUIT AND NONRECIPROCAL CIRCUIT

The circuit is called a reciprocal circuit when

$$Z_{ij} = Z_{ji}, \qquad Y_{ij} = Y_{ji} \tag{23}$$

and it is called a nonreciprocal circuit when

$$Z_{ij} \neq Z_{ji}, \qquad Y_{ij} \neq Y_{ji} \tag{24}$$

We consider the electric and magnetic fields corresponding to the two cases in Fig. 17:

First case

$$I_i = 1, \qquad K_k = 0 \quad (k \neq i) \tag{25a}$$

second case

$$I_j = 1, \qquad I_k = 0 \quad (k \neq j) \tag{25b}$$

Expressing the electric and magnetic fields of the first and second cases by $\boldsymbol{E}_i$ and $\boldsymbol{H}_i$, and $\boldsymbol{E}_j$ and $\boldsymbol{H}_j$, respectively, Z_{ij} and Z_{ji} can be calculated from

$$Z_{ij} = j\omega \iiint_\tau (\boldsymbol{H}_i \cdot \hat{\mu} \boldsymbol{H}_j + \boldsymbol{E}_j \cdot \hat{\varepsilon} \boldsymbol{E}_i)\, d\tau \tag{26a}$$

$$\begin{aligned} Z_{ji} &= j\omega \iiint_\tau (\boldsymbol{H}_j \cdot \hat{\mu} \boldsymbol{H}_i + \boldsymbol{E}_i \cdot \hat{\varepsilon} \boldsymbol{E}_j)\, d\tau \\ &= j\omega \iiint_\tau (\boldsymbol{H}_i \cdot \tilde{\mu}^t \boldsymbol{H}_j + \boldsymbol{E}_j \cdot \tilde{\varepsilon}^t \boldsymbol{E}_i)\, d\tau \end{aligned} \tag{26b}$$

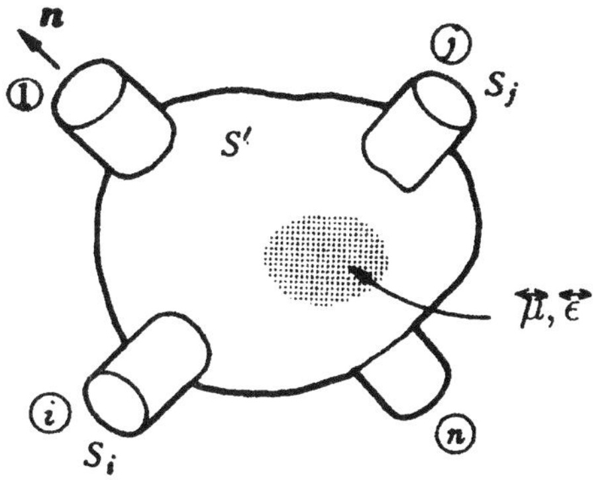

Figure 17. Circuit including medium with $\overleftrightarrow{\mu}$, $\overleftrightarrow{\varepsilon}$.

$$Z_{ij} - Z_{ji} = j\omega \iiint_{\tau} \boldsymbol{H}_i(\hat{\mu} - \hat{\mu}')\boldsymbol{H}_j + \boldsymbol{E}_i(\hat{\varepsilon} - \hat{\varepsilon}')\boldsymbol{E}_i \, d\tau \tag{27a}$$

$$\hat{\mu} = \begin{bmatrix} \mu_{11} & \mu_{12} & \mu_{13} \\ \mu_{21} & \mu_{22} & \mu_{23} \\ \mu_{31} & \mu_{32} & \mu_{33} \end{bmatrix}, \qquad \boldsymbol{B} = \mu \boldsymbol{H} \tag{27b}$$

$$\hat{\varepsilon} = \begin{bmatrix} \varepsilon_{11} & \varepsilon_{12} & \varepsilon_{13} \\ \varepsilon_{21} & \varepsilon_{22} & \varepsilon_{23} \\ \varepsilon_{31} & \varepsilon_{32} & \varepsilon_{33} \end{bmatrix}, \qquad \boldsymbol{D} = \varepsilon \boldsymbol{E} \tag{27c}$$

where $\hat{\mu}$ and $\hat{\varepsilon}$ are tensor permeability and tensor dielectric constant, respectively, $\boldsymbol{B}$ and $\boldsymbol{D}$ are magnetic flux density and electric flux density, respectively.

It is understood from Eq. (27a) that nonreciprocity is caused by

$$\hat{\mu} \neq \hat{\mu}' \tag{27d}$$

$$\hat{\varepsilon} \neq \hat{\varepsilon}' \tag{27e}$$

In other words, the condition on Eq. (27d) or (27e) is necessary for a nonreciprocal circuit.

As an example, ferrite material magnetized by a DC magnetic field along the Z-axis has a tensor permeability of

$$\hat{\mu} = \begin{bmatrix} \mu & -j\kappa & 0 \\ j\kappa & \mu & 0 \\ 0 & 0 & \mu_0 \end{bmatrix}, \qquad \hat{\varepsilon} = \varepsilon \tag{28}$$

Several nonreciprocal circuits such as isolators and circulators use ferrite material with a DC magnetic field.

4 LUMPED ELEMENT CIRCUIT AND DISTRIBUTED CIRCUIT

4.1 Lumped Element

The circuits constructed by condensers, inductors, and resistors are called lumped element circuits. On the other hand, the circuits constructed by distributed lines such as coaxial lines and microstrip lines are called distributed circuits. The distinction between a lumped element circuit and a distributed circuit is based on the concept of the construction.

When we consider the circuits from the viewpoint of an electromagnetic concept, a clarification is needed. For example, we can use a short coaxial line shortened at the end as an inductor. On the other hand, the lumped element inductance becomes a capacitance at higher frequency.

In the case of $\omega \neq 0$, however, a capacitance has an inductive energy corresponding to the magnetic field produced by Ampere's law for the displacement current flowing capacity, and an inductance has a capacitive energy corresponding to the electric field produced by Faraday's law. Therefore, only the existence of electric energy in a capacity and only the existence of magnetic energy in a inductance are not possible except if $\omega = 0$.

Then, we can define the inductor and the capacitor as follows: In an inductor,

$$\tilde{W}_m \gg \tilde{W}_e$$

$$\eta_e = \frac{\tilde{W}_e}{\tilde{W}_m} \ll 1 \tag{29}$$

In a capacitor,

$$\tilde{W}_e \gg \tilde{W}_m$$

$$\eta_m = \frac{\tilde{W}_m}{\tilde{W}_e} \ll 1 \tag{30}$$

When we consider the distributed lines as shown in Figs. 18a and 18b.

In the case of Figs. 18a and 18b, we can calculate the values of η_e and η_m as follows:

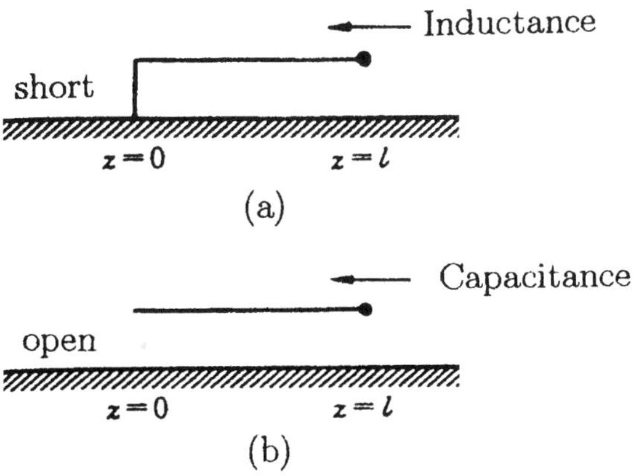

Figure 18. Examples of inductance and capacitance by using distributing lines with a short length.

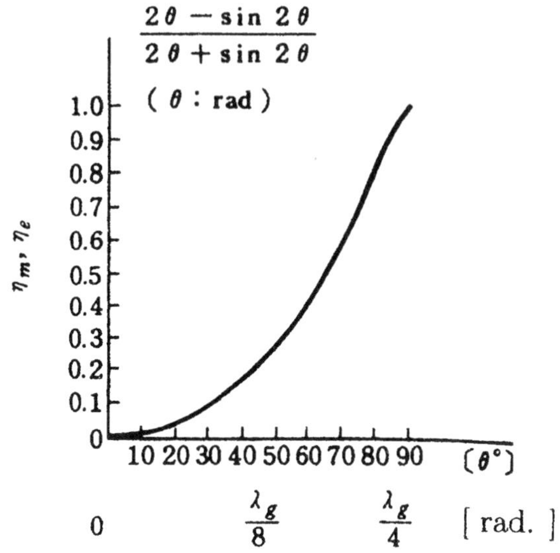

Figure 19. Ratios of $\tilde{W}_{e,m}/\tilde{W}_{m,e} = \eta_{e,m}$ in the case of Figs. 18a and 18b corresponding to $\theta = 2\pi l/\lambda_g$ (radians) $= 360l/\lambda_g$ (degrees).

$$\eta_{e,m} = \frac{2\theta - \sin 2\theta}{2\theta + \sin 2\theta}, \qquad \theta = \beta l = 2\pi \frac{l}{\lambda_g} \tag{31}$$

The values of $\eta_{e,m}$ are shown in Fig. 19. For example, in the case of Fig. 18a, to make an inductor, the values of η_e is about 0.2 for a length of $\lambda_g/8$, and η_e becomes much smaller for a shorter length. This means the shorter length makes it a lumped element.

4.2 Approximate Equivalent Lumped Element Network of a One-Port (Two-Terminal) Distributed Circuit at Used Frequency

In this section, we obtain the equivalent lumped element network of 1 port distributed circuit as shown in Fig. 20a.

When we consider the L_p, C_p parallel connected network as shown in Fig. 20b, the admittance y and $\partial y/\partial\omega$ should take the same values in the case of Figs. 20a and 20b. Values for y and $\partial y/\partial\omega$ of Fig. 20b are obtained from Eq. (32); C_p and L_p are obtained from Eqs. (33).

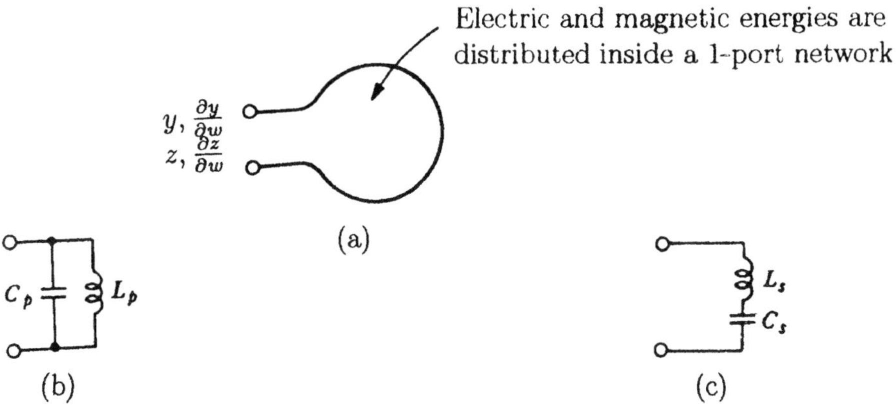

Figure 20. Equivalent lumped element network of a one-port distributing circuit. (a) one port circuit; (b) parallel equivalent network of "a"; (c) series equivalent network of "a."

$$y = j\omega C_p - j\frac{1}{\omega L_p}$$

$$\omega\frac{\partial y}{\partial \omega} = j\omega C_p + j\frac{1}{\omega L_p} \tag{32}$$

$$j\omega C_p = \frac{1}{2}\left(y + \omega\frac{\partial y}{\partial \omega}\right) \tag{33a}$$

$$j\omega L_p = \frac{2}{y - \omega(\partial y/\partial \omega)} \tag{33b}$$

We, therefore, can obtain the values of C_p and L_p of Fig. 20b from the values of y and $\partial y/\partial\omega$ of a given one-port circuit by Eqs. (33a) and (33b). As understood by the introduction to Eq. (33), the given one-port circuit is available as a general one-port circuit not only as a distributed circuit but also as a more complicated lumped element circuit, and as a combination circuit.

Next, we can also get the L_s,C_s series connected equivalent network as shown in Fig. 20c in the same way. The values of L_s and C_s can be obtained by $\dot{Z}$ and $\partial\dot{Z}/\partial\omega$ of the given one-port circuit as

$$j\omega L_s = \frac{1}{2}\left(z + \omega\frac{\partial z}{\partial \omega}\right) \tag{34a}$$

$$j\omega C_s = \frac{2}{z - \omega(\partial z/\partial\omega)} \tag{34b}$$

Equations (33) and (34) can be also obtained from Eqs. (19d), (22a), (18g), and (22b) in relation to the reactive energies as follows. Equations (19d) and (22a) are written in Eqs. (35a) and (35b):

$$y = 2j\omega(\tilde{W}_{ep} - \tilde{W}_{mp}) \tag{35a}$$

$$\omega \frac{\partial y}{\partial \omega} = 2j\omega(\tilde{W}_{ep} + \tilde{W}_{mp}) \tag{35b}$$

From the above relations, we get

$$2j\omega\tilde{W}_{ep} = \frac{y + \omega(\partial y/\partial\omega)}{2}, \qquad -2j\omega\tilde{W}_{mp} = \frac{y - \omega(\partial y/\partial\omega)}{2}$$

Comparing these with Eq. (33), we get $\omega C_p = 2\omega\tilde{W}_{ep}$ and $1/\omega L_p = 2\omega\tilde{W}_{mp}$.

It is understood that C_p and L_p represent the time average of the electric and magnetic energies, respectively, in the case of a constant voltage excitation of 1 V. Similarly, denoting the time average of the magnetic and electrical energies in the case of a constant current excitation of 1 A by $\tilde{W}_{ms}$ and $\tilde{W}_{es}$, respectively, we have Eqs. (36a) and (36b), which are derived from Eqs. (18g) and (22b), respectively:

$$z = 2j\omega(\tilde{W}_{ms} - \tilde{W}_{es}) \tag{36a}$$

$$\omega \frac{\partial z}{\partial \omega} = 2j\omega(\tilde{W}_{ms} + \tilde{W}_{es}) \tag{36b}$$

Comparing these equations with Eq. (33), we get the following results.

$$\omega L_s = 2\omega\tilde{W}_{ms}, \qquad \frac{1}{\omega C_s} = 2\omega\tilde{W}_{es} \tag{36c}$$

It is also understood that L_s and C_s are represented with the time average of electric and magnetic energies and with constant current excitation of 1 ampere.

We can, however, select the equivalent network considering the linearity of either $\partial y/\partial\omega$ in Fig. 20(b) or $\partial Z/\partial\omega$ in Fig. 20(c).

For example, the susceptibility and the reactance of $\lambda_g/4$ line with shortened, and open end, respectively, become as shown in Fig. 21(a), and the reactance of $\lambda_g/2$ line with shortened end becomes as shown in Fig. 21(b). As understood from the figures, the linear portion of a frequency range is much wider at the frequency range where B and X equal zero. Therefore, the parallel resonant network with construction of Fig. 20(b) is frequently

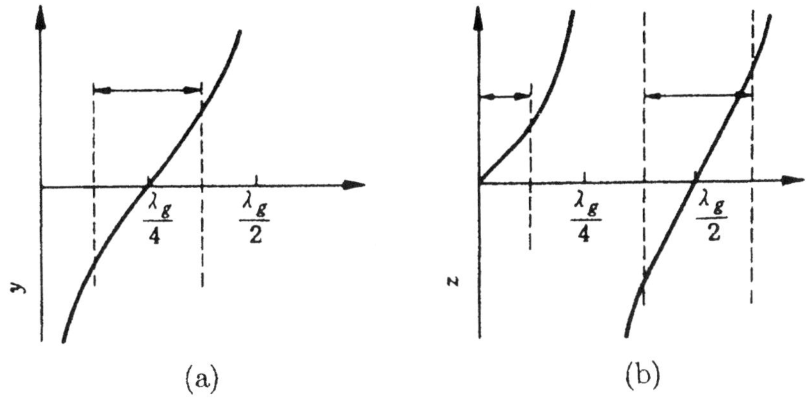

Figure 21. (a) Susceptibility and reactance of $\lambda_g/4$ line with shortened and open end; (b) reactance of $\lambda_g/2$ line with shortened end.

used in the case of $\lambda_g/4$ (line with shortened end), and the series resonant network with construction of Fig. 20(c) is frequently used in the case of $\lambda_g/4$ (line with open end), where the constants $C_pL_pC_s$ and L_s must satisfy the following relations:

$$\omega_r = \frac{1}{\sqrt{L_pC_p}} \qquad \text{[in Fig. 20(b)]} \tag{37a}$$

$$\omega_r = \frac{1}{\sqrt{L_sC_s}} \qquad \text{[in Fig. 20(c)]} \tag{37b}$$

The equivalent networks normally used in the filter designs, and the estimation of frequency performances of several devices, are described in the later chapters.

APPENDIX: NOTES TO CHAPTER 2

Note 1

Matrix $[A]$ is called the Hermitian matrix when $[\bar{A}]^* = [A]$. In this case, $A_{ij} = A_{ji}^*$, where A_{ij} is the ith row and jth column elements of $[A]$.

Note 2

The principal minor of the matrix is non-negative. This matter is introduced from the matter that the consumed power in the network is not negative because of the passive network.

Note 3

Principal minor of the matrix is positive. This matter is caused by the positive consumed power in a lossy network.

Note 4

This matter is introduced from the fact that the time average of the total reactive energies contained in a network is positive.

Note 5

In the n-row, n-column matrix, we have

$$\boldsymbol{M}\boldsymbol{U}^{(s)} = \lambda^{(s)}\boldsymbol{U}^{(s)} \tag{1}$$

Equation (1) can be expressed by

$$MU = U\Lambda, \qquad \Lambda = \begin{bmatrix} \lambda_1 & & 0 \\ & \ddots & \\ 0 & & \lambda_n \end{bmatrix}$$

Then, we get

$$M = U\Lambda U^{-1}$$

Note 6

Because we have the relations

$$[M]\boldsymbol{E}^i = \lambda^i\boldsymbol{E}^i, \qquad C_k[I]\boldsymbol{E}^i = C_k\boldsymbol{E}^i$$

adding both equations, we get

$$([M] + C_k[I])\boldsymbol{E}^i = (\lambda^i + C_k)\boldsymbol{E}^i \tag{1}$$

Multiplying both sides of Eq. (1) by $[M] + C_k[I]^{-1}$, we get

$$\boldsymbol{E}^i = (\lambda + C_k)([M] + C_k[I])^{-1}\boldsymbol{E}^i$$

$$([M] + C_k[I])^{-1}\boldsymbol{E}^s = (\lambda^s + C_k)^{-1}\boldsymbol{E}^s \tag{2}$$

Substituting Eqs. (1) and (2) into Eq. (13a) of the chapter, we get

$$f([M])\boldsymbol{E}^i = C_0([M] - C_1[I])([M] - C_2[I]) \cdots([M] - C_{-1}[I])^{-1}([M] - C_{-2}[I])^{-1}\cdots\boldsymbol{E}^i.$$

Substituting Eq. (2) into the last term of the above equation and repeating, we get

$$f([M])\boldsymbol{E}^i = C_0(\lambda^i - C_1)(\lambda^i - C_2)\cdots(\lambda^i - C_{-1})^{-1}(\lambda^i - C_{-2})^{-1}\cdots\boldsymbol{E}^i.$$

Note 7

Denoting the sth eigenvalues and eigenvectors by $p^{(s)}$ and $\boldsymbol{u}^{(s)}$, respectively, we get

$$[P]\boldsymbol{u}^{(s)} = p^{(s)}\boldsymbol{u}^{(s)} \tag{1}$$

Multiplying by $[Q]$ both sides of Eq. (1), and using Eq. (15a) of the chapter, we get

$$[P]([Q]\boldsymbol{u}^{(s)}) = p^{(s)}([Q]\boldsymbol{u}^{(s)}) \tag{2}$$

Therefore, $[Q]\boldsymbol{u}^{(s)}$ is proportional to $\boldsymbol{u}^{(s)}$. Substituting the proportional constant $q^{(s)}$,

$$[Q]\boldsymbol{u}^{(s)} = q^{(s)}\boldsymbol{u}^{(s)} \tag{3}$$

is obtained.

Note 8

In the n-port network, we denote by S_m the sectional plane of the mth port, by S the surface conductor surrounding the n-port network except ports, and by $\boldsymbol{i}_n$ the normal unit vector toward the network on S_m.

In Maxwell's equation,

$$\nabla \times \boldsymbol{E} = -j\omega\hat{\mu}\boldsymbol{H} \tag{1}$$

$$\nabla \times \boldsymbol{H} = (\sigma + j\omega\hat{\varepsilon})\boldsymbol{E} \tag{2}$$

by the mathematical process

$$\iiint_\tau [\boldsymbol{E}\cdot(1) - \boldsymbol{H}^*\cdot(2)]\, d\tau$$

and using the Gauss theorem, we get

$$\begin{aligned}\sum_{i=1}^{n} \iint_{S_i} V_i I_i^* \boldsymbol{e}_i \times \boldsymbol{h}_i^* \cdot \boldsymbol{i}_i \, dS = \sum_{i=1}^{n} V_i I_i^* \\ = \iiint_\tau \sigma|\boldsymbol{E}|^2\, d\tau \\ + j\omega \iiint_\tau (\boldsymbol{H}^*\cdot\hat{\mu}\boldsymbol{H} - \boldsymbol{E}\cdot\hat{\varepsilon}^*\boldsymbol{E}^*)\, d\tau\end{aligned} \tag{3}$$

Setting $\hat{\mu} = \hat{\mu}' - j\hat{\mu}''$ and $\hat{\varepsilon} = \hat{\varepsilon}' - j''\hat{\varepsilon}''$, we get

$$j\omega \iiint_\tau \boldsymbol{H}^* \cdot \hat{\mu}\boldsymbol{H}\, d\tau = j\omega \iiint_\tau \boldsymbol{H}^* \cdot \hat{\mu}'\boldsymbol{H}\, d\tau + \omega \iiint_\tau \boldsymbol{H}^* \cdot \hat{\mu}''\boldsymbol{H}\, d\tau$$

$$= 2j\omega\tilde{W}_m + P_m \tag{4}$$

$$-j\omega \iiint_\tau \boldsymbol{E} \cdot \hat{\varepsilon}^*\boldsymbol{E}^*\, d\tau = -j\omega \iiint_\tau \boldsymbol{E} \cdot \hat{\varepsilon}'\boldsymbol{E}^*\, d\tau + \omega \iiint_\tau \boldsymbol{E} \cdot \hat{\varepsilon}''^*\boldsymbol{E}^*\, d\tau$$

$$= -2j\omega\tilde{W}_e + P_e \tag{5}$$

Substituting Eqs. (4) and (5) into Eq. (3) and setting $P_c + P_m + P_e = P_R$, we get

$$\sum_{i=1}^{n} V_i^* I_i = P_R + 2j\omega(\tilde{W}_m - \tilde{W}_e) \tag{6}$$

Note 9

In Maxwell's equations

$$\nabla \times \boldsymbol{H} = j\omega\varepsilon\boldsymbol{E} \tag{1}$$

$$\nabla \times \boldsymbol{E} = -j\omega\mu\mathbf{H} \tag{2}$$

taking the variation, we get

$$\nabla \times \delta\boldsymbol{H} - j\varepsilon(\omega \cdot \delta\boldsymbol{E} + \boldsymbol{E} \cdot \delta\omega) = 0 \tag{3}$$

$$\nabla \times \delta\boldsymbol{E} + j\mu(\omega \cdot \delta\boldsymbol{H} + \boldsymbol{H} \cdot \delta\omega) = 0 \tag{4}$$

$$\nabla \cdot \boldsymbol{E} \times \delta\boldsymbol{H}^* = \delta\boldsymbol{H}^* \cdot \nabla \times \boldsymbol{E} - \boldsymbol{E} \cdot \nabla \times \delta\boldsymbol{H}^* \tag{5}$$

$$\nabla \cdot \delta\boldsymbol{E} \times \boldsymbol{H}^* = \boldsymbol{H}^* \cdot \nabla \times \delta\boldsymbol{E} - \delta\boldsymbol{E} \cdot \nabla \times \boldsymbol{H}^* \tag{6}$$

Substituting Eqs. (1)–(4) into Eqs. (5) and (6), integrating, and using the Gauss theorem in the vector field, we get

$$\sum_{m=1}^{n} \iint_{S_m} (\boldsymbol{E} \times \delta\boldsymbol{H}^* - \delta\boldsymbol{E} \times \boldsymbol{H}^*) \cdot (-\boldsymbol{n})\, dS$$

$$= j\delta\omega \iiint_\tau (\boldsymbol{E} \cdot \varepsilon^*\boldsymbol{E}^* + \boldsymbol{H}^* \cdot \mu\boldsymbol{H}\, d\tau$$

$$+ j\omega \iiint_\tau \{(\boldsymbol{H}^* \cdot \mu\delta\boldsymbol{H} - \delta\boldsymbol{H}^* \cdot \mu\boldsymbol{H})$$

$$+ (\boldsymbol{E} \cdot \varepsilon^*\delta\boldsymbol{E} - \delta\boldsymbol{E} \cdot \varepsilon^*\boldsymbol{E})\}\, d\tau \tag{7}$$

Because we have the relations

$$E_m = V_m e_{tm}, \qquad H_m = I_m H_{tm} \tag{8}$$

$$y^{(s)} V^i_m = I^{(s)}_m \tag{9}$$

$$\delta y^{(s)} \cdot V^{(s)}_m + y^{(s)} \cdot \delta V^{(s)}_m = \delta I_m \tag{10}$$

the left-hand side of Eq. (7) becomes

$$\delta y^{(s)} \sum_{m=1}^{n} |V^{(s)}_m|^2 + y^{(s)} \sum_{m=1}^{n} (V^{(s)}_m \delta V^{(s)*}_m - V^{(s)*}_m \delta V^{(s)}_m) \tag{11}$$

In the case of constant voltage excitation, because $\delta V_m = 0$, the first term of Eq. (11) and the first term of right-hand side of Eq. (7) is left, because only those terms are imaginary. This results in Eq. (20a) of the chapter.

Note 10

From the Maxwell equations corresponding to the conditions of Eqs. (25a) and (25b),

$$\nabla \times E_j = -j\omega\hat{\mu}H_j, \qquad E_i \cdot \nabla \times E_j = -j\omega H_i \cdot \hat{\mu}H_j \tag{1}$$

$$\nabla \times H_i = j\omega\hat{\varepsilon}E_i, \qquad E_j \cdot \nabla \times H_i = j\omega E_j \cdot \hat{\varepsilon}E_i \tag{2}$$

By Eqs. (1) and (2), we get

$$H_i \cdot \nabla \times E_j - E_j \cdot \nabla \times H_i = \nabla \cdot E_j \times H_i = -j\omega(H_i \cdot \hat{\mu}H_j + E_j \cdot \hat{\varepsilon}E_i) \tag{3}$$

By volume integrating Eq. (3) we get

$$\begin{aligned}\iiint_\tau \nabla \cdot E_j \times H_i \, d\tau &= \iint_S E_j \times H_i \cdot n \, dS \\ &= -j\omega \iiint_\tau (H_i \cdot \hat{\mu}H_j + E_j \cdot \hat{\varepsilon}E_i) \, d\tau \end{aligned} \tag{4}$$

By using the Gauss theorem and Eq. (18a), we get Eq. (26a) of the chapter. Interchanging i and j, we get Eq. (26b).

3

Basic Network Characteristics of a Transmission Line

1 SINGLE TRANSMISSION LINE

1.1 Realizing Pure Reactance

When we consider the lossless line with length l terminated by impedance $\dot{Z}_L$, as shown in Fig. 1, the input impedance $\dot{Z}_{\text{in}}$ can be obtained from Eq. (18b) of Chapter 1 as Eq. (1a):

$$\dot{Z}_{\text{in}} = \frac{\dot{Z}_L \cos\theta + jZ_c \sin\theta}{j(\dot{Z}_L/Z_c)\sin\theta + \cos\theta}$$

$$\theta = \beta l$$

$$\beta = \frac{2\pi}{\lambda_g} \tag{1a}$$

Expressing Eq. (1a) by normalized impedances of

$$\dot{Z}'_L = \frac{\dot{Z}_L}{Z_c} \quad \text{and} \quad \dot{Z}'_{\text{in}} = \frac{\dot{Z}_{\text{in}}}{Z_c}$$

we obtain

$$\dot{Z}'_{\text{in}} = \frac{\dot{Z}'_L + j\tan\theta}{j\dot{Z}'_L \tan\theta + 1} \tag{1b}$$

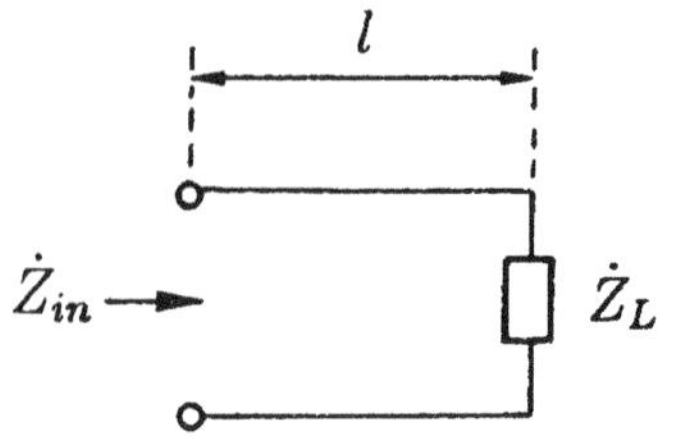

Figure 1. Lossless line of length l terminated by impedance $\dot{Z}_L$.

In the cases of a shortened end ($\dot{Z}'_L = 0$) and an open end ($\dot{Z}'_L = \infty$), we get Eqs. (1c) and (1d), respectively:

$$\dot{Z}'_{\text{in}} = j \tan \theta, \qquad \dot{Z}_{\text{in}} = jZ_c \tan \theta \tag{1c}$$

$$\dot{Z}'_{\text{in}} = -j \cos \theta, \qquad \dot{Z}_{\text{in}} = -jZ_c \cos \theta \tag{1d}$$

Therefore, we can realize an inductance L_{in} and a capacitance C_{in} as follows:

$$L_{\text{in}} = \frac{Z_c}{\omega} \tan \theta, \quad 0 < l < \frac{\lambda_g}{4} \tag{1e}$$

$$C_{\text{in}} = \frac{1}{\omega Z_c} \tan \theta, \quad 0 < l < \frac{\lambda_g}{4} \tag{1f}$$

When $l \ll \lambda_g/4$, we get

$$L_{\text{in}} = \frac{Z_c l}{f\lambda_g} = \frac{Z_c l}{v_p} \tag{1g}$$

$$C_{\text{in}} \simeq \frac{l}{f\lambda_g Z_c} = \frac{l}{v_p Z_c} \tag{1h}$$

Equations (1g) and (1h) are naturally understood from Eq. (4e) of Chapter 1, where L and C are the values of 1 m of line, that is, L (H/m) and C (F/m).

The values of L and C above, however, were obtained only for the concerned frequency. The real values should be obtained by taking into account of the values of $\partial y/\partial\omega$ and $\partial z/\partial\omega$ together with the admittance y and the impedance z as described in Eqs. (33a), (33b), (34a), and (34b) in Chapter 2.

In the case of a line with a shortened end in Fig. 2, we get a network equivalent to Fig. 2a and the values of L'_{in} and C'_{in} are obtained as Eq. (1i) from Eq. (34a) and Eq. (34b) of Chapter 2:

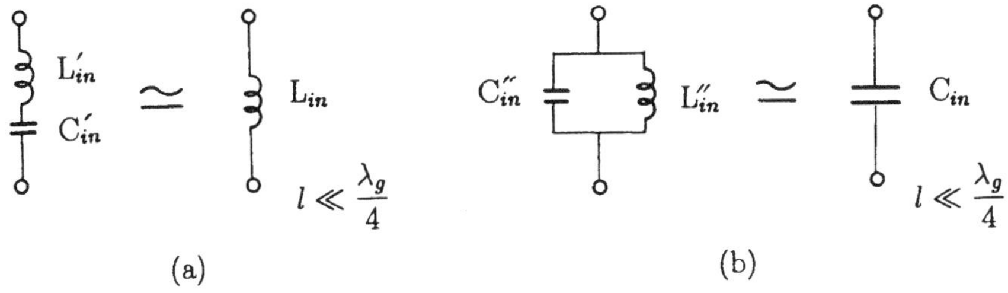

Figure 2. Equivalent network taking into account the values of $\partial Z/\partial\omega$ and $\partial y/\partial\omega$. (a) Equivalent network of the line with a shortened end; (b) equivalent network of the line with an open end.

$$L'_{\text{in}} = \frac{Z_c}{2\omega}\tan\theta\left(1+\frac{2\theta}{\sin\theta}\right)$$

$$C'_{\text{in}} = \frac{2}{\omega Z_c \tan\theta}\,\frac{1}{(2\theta/\sin 2\theta)-1}$$

$$\theta = \frac{2\pi}{\lambda_g}\,l \tag{1i}$$

For $l \ll \lambda_g/4$, we get

$$L'_{\text{in}} \simeq L_{\text{in}}, \qquad C'_{\text{in}} \simeq \infty \tag{1j}$$

which coincide with the values of Eqs. (1e) and (1g).

In the case of the line with an open end in Fig. 1, we get the equivalent network of Fig. 2b and the values of Eqs. (1k) and (1l) from Eqs. (33a) and (33b) in Chapter 2:

$$C''_{\text{in}} = \frac{Y_c\tan\theta}{2\omega}\left(1+\frac{2\theta}{\sin 2\theta}\right)$$

$$L''_{\text{in}} = \frac{2}{\omega Y_c\tan\theta}\,\frac{1}{(2\theta/\sin 2\theta)-1}$$

$$Y_c = \frac{1}{Z_c} \tag{1k}$$

In the case of $l \ll \lambda_g/4$, we get

$$C''_{\text{in}} \simeq C_{\text{in}}, \qquad L''_{\text{in}} \simeq \infty \tag{1l}$$

of which coincide with the values of Eqs. (1f) and (1h).

The inductive element and capacitive element described above are used in the matching circuit described in a later chapter.

1.2 Realizing the Impedance Transformer by Using the $\lambda_g/4$ Line

Consideration of Center Frequency

As mentioned in Chapter 1, the reflection coefficient is changed by moving the position on a line, which results in a change of the impedance. Maximum and minimum impedances appear on the line every quarter-wavelength, alternatively, as mentioned in Chapter 1. This leads to the realization of the impedance transformer with quarter-wave lines.

Now, substituting the values of $\pi/2$ into βl of Eq. (18b) of Chapter 1, we get

$$[F] = \begin{bmatrix} 0 & jZ_c \\ \dfrac{j}{Z_c} & 0 \end{bmatrix} \tag{2a}$$

Denoting the input impedance by R_1 and using Eq. (2a), we obtain directly the relation

$$R_1 = \frac{Z_c^2}{R_2} \tag{2b}$$

The values of R_1 are the same values of input impedance of Fig. 3, where n is calculated from

$$n = \frac{Z_c}{R_z} \tag{2c}$$

The explanation mentioned above only takes into account the coincidence of the impedance. We, however, must obtain the equivalent network to also meet the phase delay of $\pi/2$ radians. Under such a consideration, we get the equivalent network in Figs. 4b and 4c corresponding to the network in Fig. 4a.

The equivalency of Figs. 4a, 4b, and 4c can be confirmed by comparing the F matrices of each equivalent network. In the case when $n = Z_c/R_2$ as in Eq. (2c), the characteristic impedances of quarter-wave lines in Figs. 4b and 4c becomes R_2 and R_1 $(=Z_c^2/R_2)$, respectively.

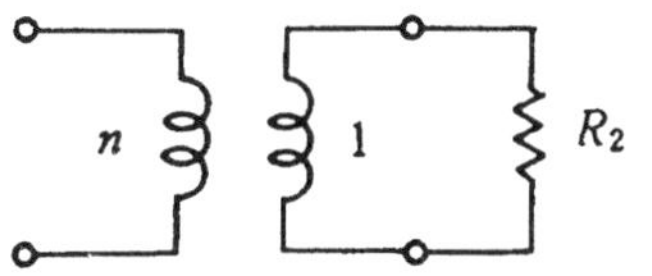

Figure 3. Ideal transformer equivalent to the $\lambda_g/4$ line.

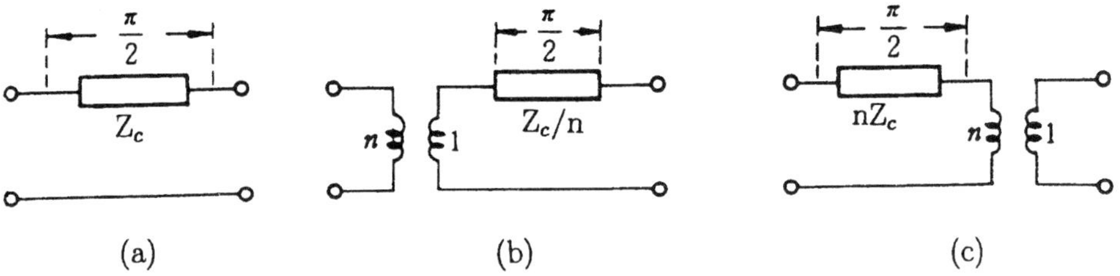

Figure 4. Equivalent network of the $\pi/2$ line with different characteristic impedances.

Equivalent Network of the $\lambda_g/4$ Line Considering the Neighboring Frequency from the Center Frequency

Denoting the center angular frequency by ω_0, we can obtain the first approximation of βl as

$$\beta l = \frac{\omega}{v_p} l = \frac{\pi}{2}\left(1 + \frac{\Delta\omega}{\omega_0}\right) \tag{3}$$

The $[F]$ matrix of Eq. (18b) in Chapter 1, however, can be transformed to the $[Y]$ matrix by Eq. (7b) in Chapter 2 and the first approximated values of the elements of the $[Y]$ matrix are obtained by using Eq. (3). The results are as follows:

$$Y = Y_1 + Y_2$$

$$Y_1 = \begin{bmatrix} y_0 & 0 \\ 0 & y_0 \end{bmatrix}, \qquad Y_2 = \begin{bmatrix} 0 & \dfrac{j}{W} \\ \dfrac{j}{W} & 0 \end{bmatrix}$$

$$y_0 = \frac{\cot k_z l}{jW} = j\frac{\pi}{2W}\frac{\Delta\omega}{\omega_0} \tag{4}$$

If we consider the LC parallel resonance circuit resonating at ω_0, the admittance at $\omega = \omega_0 + \Delta\omega$ takes the values of $2j\Delta\omega C$.

Comparing these values with y_0 of Eq. (4), we can realize the y_0 of Eq. (4) by the LC parallel tuned network, the values of which are shown in Eq. (5):

$$C = \frac{\pi}{4Z_z\omega_0}, \qquad L = \frac{4}{\pi}\frac{Z_c}{\omega_0} \tag{5}$$

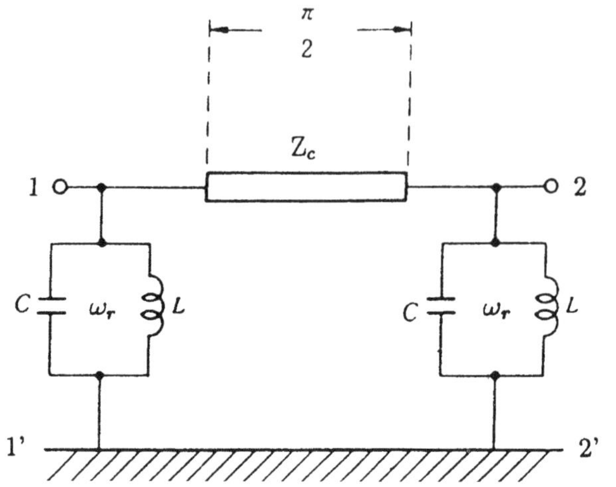

Figure 5. Equivalent network of the $\lambda_g/4$ distributed constant line.

Therefore, we get the equivalent network of a $\lambda_g/4$ line as shown in Fig. 5.

Frequency Performance of a Single $\lambda_g/4$ Transformer

When we connect the load impedance R at terminals 2 and 2′ of Fig. 5, we can obtain the input admittance, Y_i, as

$$Y_i = \Delta y + \frac{1}{W^2(1/R + \Delta y)}$$
$$\simeq \frac{R}{W^2}[(1 - \Delta y)R] + \Delta y = \frac{1}{Z_0} + \Delta y\left(1 - \frac{R}{Z_0}\right) \tag{6a}$$

where $\Delta y = 2j\Delta\omega C$

The voltage standing wave ratio (VSWR), however, takes the values

$$\text{VSWR} = \frac{1 + |\dot{\Gamma}|}{1 - |\dot{\Gamma}|},$$
$$\dot{\Gamma} = \frac{1/Z_0 - Y_i}{1/Z_0 + Y_i} \simeq \frac{-\Delta y(1 - R/Z_0)}{2/Z_0} = -j\omega C(Z_0 - R)\frac{\Delta\omega}{\omega} \tag{6b}$$

where Z_0 is the characteristic impedance of the line connected at terminals 1 and 1′.

Substituting Eq. (6a) into Eq. (6b) and using the relations in Eq. (5), we get the values of VSWR in the case of $|\dot{\Gamma}| \ll 1$:

$$\text{VSWR} \simeq 1 + 2|\dot{\Gamma}| = 1 + 2\omega C|Z_0 - R|\frac{\Delta\omega}{\omega}$$
$$= 1 + \frac{\pi}{2}\left|\sqrt{\frac{Z_0}{R}} - \sqrt{\frac{R}{Z_0}}\right|\frac{\Delta\omega}{\omega} \tag{6c}$$

It is understood that the bandwidth ratio ($=\Delta\omega/\omega$) becomes narrower in the case of a larger difference between Z_0 and R.

As an example, we will obtain the bandwidth ratio to be less than 1.2 of the VSWR for the case where $R = 200\ \Omega$, $Z_0 = 50\ \Omega$, and $Z_c = 100\ \Omega$. Substituting the given values into Eq. (6c),

$$\left|\sqrt{\frac{Z_0}{R}} = \sqrt{\frac{R}{Z_0}}\right| = \left|\sqrt{\frac{200}{50}} - \sqrt{\frac{50}{200}}\right| = 2 - \frac{1}{2} = \frac{3}{2}$$

Then

$$\text{VSWR} = 1 + \frac{\pi}{2}\left|\sqrt{\frac{Z_0}{R}} - \sqrt{\frac{R}{Z_0}}\right|\frac{\Delta\omega}{\omega} = 1.2$$

Therefore,

$$\frac{\Delta\omega}{\omega} = \frac{1.2 - 1}{(\pi/2)(3/2)} = 0.0849 = 8.49$$

The $\lambda_g/4$ transformers mentioned are used for the power divider, cominer, branch-type directional couplers, and matching circuits. These devices are explained in Chapter 6.

1.3 Realization of Resonant Circuit

Parallel Resonant Circuit by the $\lambda_g/4$ Line with a Shortened End

When the terminals 2 and 2′ of Fig. 5 are shortened, we can obtain the *LC* parallel resonant circuit, because the impedance of a shortened end line with an electrical angle of $\pi/2$ takes infinite values. Therefore, we obtain the resonator equivalent network of the parallel resonant *LC* circuit, where *L* and *C* takes the values from Eqs. (5), which is shown in Fig. 7 and Eqs. (7).

The frequency performance of a L_pC_c parallel circuit with the values of Eqs. (5) and the $\lambda_g/4$ line are compared in Fig. 8. It is understood that both performances are coincident in the range $\Delta\omega \ll \omega_0$.

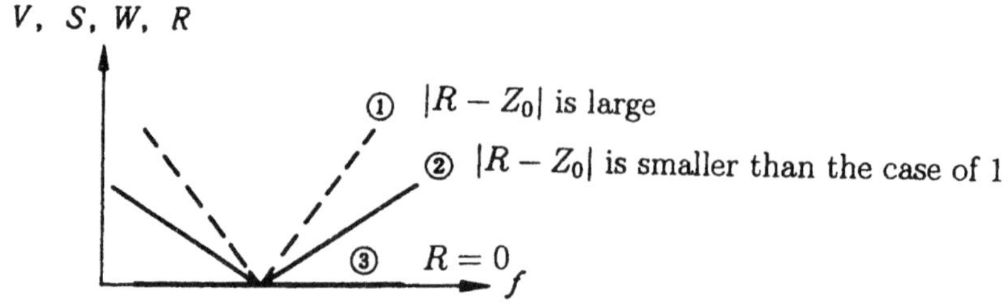

Figure 6. Comparison of VSWR among several load impedances.

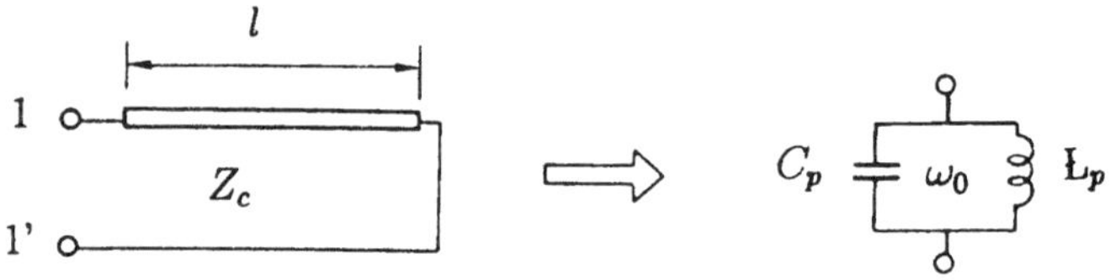

$$\left.\begin{aligned} C_p &= \frac{\pi}{4Z_c\omega_0} \\ L_p &= \frac{4Z_c}{\pi\omega_0} \end{aligned}\right\} \quad (7)$$

Figure 7. Shortened end $\lambda_g/4$ line makes the *LC* parallel resonant circuit.

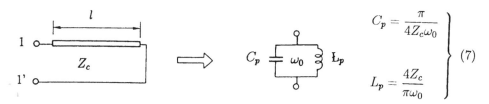

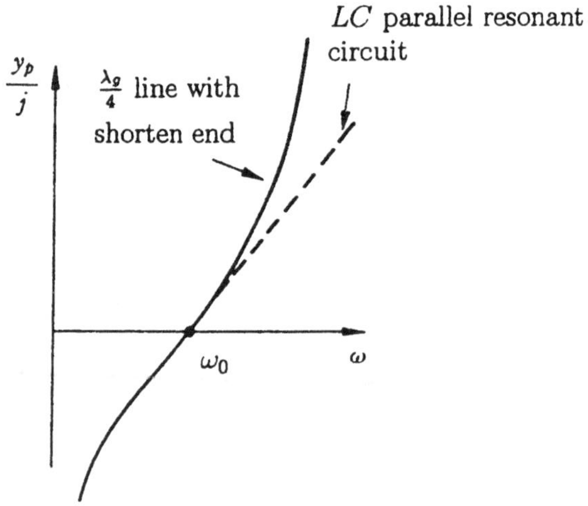

Figure 8. Comparison of frequency performances between the $\lambda_g/4$ line and the *LC* parallel resonant circuit.

Series Resonant Circuit by the $\lambda_g/2$ Line with a Shortened End

Because the impedance of a shortened end line as shown in Fig. 9 takes the values $jZ_s \tan \beta l$, one obtains a very small impedance Δz when $\beta l \simeq \pi$. It can be obtained by using

$$\beta l \simeq \pi \left(1 + \frac{\Delta\omega}{\omega_0}\right) \tag{8a}$$

as

$$\Delta z = jZ_c \frac{\pi}{\omega} \Delta\omega \tag{8b}$$

If we consider the L_sC_s series resonant circuit as shown in Fig. 9b, it has the values

$$2j\Delta\omega L_s \tag{8c}$$

Comparing Eqs. (8b) and (8c), when the values for L_s and C_s are obtained from Eq. (8d), the performance of Fig. 9b can be coincident in the frequency range around ω_0, as shown in Fig. 9c:

$$L_s = \frac{\pi Z_c}{2\omega}, \qquad C_s = \frac{1}{\omega^2 L} = \frac{2}{\pi\omega Z_c} \tag{8d}$$

In the same way, the condition of zero impedance at ω_r can be achieved by using a $\lambda_g/4$ line with an open end, as shown in Fig. 10a. In this case, the values of L and C of the equivalent LC series resonant circuit, as shown in Fig. 10b, are obtained from Eq. (8e), which are the half-values of L_s of Eq. (8d). This matter is quite understandable because if the reactive energies becomes half of those in Fig. 9a, then L_s is halved based on Eq. (36c) in Chapter 2.

$$L_s = \frac{\pi Z_c}{4\omega_r}, \qquad C_s = \frac{1}{\omega_r^2 L} = \frac{4}{\pi\omega_r Z_c} \tag{8e}$$

The resonant circuits are used for filters described in a later chapter.

2 COUPLED TRANSMISSION LINES

2.1 Telegrapher Equation of a Coupled Multiple Transmission Line and the Characteristics

Single Transmission Line

First, we describe the telegrapher equation of the single transmission line. When we consider the one conductor of Fig. 3a in Chapter 2 that is placed

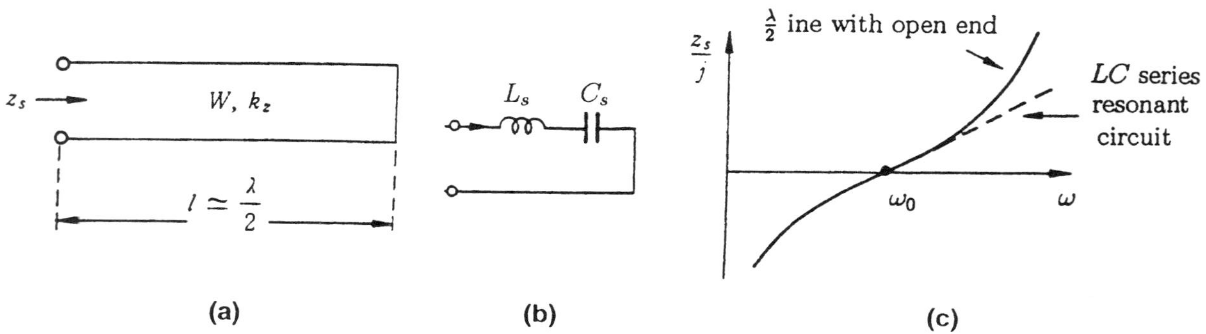

Figure 9. L_sC_s series resonant circuit (a), $\lambda_g/2$ line (b), and their frequency performance of impedance (c). (a) Resonant circuit with a $\lambda_g/2$ line; (b) LC series resonant circuit equivalent to (a); (c) comparison of the frequency performance of (a) and (b).

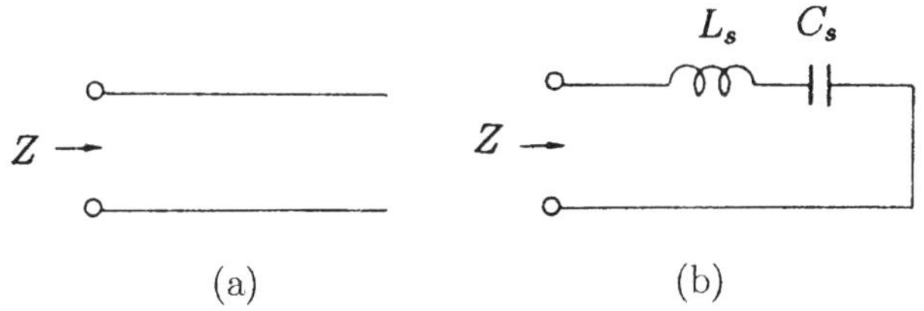

Figure 10. The $\lambda_g/4$ line with an open end (a) and the equivalent *LC* series resonant circuit (b).

parallel to ground, we get the sectional view and the equivalent network of Fig. 11. In this case, we can easily obtain the differential equations (9a) and (9b) from Fig. 11:

$$-\frac{d\dot{V}}{dz} = \dot{Z}\dot{I}, \qquad \dot{Z} = G + j\omega L \tag{9a}$$

$$-\frac{d\dot{I}}{dz} = \dot{Y}\dot{V}, \qquad \dot{Y} = G + j\omega C \tag{9b}$$

In Eqs. (9a) and (9b), V and I are the complex voltage and current, respectively, associated with a steady-state condition with sinusoidal time variation. Taking the second derivatives of Eqs. (9a) and (9b), we obtain

$$\frac{d^2\dot{V}}{dz^2} = \dot{Z}\dot{Y}\dot{V} = \gamma^2\dot{V} \tag{10a}$$

$$\frac{d^2\dot{I}}{dz^2} = \dot{Z}\dot{Y}\dot{I} = \gamma^2\dot{I} \tag{10b}$$

The general solutions of Eqs. (10) becomes

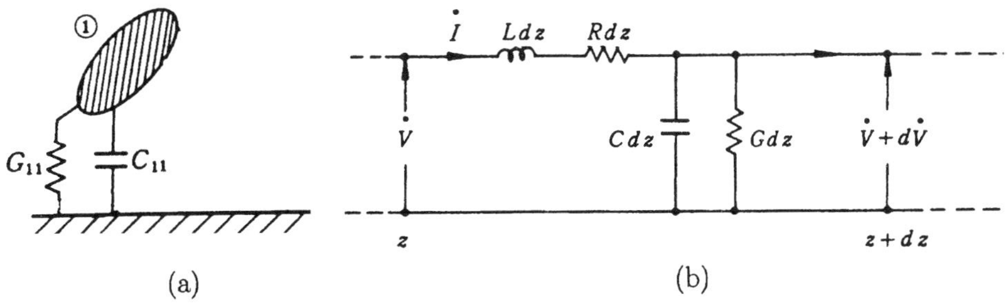

Figure 11. (a) Sectional view and (b) equivalent network of a single transmission line and *L*, *C*, *R*, *G* are 1 m.

$$\dot{V} = \dot{V}_{+}e^{-\gamma z} + \dot{V}_{-}e^{\gamma z} \tag{11a}$$

$$\dot{I} = \frac{\dot{V}_{+}}{Z_c}e^{-\gamma z} - \frac{\dot{V}_{-}}{Z_c}e^{\gamma z} \tag{11b}$$

where

$$Z_c = \sqrt{\frac{\dot{Z}}{\dot{Y}}} = \sqrt{\frac{R + j\omega L}{G + j\omega C}} \tag{12a}$$

$$\gamma = \sqrt{\dot{Z}\dot{Y}} = \alpha + j\beta \tag{12b}$$

$$\alpha = [\tfrac{1}{3}\sqrt{(R^2 + \omega^2L^2)(G^2 + \omega^2C^2)} + \tfrac{1}{2}(RG - \omega^2LC)]^{1/2} \tag{12c}$$

$$\beta = [\tfrac{1}{2}\sqrt{(R^2 + \omega^2L^2)(G^2 + \omega^2C^2)} - \tfrac{1}{2}(RG - \omega^2LC)]^{1/2} \tag{12d}$$

In the lossless line,

$$\gamma = j\beta = j\omega\sqrt{LC} = j\,\frac{2\pi}{\lambda} \tag{12e}$$

$$Z_c = \sqrt{\frac{L}{C}} \tag{12f}$$

Equations (11a), (11b), (12e), and (12f) are the results of Eqs. (6b), (4b), and (4a) of Chapter 1.

When we consider the line with the length of l as shown in Fig. 12, we can introduce Eqs. (13a) and (13b) in the lossless case from Eqs. (11a) and (11b):

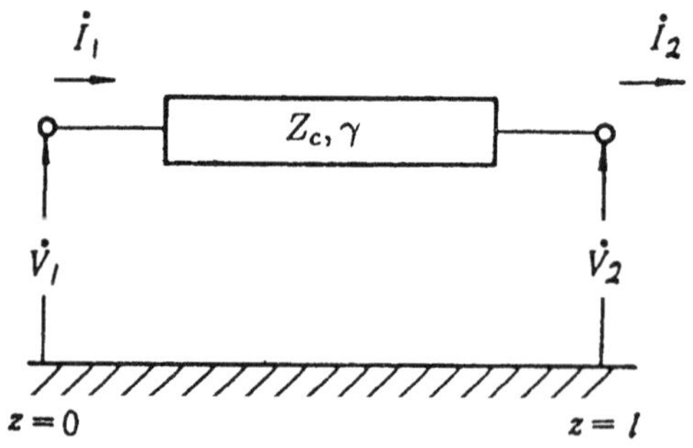

Figure 12. Voltage and current at $Z = 0$ and $Z = 1$ of the line with length l.

$$\dot{V}_1 = \dot{V}_2 \cosh \gamma l + \dot{I}_2 Z_c \sinh \gamma l$$

$$\dot{I}_1 = j\dot{V}_2 \frac{1}{Z_c} \sinh \gamma l + \dot{I}_2 \cosh \gamma l \tag{13a}$$

$$\dot{V}_1 = \dot{V}_2 \cos \beta l + j\dot{I}_2 Z_c \sin \beta l$$

$$\dot{I}_1 = j\dot{V}_2 \frac{1}{Z_c} \sinh \beta l + \dot{I}_2 \cos \beta l \tag{13b}$$

Coupled Multiple Line

Diagonalization of the Telegrapher Equation In the case of a coupled multiple line as shown in Fig. 13, we have Eqs. (14a) and (14b):

$$-\frac{dV_1}{dz} = Z_{11}I_1 + Z_{12}I_2 + \cdots + Z_{1n}I_n$$

$$-\frac{dV_2}{dz} = Z_{21}I_1 + Z_{22}I_2 + \cdots + Z_{2n}I_n$$

$$\vdots$$

$$-\frac{dV_n}{dz} = Z_{n1}I_1 + Z_{n2}I_2 + \cdots + Z_{nn}I_n \tag{14a}$$

$$-\frac{dI_1}{dz} = Y_{11}V_1 + Y_{12}V_2 + \cdots + Y_{1n}V_n$$

$$-\frac{dI_2}{dz} = Y_{21}V_1 + Y_{22}V_2 + \cdots + Y_{2n}V_n$$

$$\vdots$$

$$-\frac{dI_n}{dz} = Y_{n1}V_1 + Y_{n2}Y_2 + \cdots + Y_{nn}V_n \tag{14b}$$

where

$$Z_{ij} = R_{ij} + jX_{ii}, \qquad R_{ij} = R_i, \qquad X_{ij} = j\omega L_{ij}, \qquad L_{ii} = L_i$$

$$Y_{ij} = G_{ij} + jB_{ij}, \qquad B_{ij} = j\omega C_{ij} \quad (i, j = 1, \ldots, n)$$

To express it more simply, we can show the relation by using vectors of voltage and currents associated to each terminal as follows:

$$-\frac{d\boldsymbol{V}}{dz} = Z\boldsymbol{I} \tag{15a}$$

$$-\frac{d\boldsymbol{I}}{dz} = Y\boldsymbol{V} \tag{15b}$$

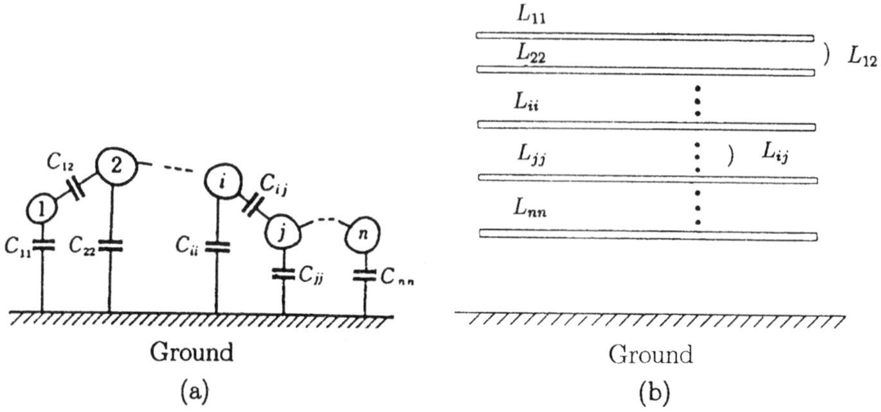

Figure 13. Capacitances and inductances of coupled n uniform lines. (a) Capacitances; (b) self-inductances and mutual inductances.

$$\boldsymbol{V} = \begin{bmatrix} V_1 \\ \vdots \\ V_n \end{bmatrix}, \quad \boldsymbol{I} = \begin{bmatrix} I_1 \\ \vdots \\ I_n \end{bmatrix}, \quad \boldsymbol{Z} = \begin{bmatrix} Z_{11} & \cdots & Z_{1n} \\ \vdots & & \vdots \\ Z_{n1} & \cdots & Z_{nn} \end{bmatrix},$$

$$\boldsymbol{Y} = \begin{bmatrix} Y_{11} & \cdots & Y_{1n} \\ \vdots & & \vdots \\ Y_{n1} & \cdots & Y_{nn} \end{bmatrix} \tag{15c}$$

Now we consider a mode with special set of voltages $V_i^{(s)}$ and special currents $I_i^{(s)}$ at each line, where i denotes the ith line and s denotes the sth set, as shown in Table 1.

In this case, we assume that there exists an incident wave with propagation constant $\gamma^{(s)}$ ($s = 1, \ldots, n$). The values of $V_i^{(s)}$, $I_i^{(s)}$, and $\gamma^{(s)}$ will be

Table 1. Voltages and Currents of Modes

	1st Mode		2nd Mode		sth Mode		nth Mode	
Line	Voltage	Current	Voltage	Current	Voltage	Current	Voltage	Current
1	V_1^1	I_1^1	V_1^2	I_1^2	V_1^s	I_1^s	V_1^n	I_1^n
2	V_2^1	I_2^1	V_2^2	I_2^2	V_2^s	I_2^s	V_2^n	I_2^n
i	V_i^1	I_i^1	V_i^2	I_i^2	V_i^s	I_i^s	V_i^n	I_i^n
n	V_n^1	I_n^1	V_n^2	I_n^2	V_n^s	I_n^s	V_n^n	I_n^n

obtained as follows. The voltage V_i and current I_i of the ith line, therefore, should be expressed by using the modes of Table 1 as

$$V_i = v^{(1)}V_i^{(1)} + v^{(2)}V_i^{(2)} + \cdots + v^{(n)}V_i^{(n)}$$
$$I_i = i^{(1)}I_i^{(1)} + i^{(2)}I_i^{(2)} + \cdots + i^{(n)}I_i^{(n)} \quad (i = 1, \ldots, n) \tag{16a}$$

Equation (16a) can be presented in matrix form by Eq. (16b)

$$\boldsymbol{V} = T_V \boldsymbol{v}$$

$$\boldsymbol{I} = T_I \boldsymbol{i}$$

$$T_V = \begin{bmatrix} V_1^1 & V_1^2 & \cdots & V_1^n \\ V_2^1 & V_2^2 & \cdots & V_2^n \\ \vdots & \vdots & & \vdots \\ V_n^1 & V_n^2 & \cdots & V_n^n \end{bmatrix}, \qquad T_I = \begin{bmatrix} I_1^1 & I_1^2 & \cdots & I_1^n \\ I_2^1 & I_2^2 & \cdots & I_2^n \\ \vdots & \vdots & & \vdots \\ I_n^1 & I_n^2 & \cdots & I_n^n \end{bmatrix}$$

$$\boldsymbol{v} = \begin{bmatrix} v^1 \\ v^2 \\ \vdots \\ v^n \end{bmatrix}, \qquad \boldsymbol{i} = \begin{bmatrix} i^1 \\ i^2 \\ \vdots \\ i^n \end{bmatrix} \tag{16b}$$

Substituting Eq. (16b) into Eqs. (15a) and (15b), we get

$$-\frac{d\boldsymbol{v}}{dz} = T_V^{-1} Z T_I \boldsymbol{i} \tag{16c}$$

$$-\frac{d\boldsymbol{i}}{dz} = T_I^{-1} Y T_V \boldsymbol{v} \tag{16d}$$

If the right-hand side of Eqs. (16c) and (16d) are diagonalized, the equations are similar to Eqs. (9a) and (9b), respectively.

Taking the second derivatives of Eqs. (16c) and (16d), we get

$$\frac{\partial^2 \boldsymbol{v}}{\partial z^2} = T_V^{-1} Z Y T_V \boldsymbol{v} \tag{16e}$$

$$\frac{\partial^2 \boldsymbol{i}}{\partial z^2} = T_I^{-1} Y Z T_I \boldsymbol{i} \tag{16f}$$

When the right-hand side of Eqs. (16e) and (16f) are diagonalized to the same diagonal matrix with the elements of the square of propagation constants $\gamma_1^2, \ldots, \gamma_n^2$, we can have n independent lines corresponding to independent modes. In this case, Eqs. (16e) and (16f) are similar to Eqs. (16a) and (16b), respectively. We, however, have the relation between T_V and T_I as follows.

The incident power P carried by n lines can be obtained from Eq. (16b) as Eq. (17a), which results in Eq. (17b):

$$P = (\boldsymbol{i}^*)^t \boldsymbol{v} = (\boldsymbol{i}^*)^t (T_I^*)^t T_V \boldsymbol{v} \tag{17a}$$

$$T_I = (T_V^{*t})^{-1} \tag{17b}$$

Taking the transpose matrix of Eq. (16f) and substituting the relation of Eq. (17b) into T_I, we obtain Eq. (16e) under the condition that T_V is real matrix (see Note 1). This means that Eq. (16f) is subordinate to Eq. (16e).

We can obtain the eigenvalues of ZY and γ_s^2 by solving the eigenequation. Substituting γ_s into the eigenequation again, we can obtain the relative values of the eigenvector, $\boldsymbol{v}_V^{(1)}$, corresponding to γ_s. Arranging $\boldsymbol{u}_V^{(1)}, \ldots, \boldsymbol{u}_V^{(n)}$ as rows of the matrix, we get T_V.

Sometimes, eigenvectors are already known; for example, when the sectional view of a multiline takes the symmetrical construction, such as plane symmetry and rotationally symmetry, which is the same as the case of the circuit described in Chapter 2, Section 2.1.

For a simpler case, we have

$$T_V = T_I = P \tag{17c}$$

where P is the orthogonal matrix.

In this case, substituting Eq. (17c) into Eqs. (16c), (16d), and (16e), we get

$$\begin{aligned} T_V^t Z T_V &= \Lambda_z = \mathrm{diag}(z^{(1)} \cdots z^{(n)}) \\ T_V^t Y T_V &= \Lambda_y = \mathrm{diag}(y^{(1)} \cdots y^{(n)}) \end{aligned} \tag{17d}$$

$$\begin{aligned} T_V^t ZY T_V &= (T_V^t Z T_V)(T_V^t Y T_V) \\ &= \Lambda_z \Lambda_y = \Lambda_\gamma^2 \\ &= \Lambda_\gamma = \mathrm{diag}(\gamma_1, \ldots, \gamma_n). \end{aligned} \tag{17e}$$

Then, we get

$$\gamma_s = \sqrt{z^{(s)} y^{(s)}} \tag{17f}$$

Substituting Eq. (17f) into Eqs. (16c) and (16d) again, and considering $v_+^{(s)} = v_0^{(s)} e^{-\gamma_s z}$ for the incident sth mode, we get

$$v_+^{(s)} = \frac{z^{(s)}}{\gamma^{(s)}} i_+^{(s)} = W_s i_+^{(s)}, \qquad W_s = \sqrt{\frac{z^{(s)}}{y^{(s)}}} \tag{17g}$$

$$i_+^{(s)} = \frac{y^{(s)}}{\gamma^{(s)}} v_+^{(s)} = Y_s v_+^{(s)}, \qquad Y_s = \frac{1}{W_s} \tag{17h}$$

When we consider the impedance of line for the sth mode, we can get the results from Eq. (16b) by the following process:

$$\boldsymbol{V}_+ = \boldsymbol{u}_V^{(s)} v_+^{(s)}, \qquad \boldsymbol{I}_+ = \boldsymbol{u}_I^{(s)} i_+^{(s)}$$

Taking account the relation in Eq. (17c), which means $\boldsymbol{u}_V^{(s)} = \boldsymbol{u}_I^{(s)}$, and using Eq. (17g), we get

$$\frac{\boldsymbol{V}_+^{(s)}}{\boldsymbol{I}_+^{(s)}} = Z_s = W_s = \sqrt{\frac{z^{(s)}}{y^{(s)}}} \tag{17i}$$

This means that the impedance of all lines for the sth mode, Z_s, takes the same values of mode impedance W_s.

Next, we explain the lossless lines. First, we explain that T_V is a real matrix and Eq. (17b) becomes

$$T_I = (T_V^t)^{-1} \tag{17j}$$

Because R_{ij} and G_{ij} of Eqs. (14a) and (14b) become zero in the lossless lines, Eqs. (16c), (16d), (16e), and (16f) become Eqs. (18a), (18b), (18c), and (18d), respectively:

$$-\frac{\partial \boldsymbol{v}}{\partial z} = j\omega T_V^{-1} L T_I \boldsymbol{i} \tag{18a}$$

$$-\frac{\partial \boldsymbol{i}}{\partial z} = j\omega T_I^{-1} C T_V \boldsymbol{v} \tag{18b}$$

$$\frac{\partial^2 \boldsymbol{v}}{\partial z^2} = -\omega^2 T_V^{-1} L C T_V \boldsymbol{v} \tag{18c}$$

$$\frac{\partial^2 \boldsymbol{i}}{\partial z^2} = -\omega^2 T_I^{-1} C L T_I \boldsymbol{i} \tag{18d}$$

where L and C are the positive real symmetric matrices (see Note 2). We can show that T_V and T_I in Eqs. (18a)–(18d) are the real matrices as follows.

We can consider two positive real symmetric matrices L^{-1} and C, because the eigenvalues of L^{-1} are also positive. We, however, can verify that there exists the real matrix T to diagonalize two real symmetrical matrices A and B at the same time in the form

$$T^t A T = \Lambda_a \tag{19a}$$

$$T^t B T = \Lambda_b \tag{19b}$$

where A and B are positive real symmetric (see Note 3).

Replacing T by T_V, A by L^{-1}, and V by C, we have the following relations under the consideration of Eq. (17c) by the reason of the real matrix of T_V:

$$T_V^t L^{-1} T_V = \{T_V^{-1} L (T_V)^{-1}\}^{-1} = \Lambda_L^{-1}$$

$$T_V^{-1} L (T_V^t)^{-1} = T_V^{-1} L T_I = \Lambda_L = \mathrm{diag}(\lambda_{L_1}, \ldots, \lambda_{L_n}) \tag{19c}$$

$$T_V^t C T_V = T_I^{-1} C T_V = \Lambda_I = \mathrm{diag}(\lambda_{c_1}, \ldots, \lambda_{c_n}) \tag{19d}$$

Equations (19c) and (19d) are the diagonal matrices of Eqs. (18a) and (18b).
Multiplying Eq. (19c) by Eq. (19d), and Eq. (19d) by Eq. (19c), we get

$$T_V^{-1} L C T_V = \Lambda_L \Lambda_C \tag{19e}$$

$$T_I^{-1} C L T_I = \Lambda_C \Lambda_L \tag{19f}$$

Equations (19e) and (19f) are included in the right-hand sides of Eqs. (18c) and (18d), respectively. Therefore, we get

$$\frac{\partial^2 \boldsymbol{v}}{\partial Z^2} = -\omega^2 \Lambda^2 \boldsymbol{v} \tag{20a}$$

$$\frac{\partial^2 \boldsymbol{i}}{\partial Z^2} = -\omega \Lambda^2 \boldsymbol{i} \tag{20b}$$

where

$$\Lambda^2 = \Lambda_L \Lambda_C = \mathrm{diag}(\lambda_1^2, \ldots, \lambda_n^2) \tag{20c}$$

To obtain the values of Λ_{LC}, we can calculate the eigenvalues directly from the eigenequation as follows:

$$|LC - \lambda_s^2 I| = 0 \tag{21a}$$

where I is the unit matrix.

Substituting λ_s into the eigenequation again, we get the eigenvectors $\boldsymbol{u}_V^{(s)}$ and $\boldsymbol{u}_I^{(s)}$, and T_V and T_I as follows:

$$\boldsymbol{u}_V^{(1)} = \begin{bmatrix} V_1^{(1)} \\ \vdots \\ V_n^{(1)} \end{bmatrix}, \qquad \boldsymbol{u}_V^{(s)} = \begin{bmatrix} V_1^{(s)} \\ \vdots \\ V_n^{(s)} \end{bmatrix}, \qquad \boldsymbol{u}_V^{(n)} = \begin{bmatrix} V_1^{(n)} \\ \vdots \\ V_n^{(n)} \end{bmatrix} \tag{21b}$$

$$T_V = [\boldsymbol{u}_V^{(1)} \cdots \boldsymbol{u}_V^{(s)} \cdots \boldsymbol{u}_V^{(n)}] \tag{21c}$$

$$T_V = (T_V^t)^{-1} = [\boldsymbol{u}_I^{(1)}, \ldots, \boldsymbol{u}_I^{(s)}, \ldots, \boldsymbol{u}_I^{(n)}] \tag{21d}$$

In the case of

$$LC = CL \tag{22a}$$

comparing Eqs. (19e) and (19f), we can set

$$T_V = T_I = (T_V^t)^{-1}, \qquad T_V^t T_V = \mathrm{I}$$

Then, we get

$$T_V = T_I = P \tag{22b}$$

where P is the orthogonal matrix.

The practical example for the case of when the eigenvalues are all different is symmetrical two lines in an inhomogeneous medium, such as symmetrical parallel microstrip lines mentioned later.

Another example for the case of so-called degenerated eigenvalues, which all values are the same, is the parallel multilines in the homogeneous medium mentioned later.

Equation (22b), however, is also based on the general mathematical theorem

> The necessary and sufficient condition to diagonalize two real symmetrical matrices A and B by an orthogonal matrix P is $AB = BA$.

Therefore, for the case

$$LC \neq CL \tag{22c}$$

$$T_V, T_I \neq P \tag{22d}$$

A practical example is two coupled asymmetrical lines in inhomogeneous medium, such as asymmetrical parallel microstrip lines.

Equivalent Network of Multicoupled Lines Because Eqs. (18a) and (18b) are diagonalized by the real matrices T_V and T_I as obtained in Eqs. (19c) and (19d), we can also define them as

$$-\frac{dv^{(s)}}{dz} = z^{(s)} i^{(s)} \tag{23a}$$

$$-\frac{di^{(s)}}{dz} = y^{(s)} v^{(s)} \tag{23b}$$

where

$$z^{(s)} = j\omega\lambda_L^{(s)}, \qquad y^{(s)} = j\omega\lambda_C^{(s)} \tag{23c}$$

Equations (23a) and (23b) are in the same form as Eqs. (9a) and (9b). Therefore, the sth mode can be expressed by Eqs. (23d) and (23e) or Eqs. (23f) and (23g), similar to Eqs. (11a), (11b), and (13b):

$$v^s = v_+^s \exp(-jk_z^{(s)} z) + v_-^s \exp(jk_z^{(s)} z) \tag{23d}$$

$$i^s = \frac{v_+^s}{W_s} \exp(-jk_z^{(s)} z) - \frac{v_-^s}{W_s} \exp(jk_z^{(s)} z) \tag{23e}$$

$$v^s = v_0^s \cos k_z^{(s)} z - jW_s i_0^s \sin k_z^{(s)} z \tag{23f}$$

$$i^s = -j\frac{v_0^s}{W_s}\sin k_z^{(s)}z + i_0^s \cos k_z^{(s)}z \tag{23g}$$

The phase constant k_z and characteristic impedance W_s are defined in Eqs. (23h) and (23i), respectively, similar to Eqs. (12e) and (12f):

$$k_z^{(s)} = \omega\sqrt{\lambda_L^{(s)}\lambda_C^{(s)}} \tag{23h}$$

$$W_s = \frac{\omega\lambda_L^{(s)}}{k_z^{(s)}} = \frac{k_z^{(s)}}{\omega\lambda_C^{(s)}} = \sqrt{\frac{\lambda_L^{(s)}}{\lambda_C^{(s)}}} \tag{23i}$$

We obtain the equivalent network of lossless multicoupled lines as shown in Fig. 14. In Fig. 14, N_T is the ideal transformer network which takes the role of T_V and T_I in Eq. (16b). When N_T satisfies T_V, T_I is automatically satisfied through the power invarient relation shown in Eq. (17j).

We can also obtain the equation of V_k^s and I_k^s ($k = 1, \ldots, n$) corresponding to Eqs. (23d)–(23g). As for these equations, we will introduce the practical examples in Section 2.2.

In Fig. 14, V_{k0} and V_{kl} ($k = 1, \ldots, n$) show the voltages of the kth line at $z = 0$ and $z = l$. I_{k0} and I_{kl} show the current of the kth line at $z = 0$ and $z = l$. $V_0^{(s)}$ and $i_0^{(s)}$ are the voltage and current, respectively, at the sth independent line corresponding to the sth mode at $z = 0$, and $v_l^{(s)}$ and $i_l^{(s)}$ are the values of $z = 1$.

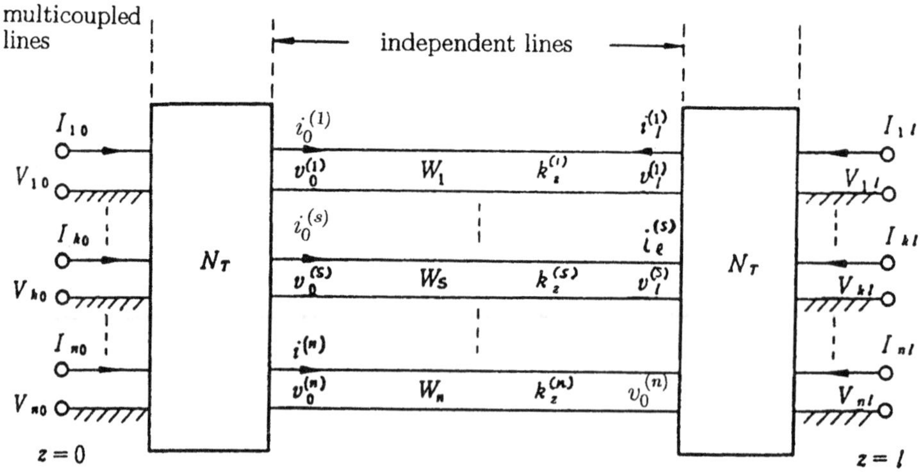

Figure 14. Equivalent network of n parallel lossless lines.

Impedance of Each Line Corresponding to Each Mode The voltages of lines corresponding to the incident wave of the sth mode, $V_+^{(s)}$, is calculated using

$$\boldsymbol{V}_+ = \boldsymbol{u}_V^{(s)} v_+^{(s)} \tag{24a}$$

The currents of lines corresponding to the sth mode, $\boldsymbol{I}_+^{(s)}$ is calculated using

$$\boldsymbol{I}_+^{(s)} = \boldsymbol{u}_I^{(s)} i_+^{(s)} = \boldsymbol{u}_I^{(s)} \frac{v_+^{(s)}}{W_s} \tag{24b}$$

Therefore, in the case of $LC \neq CL$, we get

$$\boldsymbol{V}_+^{(s)} \neq W_s \boldsymbol{I}_+^{(s)} \tag{24c}$$

because $\boldsymbol{u}_V^{(s)} \neq \boldsymbol{u}_I^{(s)}$, as $T_V \neq T_I$. This result shows that the impedance at each line corresponding to the incident wave of the sth mode is different. This result appears in the case of unsymmetrical parallel microstrip lines described later.

In the case of $LC = CL$, as we have the relation $\boldsymbol{u}_V^{(s)} = \boldsymbol{u}_I^{(s)}$ from $T_V = T_I$, we get

$$\boldsymbol{V}_+^{(s)} = W_s \boldsymbol{I}_+^{(s)} \tag{24d}$$

This result shows that the all impedances at lines corresponding to the incident sth mode take the same values of W_s.

The fact mentioned above is true for the case of LC the eigenvalues of which are different.

When LC has degenerate eigenvalues, the combination of the eigenvectors corresponding to the same eigenvalues can also become the eigenvector corresponding to the same eigenvalues. By using such eigenvectors, originally even the orthogonal matrix, T_V, changes to a nonorthogonal matrix, which results in $T_V \neq T_I$. In this case, the impedance between line and ground is different, even for the same mode.

2.2 Basic Characteristics of Coupled Lines

Symmetric Coupled Lossless Two Lines in a Homogeneous or Inhomogeneous Medium

As symmetrical coupled lines, we have coupled triplate lines in homogeneous medium, coupled microstrip lines, and coupled parallel lines in inhomogeneous medium. We show the sectional views of them in Fig. 15. The inhomogeneous medium means that there are more than two media, for example, such as a dielectric material and air.

Considering the symmetric structure of the sectional view, we have the relation in Eq. (16d) in Chapter 2, and the eigenvector of the Z and Y

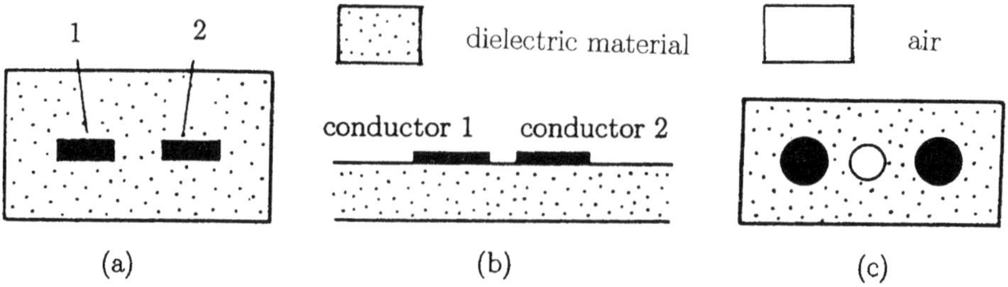

Figure 15. Sectional views of practical examples of symmetrical coupled lines. (a) Coupled triplate lines; (b) coupled microstrip lines; (c) coupled parallel lines in an inhomogeneous medium.

matrices is the same vector as in Eq. (16f) in Chapter 2, that is, as the same as

$$\boldsymbol{U}_V^{(1)} = \boldsymbol{U}_I^{(1)} = \begin{bmatrix} \dfrac{1}{\sqrt{2}} \\ \dfrac{1}{\sqrt{2}} \end{bmatrix} \tag{25a}$$

$$\boldsymbol{U}_V^{(2)} = \boldsymbol{U}_I^{(2)} = \begin{bmatrix} \dfrac{1}{\sqrt{2}} \\ \dfrac{-1}{\sqrt{2}} \end{bmatrix} \tag{25b}$$

therefore, $\boldsymbol{T}_V$ and $\boldsymbol{T}_I$ in matrix form are

$$\boldsymbol{T}_V = \boldsymbol{T}_I = \frac{1}{\sqrt{2}} \begin{bmatrix} 1 & 1 \\ 1 & -1 \end{bmatrix} \tag{26}$$

$\boldsymbol{T}_V$ and $\boldsymbol{T}_I$ are the orthogonal matrices.

$\boldsymbol{U}_V^{(1)}$ and $\boldsymbol{U}_I^{(1)}$ are the vectors for the so-called even mode, and $\boldsymbol{U}_V^{(2)}$ and $\boldsymbol{U}_I^{(2)}$ are the vectors for the so-called odd mode. The electromagnetic fields for even and odd modes are shown in Figs. 17a and 17b, respectively. It is understood that the symmetrical plane S becomes the magnetic wall and electrical wall for the even mode and the odd mode, respectively, as shown in Fig. 17.

The equivalent network of a symmetrical coupled line is shown in Fig. 16a and 16b.

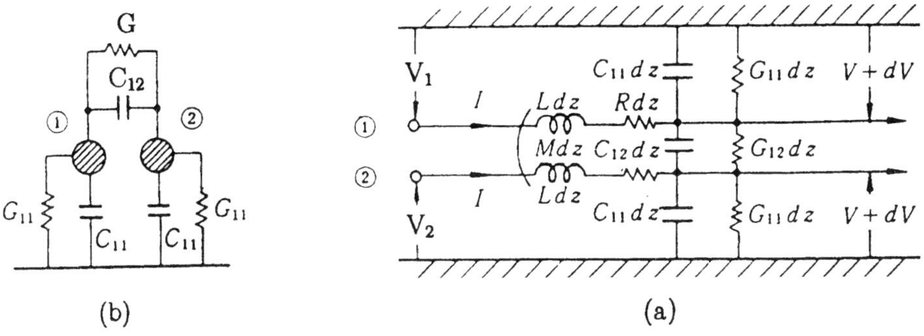

Figure 16. Distributed constant of coupled lines. (a) Side view; (b) sectional view.

However, because the matrix $Z \cdot \Delta z$ is the impedance matrix viewed from $z = 0$ when the lines are shortened at $z = \Delta z$, as shown in Fig. 18a, the impedance of each line for even and odd modes takes the values $z^{(e)}$ and $z^{(o)}$, respectively:

$$z^{(e)} = R + j\omega(L + M) \tag{27a}$$

$$z^{(o)} = R + j\omega(L - M) \tag{27b}$$

as shown in Fig. 18b, where $Z^{(e)}$ and $Z^{(o)}$ are the eigenvalues of the Z matrix. Because $Y \cdot \Delta Z$ is the admittance matrix viewed from $Z = 0$ when lines are opened at $Z = \Delta Z$, as shown in Fig. 18c, the admittance of each line for even and odd modes takes the values of $y^{(e)}$ and $y^{(o)}$:

$$y^{(e)} = G_o - G_{12} + j\omega(C_o - C_{12}) \tag{28a}$$

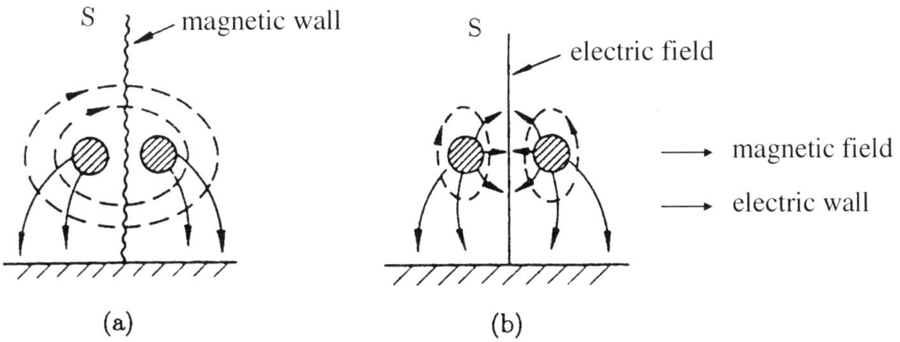

Figure 17. Electromagnetic fields of even (a) and odd (b) modes.

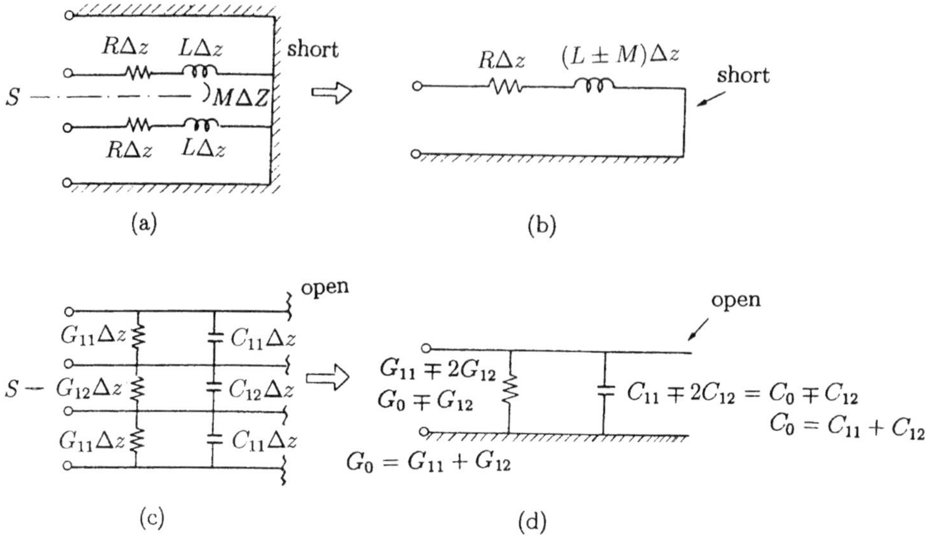

Figure 18. Explanation of eigenvalues of Z and Y matrices. (a) Network with $Z\cdot\Delta z$ matrix; (b) equivalent network of eigenvalues of $Z\cdot\Delta z$ matrix; (c) network with $Y\cdot\Delta z$ matrix; (d) equivalent network of eigenvalues of $Z\cdot\Delta z$ matrix.

$$y^{(o)} = G_o + G_{12} + j\omega(C_o + C_{12}) \tag{28b}$$

where

$$G_o = G_{11} + G_{12}, \qquad C_o = C_{11} + C_{12}$$

as shown in Fig. 18d, where $y^{(e)}$ and $y^{(o)}$ are the eigenvalues of the Y matrix. Therefore, the propagation constant $\gamma_{e,o}$ and the characteristic impedance of even and odd modes $Z_{e,o}$ are calculated using Eqs. (29a) and (29b) following Eqs. (17f) and (17i):

$$\gamma_{e,o} = \sqrt{z^{(e)(o)}y^{(e)(o)}}$$

$$= \{[R + j\omega(L \pm M)][(G_o \mp G_{12}) + j\omega(C_o \pm C_{12})]\}^{1/2} \tag{29a}$$

$$z_{e,o} = \left(\frac{R + j\omega(L \pm M)}{(G_o \mp G_{12}) + j\omega(C_o \mp C_{12})}\right)^{1/2} \tag{29b}$$

where the upper and lower signs correspond to the even and odd modes, respectively.

In the lossless lines, setting $R = G_o = G_{12} = 0$, we get

$$\frac{\gamma_e}{j} = k_{ze} = \omega\sqrt{(L + M)(C_o - C_{12})} = \omega\sqrt{LC_o(1 + k_l)(1 - k_c)} \tag{30a}$$

$$\frac{\gamma_0}{j} = k_{zo} = \omega\sqrt{(L - M)(C_o + C_{12})} = \omega\sqrt{LC_o(1 - k_l)(1 + k_c)} \tag{30b}$$

$$Z_e = W_e = \sqrt{\frac{L + M}{C_o - C_{12}}} = \sqrt{\frac{L}{C_o}}\sqrt{\frac{1 + k_l}{1 - k_c}} \tag{30c}$$

$$Z_o = W_o = \sqrt{\frac{L - M}{C_o + C_{12}}} = \sqrt{\frac{L}{C_o}}\sqrt{\frac{1 - k_l}{1 + k_c}} \tag{30d}$$

where

$$k_l = \frac{M}{L} \qquad k_c = \frac{C_{12}}{C_o} \tag{30e}$$

When the coupled symmetrical lines are in the homogeneous medium, we have

$$k_{ze} = k_{zo} \tag{31a}$$

Substituting Eq. (31a) into Eqs. (30a)–(30d), we get

$$k_l = k_c = k \tag{31b}$$

$$k_z = \omega\sqrt{LC_o(1 - k^2)} \tag{31c}$$

$$Z_e = W_e = \sqrt{\frac{L}{C_o}\left(\frac{1 + k}{1 - k}\right)} \tag{31d}$$

$$Z_o = W_o = \sqrt{\frac{L}{C_o}\left(\frac{1 - k}{1 + k}\right)} \tag{31e}$$

From Eqs. (30a)–(30e), it is understood that the voltages and currents of conductive lines are superposition of the even mode with the constant in Eqs. (30a) and (30c) and of the odd mode with the constant in Eqs. (30b) and (30d). Therefore, denoting the voltage and current corresponding to the even and odd modes at $z = 0$ by $V_{i\pm}^{(e)}$ and $I_{i\pm}^{(e)}$, and $V_{i\pm}^{(o)}$ and $I_{i\pm}^{(o)}$ ($i = 1, 2$), respectively, where the plus and minus sign show the incident and reflected waves, respectively. We can decompose V_{i+} and I_{i+}, and V_{i-} and I_{i-} to those shown in Figs. 19a and 19b. It is also expressed by Eqs. (32a), (32b), (32c), and (32d):

$$V_1 = V_+^{(e)}e^{-\gamma_e z} + V_-^{(e)}e^{\gamma_e z} + V_+^{(o)}e^{-\gamma_o z} + V_-^{(o)}e^{\gamma_o z} \tag{32a}$$

$$V_2 = V_+^{(e)}e^{-\gamma_e z} + V_-^{(e)}e^{\gamma_e z} - V_+^{(o)}e^{-\gamma_o z} - V_-^{(o)}e^{\gamma_o z} \tag{32b}$$

$$I_1 = \frac{V_+^{(e)}}{Z_e}e^{-\gamma_e z} - \frac{V_-^{(e)}}{Z_e}e^{\gamma_e z} + \frac{V_+^{(o)}}{Z_o}e^{-\gamma_o z} - \frac{V_-^{(o)}}{Z_o}e^{\gamma_o z} \tag{32c}$$

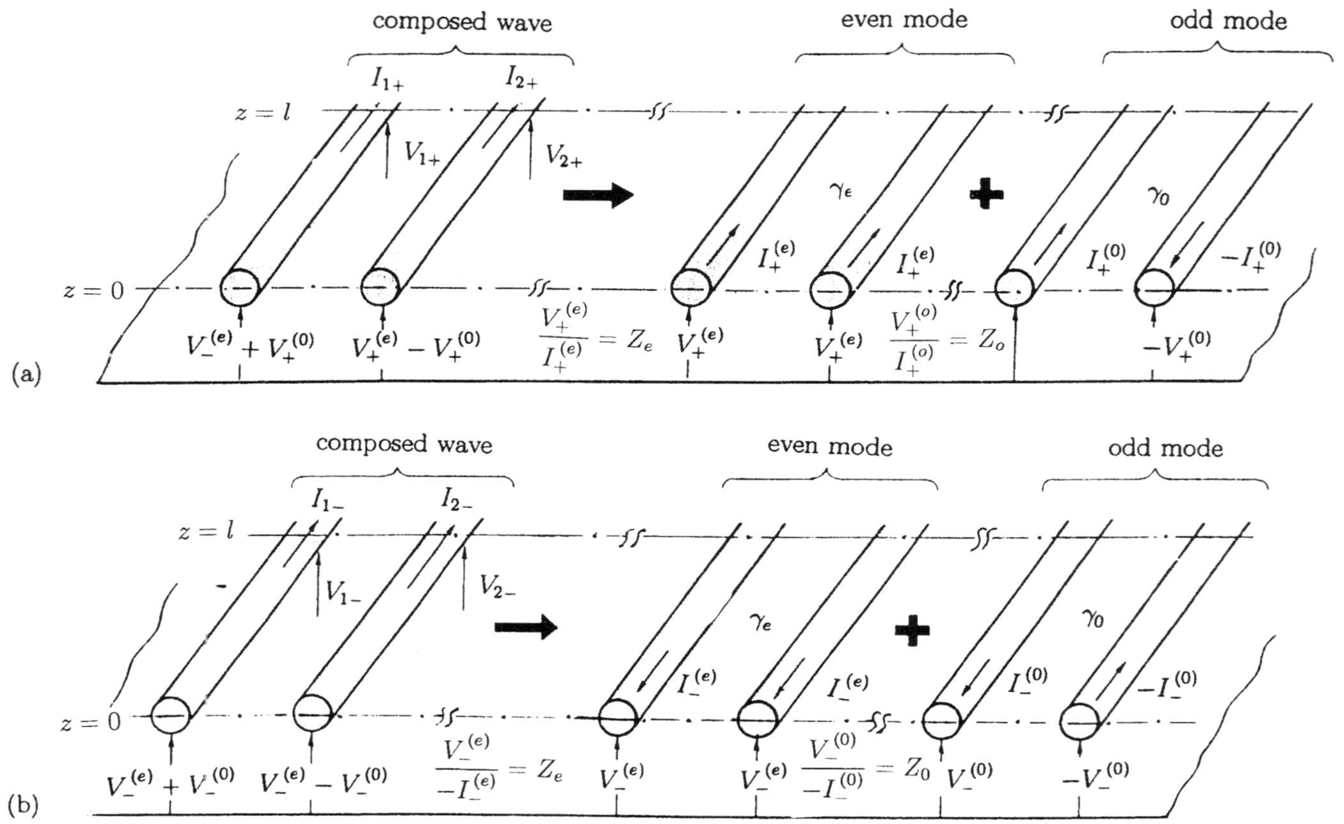

Figure 19. (a) Decomposition to even mode and odd modes of an incident wave; (b) decomposition to even and odd modes of a reflected wave.

$$I_2 = \frac{V_+^{(e)}}{Z_e} e^{-\gamma_e z} - \frac{V_-^{(e)}}{Z_e} e^{\gamma_e z} - \frac{V_+^{(o)}}{Z_o} e^{-\gamma_o z} + \frac{V_-^{(o)}}{Z_o} e^{\gamma_o z} \tag{32d}$$

In the above equations, if we use the values of T_V and T_I of Eq. (26) and the mode vectors $\boldsymbol{v}^{(e)}$ and $\boldsymbol{v}^{(o)}$ of Eq. (24a),

$$V_\pm^{(e)} = \frac{v_\pm^{(e)}}{\sqrt{2}}, \qquad V_\pm^{(o)} = \frac{v_\pm^{(o)}}{\sqrt{2}} \tag{32e}$$

Equations (32a)–(32d) correspond to Eqs. (11a) and (11b).

Multicoupled Lossless Line in a Lossless Homogeneous Medium

General Theory As described in Section 3.2 in Chapter 1, there exist TEM modes in parallel multicoupled lines in a lossless homogeneous medium. Denoting the sth mode voltages and currents by $\boldsymbol{V}_\pm^{(s)}$ and $\boldsymbol{I}_\pm^{(s)}$, where the plus and minus signs show the incident and reflected waves, $\boldsymbol{V}_\pm^{(s)}$ and $\boldsymbol{I}_\pm^{(s)}$, travel with same phase constant, k_z. They are expressed by

$$k_z = \omega\sqrt{\mu\varepsilon} \tag{33a}$$

$$\boldsymbol{V}_\pm^{(s)} = \boldsymbol{V}_{o\pm}^{(s)} e^{\mp jk_z z}, \quad \boldsymbol{I}_\pm = \boldsymbol{I}_{o\pm}^{(s)} e^{\mp jk_z z} \tag{33b}$$

for $s = 1, \ldots, n$ and k_z is the phase constant.

Substituting Eq. (33b) into Eqs. (15a) and (15b), and using the relation $Z = j\omega L$, $Y = j\omega C$ described in Note 2, we get the following results:

$$\frac{d^2\boldsymbol{V}}{dz^2} = ZY\boldsymbol{V} = -\omega^2 LC\boldsymbol{V} = -k_z^2\boldsymbol{V}$$
$$\frac{d^2\boldsymbol{I}}{dz^2} = YZ\boldsymbol{I} = -\omega^2 CL\boldsymbol{I} = -k_z^2\boldsymbol{I} \tag{33c}$$

and

$$LC = CL = \frac{k_z^2}{\omega^2}\,\mathrm{I} \tag{33d}$$

$$L = \frac{k_z^2}{\omega^2} C^{-1} \tag{33e}$$

where I is the unit matrix.

Because LC is commutative as shown in Eq. (33d), we can diagonalize the orthogonal matrix P as

$$P^t LP = \Lambda_L = \mathrm{diag}(\lambda_{L_1}, \ldots, \lambda_{L_n}) \tag{33f}$$

$$P^t CP = \Lambda_C = \mathrm{diag}(\lambda_{C_1}, \ldots, \lambda_{C_n}) \tag{33g}$$

L and C are the matrices of a real existing network. Then, λ_{L_s} and λ_{C_s} ($s = 1, \ldots, n$) are positive real numbers. Because

$$P^tLCP = P^tLP \cdot P^tCP = \Lambda_L \Lambda_C = \frac{k_z^2}{\omega^2}\,\mathrm{I} \tag{33h}$$

then

$$\lambda_{L_s}\lambda_{C_s} = \frac{k_z^2}{\omega^2} = \frac{1}{v_p^2} = \mu\varepsilon, \quad \mathrm{s} = 1, \ldots, n \tag{33i}$$

where v_p is a phase velocity, μ is a permeability, and ε is a dielectric constant.

Substituting Eqs. (33f) and (33g) into Eqs. (18a) and (18b), we get Eqs. (34a)–(34c):

$$\boldsymbol{v}_\pm = \pm\frac{\omega}{k_z} P^{-1}LP\boldsymbol{i}_\pm = \pm\frac{\omega}{k_z} \Lambda_L \boldsymbol{i}_\pm = \pm\Lambda_W \boldsymbol{i}_\pm \tag{34a}$$

$$\boldsymbol{i}_\pm = \pm\frac{\omega}{k_z} P^{-1}LP\boldsymbol{v}_\pm = \pm\frac{\omega}{k_z} \Lambda_c \boldsymbol{v}_\pm = \pm\Lambda_Y \boldsymbol{v}_\pm \tag{34b}$$

$$\Lambda_W = \begin{bmatrix} W_1 & & 0 \\ & \ddots & \\ 0 & & W_n \end{bmatrix}, \qquad \Lambda_Y = \Lambda_W^{-1} = \begin{bmatrix} Y_{c1} & & 0 \\ & \ddots & \\ 0 & & Y_{cn} \end{bmatrix}$$

$$Z_s = W_s = \frac{\omega\lambda_L^{(s)}}{k_z} = \frac{k_z}{\omega\lambda_c^{(s)}} = \sqrt{\frac{\lambda_L^{(s)}}{\lambda_c^{(s)}}} = v_p\lambda_L^{(s)}$$

$$v_p = \frac{1}{\sqrt{\mu\varepsilon}} \tag{34c}$$

Considering the only sth mode voltages $\boldsymbol{V}_\pm^{(s)}$ and currents $\boldsymbol{I}_\pm^{(s)}$ on the lines, we have the same relation as Eq. (24d); that is,

$$\boldsymbol{V}_\pm = \pm W_s \boldsymbol{I}_\pm \tag{34d}$$

It is understood that the impedance between a line and a ground takes the same values of W_s for the sth incident mode, and $-W_s$ for the sth reflected wave. This matter is correct only for the case when the orthogonal matrix P is used for T_v in Eq. (18a). In this case, however, all eigenvalues, i.e., phase constants, takes the same values.

The relative values of an arbitrary row of T_V is changeable, which results in taking a different impedance, as shown in later example.

In the expressions

$$\boldsymbol{v} = \boldsymbol{v}_+ e^{-jk_z z} + \boldsymbol{v}_- e^{jk_z z} \tag{35a}$$

$$\boldsymbol{i} = \boldsymbol{i}_{+}e^{-jk_z z} + \boldsymbol{i}_{-}e^{jk_z z} \tag{35b}$$

by setting

$$\boldsymbol{v}_{+} + \boldsymbol{v}_{-} = \boldsymbol{a}, \qquad \boldsymbol{v}_{-} - \boldsymbol{v}_{+} = \boldsymbol{b} \tag{35c}$$

and substituting Eq. (34a) into Eq. (35b), we get

$$\boldsymbol{v} = \boldsymbol{a} \cos k_z z + \boldsymbol{b} j \sin k_z z \tag{35d}$$

$$\boldsymbol{i} = -\Lambda_W^{-1}(\boldsymbol{b} \cos k_z z + \boldsymbol{a} j \sin k_z z) \tag{35e}$$

Representing the voltages and the currents at $z = 0$ by $\boldsymbol{v}_0$ and $\boldsymbol{i}_0$, because $\boldsymbol{a} = \boldsymbol{v}_0$ and $\boldsymbol{b} = -\Lambda_W \boldsymbol{i}_0$, we get

$$\boldsymbol{v} = \boldsymbol{v}_0 \cos k_z z - \Lambda_W \boldsymbol{i}_0 j \sin k_z z \tag{35f}$$

$$\boldsymbol{i} = \Lambda_W^{-1} \boldsymbol{v}_0 j \sin k_z z + \boldsymbol{i}_0 \cos k_z z \tag{35g}$$

Representing the voltages and the currents at $z = l$ by $\boldsymbol{v}_l$ and $\boldsymbol{i}_l$, we get

$$\begin{aligned} \boldsymbol{v}_l &= \frac{1}{\sqrt{1 - u_0^2}} (\boldsymbol{v}_0 - u_0 \Lambda_W \boldsymbol{i}_0) \\ &= \frac{1}{\sqrt{1 - u_0^2}} (-u_0 \Lambda_W^{-1} \boldsymbol{v}_0 + \boldsymbol{i}_0) \\ u_0 &= j \tan k_z l \end{aligned} \tag{36a}$$

From Eqs. (36a) and (35g), we get

$$\begin{bmatrix} \boldsymbol{v}_l \\ \boldsymbol{i}_l \end{bmatrix} = \frac{1}{\sqrt{1 - u_0^2}} \begin{bmatrix} U & -u_0 \Lambda_W \\ -u_0 \Lambda_W^{-1} & U \end{bmatrix} \begin{bmatrix} \boldsymbol{v}_0 \\ \boldsymbol{i}_0 \end{bmatrix}, \qquad U = \begin{bmatrix} 1 & 0 \\ 0 & 1 \end{bmatrix} \tag{36b}$$

From Eq. (36b), we get

$$\begin{bmatrix} \boldsymbol{v}_0 \\ \boldsymbol{i}_0 \end{bmatrix} = \frac{1}{\sqrt{1 - u_0^2}} \begin{bmatrix} U & u_0 \Lambda_W \\ u_0 \Lambda_W^{-1} & U \end{bmatrix} \begin{bmatrix} \boldsymbol{v}_l \\ \boldsymbol{i}_l \end{bmatrix} \tag{36c}$$

This equation corresponds to Eq. (13b) in a single transmission line; Eq. (36c) is the expression by mode voltage and current vectors.

As we have the relations

$$P\boldsymbol{v}_{l,0} = \boldsymbol{V}_{l,0}, \qquad \boldsymbol{v}_{l,0} = P^{-1}\boldsymbol{V}_{l,0} \quad \text{[from Eqs. (16b) and (22b)]}$$

$$P\Lambda_W P^{-1} = \frac{\omega}{k_z} L = \frac{L}{\sqrt{\mu\varepsilon}} \quad \text{[from Eqs. (34a) and (34c)]}$$

we get the expression in Eq. (37) from Eq. (36c):

$$\begin{bmatrix} \boldsymbol{V}_0 \\ \boldsymbol{I}_0 \end{bmatrix} = F \begin{bmatrix} \boldsymbol{V}_l \\ \boldsymbol{I}_l \end{bmatrix}, \qquad F = \frac{1}{\sqrt{1 - u_0^2}} \begin{bmatrix} U & u_0 \dfrac{L}{\sqrt{\mu\varepsilon}} \\ u_0 \dfrac{C}{\sqrt{\mu\varepsilon}} & U \end{bmatrix} \tag{37}$$

Equation (37) corresponds to Eq. (13b) in the case of a single transmission line. To examine this, we set $u_0 = j \tan k_z l$ and

$$\frac{L}{\sqrt{\mu\varepsilon}} \to Z_c \quad \text{in a single line}$$

$$\frac{C}{\sqrt{\mu\varepsilon}} \to \frac{1}{Z_c} \quad \text{in a single line}$$

$$\begin{bmatrix} V_1 \\ I_1 \end{bmatrix} = F \begin{bmatrix} V_2 \\ I_2 \end{bmatrix}, \qquad F = \begin{bmatrix} \cos k_z l & jZ_c \sin k_z l \\ j \dfrac{1}{Z_c} \sin k_z l & \cos k_z l \end{bmatrix}$$

which is the F matrix expression of a single transmission line of Eq. (13b).

Two Coupled Lossless Lines

Relations Between Line Constants and Propagation Constants In the case of $n = 2$, the L and C matrices can be expressed by Eqs. (38a) and (38b), respectively:

$$L = \begin{bmatrix} L_1 & M \\ M & L_2 \end{bmatrix} = L_0 \begin{bmatrix} n_l & k_l \\ k_l & \dfrac{1}{n_l} \end{bmatrix}$$

$$L_0 = \sqrt{L_1 L_2}, \qquad k_l = \frac{M}{\sqrt{L_1 L_2}}, \qquad n_l = \sqrt{\frac{L_1}{L_2}} \tag{38a}$$

$$C = \begin{bmatrix} C_1 & -C_{12} \\ -C_{12} & C_2 \end{bmatrix} = C_0 \begin{bmatrix} n_c & -k_c \\ -k_c & \dfrac{1}{n_c} \end{bmatrix}$$

$$C_0 = \sqrt{C_1 C_2}, \qquad C_1 = C_{11} + C_{12}, \qquad C_2 = C_{22} + C_{12}$$

$$k_c = \frac{C_{12}}{\sqrt{C_1 C_2}}, \qquad n_c = \sqrt{\frac{C_1}{C_2}} \tag{38b}$$

When we apply the condition of Eq. (33e) to Eqs. (38a) and (38d), we get

$$\frac{1}{v_p^2}\frac{1}{L_0(1-k_l^2)}\begin{bmatrix}\frac{1}{n_l} & -k_l\\ -k_l & n_l\end{bmatrix} = C_0\begin{bmatrix} n_c & -k_c\\ -k_c & \frac{1}{n_c}\end{bmatrix}$$

To satisfy the above equation, we get Eqs. (39a)–(39d). Calculating the matrix $[LC]$ by using Eqs. (39a) and (39b), we get (39e):

$$k_l = k_c = k \tag{39a}$$

$$n_l n_c = 1 \tag{39b}$$

$$\frac{M}{\sqrt{L_1L_2}} = \frac{C_{12}}{\sqrt{C_1C_2}} \tag{39c}$$

$$\frac{L_1}{L_2} = \frac{C_2}{C_1} \tag{39d}$$

$$k_c = \omega\sqrt{L_0C_0(1-k^2)} = \frac{\omega}{v_p} = \omega\sqrt{\mu\varepsilon} \tag{39e}$$

In the case of symmetrical coupled lines in an inhomogenous medium, we have the following relations from Eqs. (38a)–(39e):

$$n_l = n_c = 1, \qquad L_1 = L_2 = L_0, \qquad C_1 = C_2 = C_0 \tag{40a}$$

$$\frac{M}{L_0} = \frac{C_{12}}{C_0} \quad (C_1 = C_{11} + C_{12})$$

$$k_z = \omega\sqrt{L_0C_0(1-k^2)} \tag{40b}$$

Eigenvectors and the Corresponding Characteristic Impedance Because k_z can be obtained from $[LC]$, the values of k_z take the same values whenever the orthogonal matrix P is used or a set of nonorthogonal matrices T_v and T_i are used for diagonalization of the matrix $[LC]$.

- First, we described in the case when we use P.

  Substituting

$$n_l = \frac{1}{n_c} = n, \qquad k_l = k_c = k$$

and solving Eq. (33f), we get

$$\frac{\lambda_{L_s}}{L_0} = \frac{1}{2}\left[\left(n+\frac{1}{n}\right) \pm \sqrt{\left(n-\frac{1}{n}\right)^2 + 4k^2}\right] \quad (s = 1, 2) \tag{41a}$$

Substituting the eigenvalues $\lambda_L^{(1)}$ and $\lambda_L^{(2)}$ into the eigenequation, we can obtain the eigenvectors $\boldsymbol{u}^1$ and $\boldsymbol{u}^2$ corresponding to $\lambda_L^{(1)}$ and $\lambda_L^{(2)}$ as follows:

$$P = [\boldsymbol{u}^1, \boldsymbol{u}^2] = \frac{1}{\sqrt{1 + R_c^2}} \begin{bmatrix} 1 & R_c \\ R_c & -1 \end{bmatrix} \tag{41b}$$

$$R_c = (-S + \sqrt{S^2 + 1}) \tag{41c}$$

$$S = \frac{n_l - 1/n_l}{2k} = \frac{1/n_c - n_c}{2k} = \frac{L_1 - L_2}{2M} = \frac{C_2 - C_1}{2C_{12}} \tag{41d}$$

Substituting Eq. (41a) into Eq. (34c), we get the characteristic impedance:

$$Z_{c,\pi} = W_{c,\pi} = \sqrt{\frac{L_0}{C_0}\left(\frac{[(n^2 + 1/n^2) + 2k^2 \pm n + 1/n]\sqrt{(n - 1/n)^2 + 4k^2}}{2(1 - k^2)}\right)} \tag{41e}$$

where Z_{c1} and Z_{c2} are the characteristic impedances of real lines 1 and 2, and W_1 and W_2 are the characteristic impedances of independent mode lines for $\lambda_L^{(1)}$ and $\lambda_L^{(2)}$.

In the symmetrical coupled line in an inhomogeneous medium, we get

$$P = [\boldsymbol{u}^1, \boldsymbol{u}^2] = \frac{1}{\sqrt{2}} \begin{bmatrix} 1 & 1 \\ 1 & -1 \end{bmatrix} \tag{42a}$$

$$Z_{e,o} = W_{e,o} = \sqrt{\frac{L_0}{C_0}\left(\frac{1 \pm k}{1 \mp k}\right)} \tag{42b}$$

In Eqs. (42a) and (42b), suffixes 1 and 2 show the even and odd modes, respectively, described in Eqs. (31d) and (31e).

Next, we will describe for the case of when T_v is not orthogonal matrix.

When the eigenvalues of $[LC]$ are all degenerated, we can choose two eigenvectors as follows:

1st eigenvector: The voltage of two lines is same.
2nd eigenvector: The current of two lines flow in opposite directions with the same magnitude.

As a 1st eigenvector, we assume T_v of Eq. (43a):

$$T_V = \begin{bmatrix} 1 & x \\ 1 & -y \end{bmatrix}, \qquad T_V^{-1} = \frac{1}{x + y} \begin{bmatrix} y & x \\ 1 & -1 \end{bmatrix} \tag{43a}$$

From Eq. (17j) of the power invariant condition, we get

$$T_I = (T_V^{-1})^t = \frac{1}{x+y}\begin{bmatrix} y & 1 \\ x & -1 \end{bmatrix}, \qquad T_I^{-1} = (x+y)\begin{bmatrix} 1 & 1 \\ x & -y \end{bmatrix} \tag{43b}$$

Calculating $T_V^{-1}LT_I$ by using L of Eq. (38a), T_V of Eq. (43a), and T_I of Eq. (43b), we get

$$T_V^{-1}LT_I = \frac{L_0}{(x+y)^2}\begin{bmatrix} \dfrac{x^2}{n_l} + 2k_l xy + n_l y^2 & x\left(k_l - \dfrac{1}{n_l}\right) - y(k_l - n_l) \\ x\left(k_l - \dfrac{1}{n_l}\right) - y(k_l - n_l) & n_l - 2k_l + \dfrac{1}{n_l} \end{bmatrix} \tag{43c}$$

To be diagonalized, we need the condition of Eq. (43d) under the consideration of Eqs. (39a) and (39b):

$$\frac{x}{y} = \frac{k_l - n_l}{k_l - 1/n_l} = \frac{k_c - 1/n_c}{k_c - n_c} = \delta \tag{43d}$$

As we only need the relative values of x and y, we use $x + y = 1$. In this case, we get the following results:

$$\frac{x}{x+y} = x = \frac{\delta}{1+\delta}, \qquad y = \frac{1}{1+\delta} \tag{44a}$$

$$T_V = \begin{bmatrix} 1 & \dfrac{\delta}{1+\delta} \\ 1 & -\dfrac{1}{1+\delta} \end{bmatrix} \tag{44b}$$

$$T_I = \begin{bmatrix} \dfrac{1}{1+\delta} & 1 \\ \dfrac{\delta}{1+\delta} & -1 \end{bmatrix} \tag{44c}$$

By using these relations, Eq. (43c) can be diagonalized as Eq. (44d), and C from Eq. (38b) can be diagonalized as Eq. (44e):

$$T_V^{-1}LT_I = \Lambda_L' = \text{diag}[\lambda_{L_1}', \lambda_{L_2}']$$

$$\lambda_{L_1}' = L_0\left(\frac{x^2}{n_l} + n_l y^2 + 2k_l xy\right)$$

$$\lambda_{L_2}' = L_0\left(n_l + \frac{1}{n_l} - 2k_l\right) \tag{44d}$$

$$T_I^{-1}CT_V = \Lambda_c' = \text{diag}[\lambda_c^{(1)}, \lambda_c^{(2)}]$$

$$\lambda_c^{(1)} = C_0\left(n_c + \frac{1}{n_c} - 2k_c\right)$$

$$\lambda_c^{(2)} = C_0\left(\frac{x^2}{n_c} + n_c y^2 + 2k_c xy\right) \tag{44e}$$

Substituting Eq. (33b) into Eq. (18a), we get

$$\boldsymbol{v}_a = \pm\Lambda_{W_c}'\boldsymbol{i}_a, \quad \Lambda_{W_c} = \begin{bmatrix} W_u & 0 \\ 0 & W_b \end{bmatrix} \tag{44f}$$

$$\Lambda_{W_c}' = \frac{\omega}{k_z}\Lambda_L' = \frac{1}{\sqrt{\mu\varepsilon}}\Lambda_L' = \sqrt{\mu\varepsilon}(\Lambda_c')^{-1} \tag{44g}$$

(See Note 3.) Substituting Eqs. (44d) and (44e) into Eq. (44g), we get the values of W_u and W_b as follows:

$$W_u = \frac{\sqrt{\mu\varepsilon}}{C_0(n_l + 1/n_c - 2k_c)} = \frac{\sqrt{\mu\varepsilon}}{C_1 + C_2 - 2C_{12}} \tag{44h}$$

$$W_b = \frac{1}{\sqrt{\mu\varepsilon}}L_0\left(n_l + \frac{1}{n_l} - 2k_l\right) = \frac{1}{\sqrt{\mu\varepsilon}}(L_1 + L_2 - 2L_{12}) \tag{44i}$$

When we correspond the real voltages to mode voltages, we can obtain the following relationships from Eqs. (44b) and (44c):

In the case of the 1st mode (the so-called unbalance mode): Setting

$$v^{(1)} = v_u \neq 0, \quad v^{(2)} = 0; \qquad i^{(1)} = i_u \neq 0, \quad i^{(2)} = 0$$

$$V_1 = V_2 = v^{(1)} = v_u, \qquad I_1^{(1)} = \frac{1}{1+\delta}i_u, \qquad I^{(2)} = \frac{\delta}{1+\delta}i_u$$

$$i_u = I_1^{(1)} + I_2^{(1)}$$

Then

$$W_u = \frac{v_u}{i_u} = \frac{V_1}{I_1^{(1)} + I_2^{(1)}} \tag{45a}$$

In the case of 2nd mode (the so-called balance mode): Setting

$$v^{(2)} = v_b \neq 0, \quad v^{(1)} = 0, \qquad i^{(2)} = i_b \neq 0, \quad i^{(1)} = 0$$

$$V_1 = \frac{\delta}{1+\delta}v_b, \qquad V_2 = \frac{1}{1+\delta}v_b$$

$$v_b = V_1 - V_2, \qquad i_b = I_1 = -I_2$$

Then,

$$W_b = \frac{v_b}{i_b} = \frac{V_1 - V_2}{I_1} \tag{45b}$$

The voltage and the current of the unbalance and balance mode are shown in Fig. 20.

In the case of an asymmetrical structure, it is clear from Fig. 20 that the impedance of each line even for the same mode is different. Characteristic impedances for 1st and 2nd modes at lines 1 and 2 can be directly obtained from Eqs. (44b) and (44c) as follows:

$$Z_{c1}^{(1)} = \frac{v^{(1)}}{i^{(1)}}(1 + \delta), \qquad Z_{c2}^{(1)} = \frac{v^{(1)}}{i^{(1)}} \frac{1 + \delta}{\delta}, \qquad Z_{c1}^{(1)} \neq Z_{c2}^{(1)} \tag{46a}$$

$$Z_{c1}^{(2)} = \frac{v^{(2)}}{i^{(2)}} \frac{\delta}{1 + \delta}, \qquad Z_{c2}^{(2)} = \frac{v^{(2)}}{i^{(2)}} \frac{1}{1 + \delta}, \qquad Z_{c1}^{(2)} \neq Z_{c2}^{(2)} \tag{46b}$$

where the superscripts (1) and (2) indicate the unbalance and balance modes, and 1 and 2 in the subscripts denotes the number of conductive lines.

F Matrix The F matrix becomes Eq. (37) for $n = 2$. Therefore, it is expressed as

$$\begin{bmatrix} V_{10} \\ V_{20} \\ I_{10} \\ I_{20} \end{bmatrix} = F \begin{bmatrix} V_{1l} \\ V_{2l} \\ I_{1l} \\ I_{2l} \end{bmatrix}$$

$$F = \begin{bmatrix} \cos k_z l & 0 & j\dfrac{L_{11} \sin k_z l}{\sqrt{\mu\varepsilon}} & j\dfrac{L_{12} \sin k_z l}{\sqrt{\mu\varepsilon}} \\ 0 & \cos k_z l & j\dfrac{L_{12} \sin k_z l}{\sqrt{\mu\varepsilon}} & j\dfrac{L_{22} \sin k_z l}{\sqrt{\mu\varepsilon}} \\ j\dfrac{C_{11} \sin k_z l}{\sqrt{\mu\varepsilon}} & j\dfrac{C_{12} \sin k_z l}{\sqrt{\mu\varepsilon}} & \cos k_z l & 0 \\ j\dfrac{C_{12} \sin k_z l}{\sqrt{\mu\varepsilon}} & j\dfrac{C_{22} \sin k_z l}{\sqrt{\mu\varepsilon}} & 0 & \cos k_z l \end{bmatrix} \tag{47}$$

In Eq. (47), L and C are related by Eq. (33e); the constants k_e, k_c, n_e, n_c, and k_z are related by Eqs. (39a)–(39e).

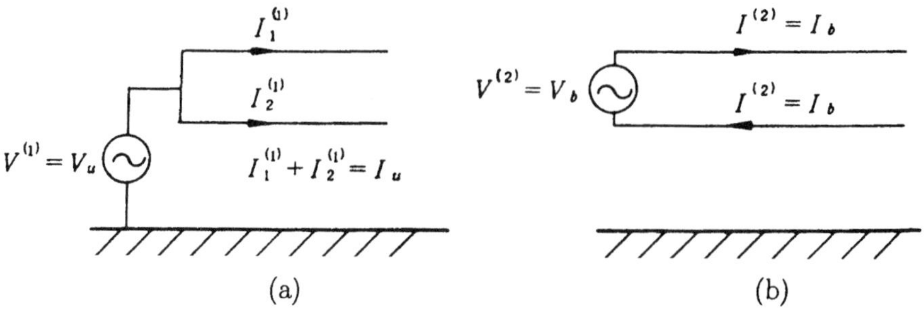

Figure 20. Description of an unbalance mode (a) and balance mode (b) of two parallel lines in a lossless homogeneous medium.

Asymmetrical Coupled Two Lossless Lines

Eigenvectors, Eigenvalues (Phase Constant), Characteristic Impedance

As examples, we have the unsymmetrical coupled lines of Fig. 21. In Fig. 22, the distributed constants of unsymmetrical coupled lines are shown. The L and C matrices can be expressed as Eqs. (38a) and (38b), respectively. The T_V and T_I diagonalize L and C in the forms

$$T_V^{-1}LT_I = \Lambda_L = \mathrm{diag}(\lambda_L^c, \lambda_L^\pi)$$
$$T_I^{-1}CT_V = \Lambda_I = \mathrm{diag}(\lambda_c^c, \lambda_c^\pi)$$

as shown in Eqs. (19c) and (19d), can be obtained from Eqs. (19e) and (17j): which are shown as follows.

As shown in Eqs. (48a) and (48b), Λ_V and Λ_I are the arbitral diagonal

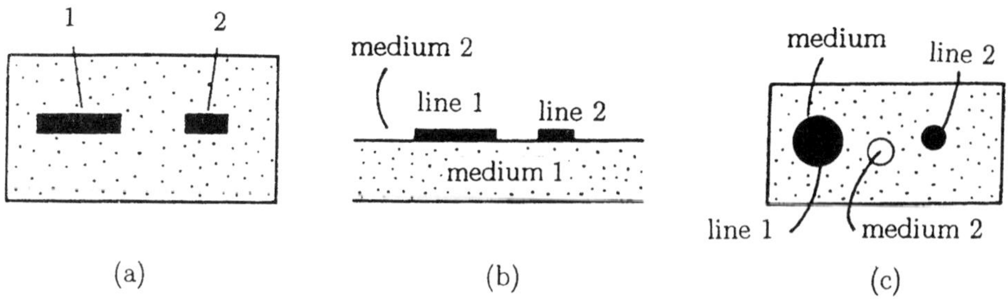

Figure 21. Asymmetrical coupled lines. (a) Asymmetrical coupled triplates; (b) asymmetrical coupled microstrip lines; (c) asymmetrical coupled parallel lines in an inhomogeneous medium.

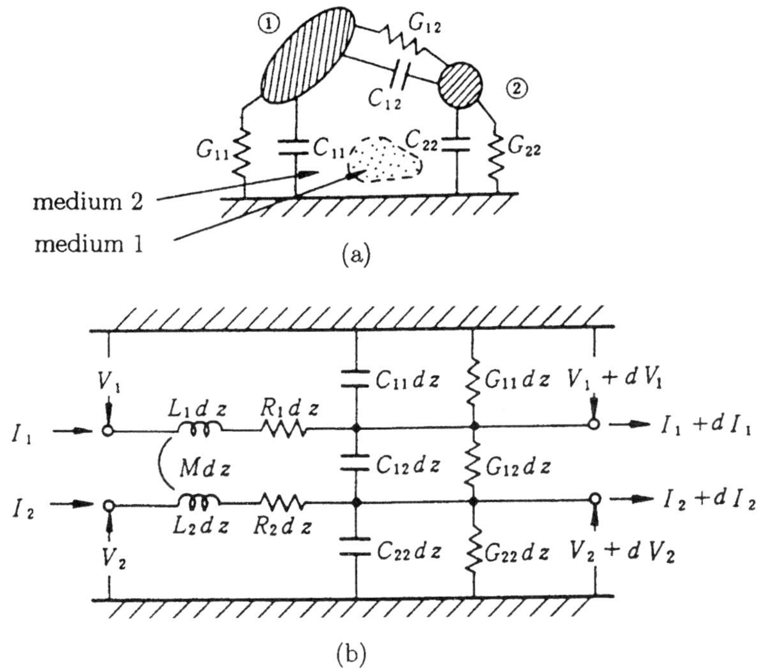

Figure 22. Distributing constants of asymmetrical coupled lines. (a) Sectional view; (b) top view.

matrices. This means that the relative values of elements of a vector is necessary.

$$T_V = \frac{1}{\sqrt{1 - R_c/R_\pi}} \begin{bmatrix} 1 & -\dfrac{1}{R_\pi} \\ R_c & -1 \end{bmatrix} \cdot \Lambda_V(\lambda_{V_1}, \ldots, \lambda_{V_n}) \tag{48a}$$

$$T_I = \frac{1}{\sqrt{1 - R_c/R_\pi}} \begin{bmatrix} 1 & R_c \\ -\dfrac{1}{R_\pi} & -1 \end{bmatrix} \cdot \Lambda_I(\lambda_{I_1}, \ldots, \lambda_{I_n}) \tag{48b}$$

The first and 2nd rows are sometimes called c and π modes, respectively [1,2]. In the case of a symmetrical structure, the c and π modes become the even and odd modes, respectively.

The eigenvalues of the phase constant $k_{zc,\pi}$ can be obtained as $k_{zc,\pi} = \omega\lambda_{L_C}^{(c)(\pi)}$, where $\lambda_{L_C}^{(c)(\pi)}$ can be obtained from Eq. (21a).

$R_{c,\pi}$ is obtained by substituting $k_{zc,\pi}$ into the eigenequation again. The characteristic impedance of an independent line corresponding c and π modes, W_c and W_π, can be obtained from Eq. (23i), where $\lambda_L^{(c)(\pi)}$ and $\lambda_C^{(c)(\pi)}$ are obtained from Eqs. (19c) and (19d) by using T_V and T_I already obtained, because $R_{c,\pi}$ is already obtained. The results are shown in the following [2,3]:

$$\begin{aligned}
k_{zc,\pi} &= \frac{\omega\sqrt{L_0C_0}}{\sqrt{2}}\left\{n_ln_c + \frac{1}{n_ln_c} - 2k_lk_c\right.\\
&\quad \left.\pm\left[\left(\frac{1}{n_ln_c} - n_ln_c\right)^2 + 4\left(\frac{k_l}{n_c} - n_lk_c\right)\left(\frac{k_l}{n_c} - n_lk_l\right)\right]^{1/2}\right\}^{1/2}\\
&= \frac{\omega}{\sqrt{2}}\{L_1C_1 + L_2C_2 - 2MC_{12}\\
&\quad \pm [(L_2C_2 - L_1C_1)^2 + 4(MC_2 - L_1C_{12})(MC_1 - L_2C_{12})]^{1/2}\}^{1/2}
\end{aligned} \tag{48c}$$

$$\begin{aligned}
R_{c,\pi} &= \left\{\frac{1}{n_ln_c} - n_ln_c'\right.\\
&\quad \left.\pm\left[\left(\frac{1}{n_ln_c} - n_ln_c\right)^2 + 4\left(\frac{k_l}{n_c} - n_lk_c\right)\left(k_ln_c - \frac{k_c}{n_l}\right)\right]^{1/2}\right\}\left[2\left(\frac{k_l}{n_c} - n_lk_l\right)\right]^{-1}\\
&= \frac{L_2C_2 - L_1C_1 \pm [(L_2C_2 - L_1C_1)^2 + 4(MC_2 - L_1C_{12})(MC_1 - L_2C_{12})]^{1/2}}{2(MC_2 - L_1C_{12})}
\end{aligned} \tag{48d}$$

$$W_{c,\pi} = \frac{\omega\lambda_L^{c,\pi}}{k_{zc,\pi}} = \frac{k_{zc,\pi}}{\omega\lambda_C^{c,\pi}} = \sqrt{\frac{\lambda_L^{c,\pi}}{\lambda_C^{c,\pi}}} \tag{48e}$$

We can obtain Eq. (48f) from Eq. (48d):

$$R_cR_\pi = \frac{k_c/n_l - k_ln_c}{k_l/n_c - n_lk_c} = \frac{L_2C_{12} - MC_1}{MC_2 - L_1C_{12}} \tag{48f}$$

To obtain the values of R_c and R_π, Eq. (48d) is only useful for the case

$$\frac{k_l}{n_c} - n_ek_c \neq 0 \tag{48g}$$

In the case of coupled line in a homogeneous medium, the left-hand side of Eq. (48g) becomes zero from Eqs. (39a) and (39b). In such a case, we must use the values of R_1 obtained in Eq. (41c) with respect to R_c. R_π must be obtained by Eq. (48h):

$$R_\pi = -\frac{1}{R_c} \tag{48h}$$

This is understandable by comparing Eq. (48a) with Eq. (41b); in other words, T_V should be the orthogonal matrix P in the case of multilines in a homogeneous medium. Therefore, from Eq. (48a), we get $-R_c - 1/R_\pi = 0$. Thus, we must have Eq. (49a) taking into consideration the commutative condition of Eq. (22a) required for the orthogonal matrix to diagonalize.

For

$$R_c R_\pi = -1 \tag{49a}$$

When $LC = CL$ or $ZY = YZ$, $T_V = P$. R_c should be obtained from Eqs. (41c) and (41d). When Eq. (48g) is satisfied, R_c can also be obtained from Eq. (48d).

$$R_c R_\pi \neq -1 \tag{49b}$$

When $LC \neq CL$ or $ZY \neq YZ$, $T_v \neq P$. R_c should be obtained from Eq. (48d).

We can summarize the matter mentioned above as follows.

1. We have two modes whose phase constants are different from each other, as shown in Eq. (48c).
2. To diagonalize L and C at the same time, the matrices T_V and T_I are not orthogonal, as shown in Eqs. (48a) and (48b).

Therefore $R_c \neq -1/R_\pi$, as shown in Eq. (49b). From this, the impedances of line 1 and line 2 are not the same, even if the wave of the same mode is incident on line 1 and line 2 in the same direction.

The matters summarized above are shown in Fig. 23 and Eqs. (50a)–(50d).

$$V_1 = (v_+^c e^{-jk_{zc}z} + v_-^c e^{jk_{zc}z}) + (v_+^\pi e^{-jk_{z\pi}z} + v_-^\pi e^{jk_{z\pi}z}) \tag{50a}$$

$$V_2 = R_c(v_+^c e^{-jk_{zc}z} + v_-^c e^{jk_{zc}z}) + R_\pi(v_+^\pi e^{-jk_{z\pi}z} + v_-^\pi e^{jk_{z\pi}z}) \tag{50b}$$

$$I_1 = Y_{c1}(v_+^c e^{-jk_{zc}z} - v_-^c e^{jk_{zc}z}) + Y_{\pi 1}(v_+^\pi e^{-jk_{z\pi}z} - v_-^\pi e^{jk_{z\pi}z}) \tag{50c}$$

$$I_2 = Y_{c2}R_c(v_+^c e^{-jk_{zc}z} - v_-^c e^{jk_{zc}z}) + Y_{\pi 2}R_\pi(v_+^\pi e^{-jk_{z\pi}z} - v_-^\pi e^{jk_{z\pi}z}) \tag{50d}$$

Notations in Fig. 23 correspond to those for Eqs. (50a)–(50d) as follows:

$$V_{\pm 1}^{(c)} = v_\pm^c, \qquad V_{\pm 2}^{(c)} = R_c v_\pm^c$$

$$V_{\pm 1}^{(\pi)} = v_\pm^\pi, \qquad V_{\pm 2}^{(\pi)} = R_\pi v_\pm^\pi$$

$$V_{\pm 1} = v_\pm^c e^{\mp jk_{zc}z} + v_+^\pi e^{\mp jk_{z\pi}z}$$

$$V_{\pm 2} = R_c v_\pm^c e^{\mp jk_{zc}z} + R_\pi v_\pm^\pi e^{\mp jk_{z\pi}z}$$

$$V_1 = V_{+1} + V_{-1}, \qquad V_2 = V_{+2} + V_{-2} \tag{50e}$$

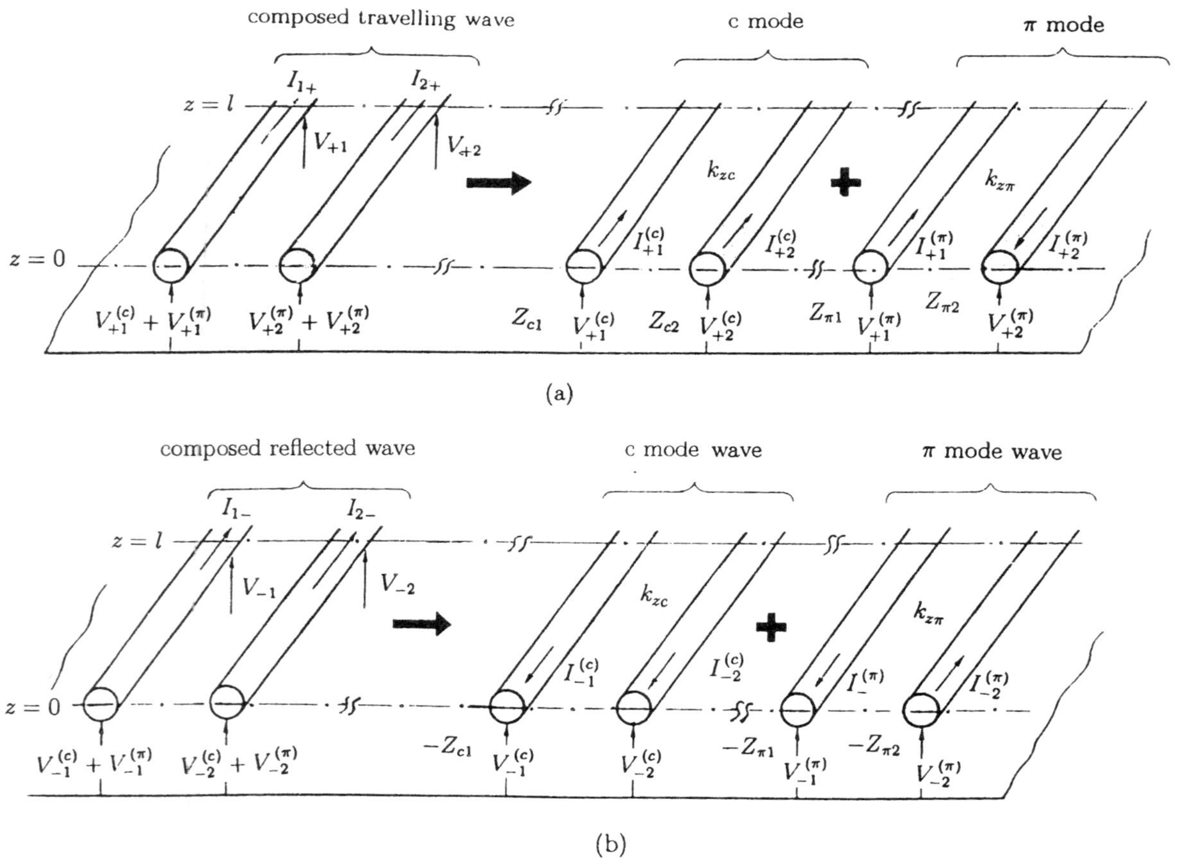

Figure 23. (a) Decomposition to the c mode and the π mode of an incident wave; (b) decomposition to the c mode and the π mode of a reflecting wave.

Y_{c1}, Y_{c2}, Z_{c1}, and Z_{c2} are as follows:

$$Y_{c1} = \frac{1}{W_c}, \qquad Y_{c2} = \frac{-1}{W_c R_c R_\pi}$$

$$Y_{\pi 1} = \frac{-R_c R_\pi}{W_\pi}, \qquad Y_{\pi 2} = \frac{1}{W_\pi}$$

$$Z_{c1,2} = \frac{1}{Y_{c1,2}}, \qquad Z_{\pi 1,2} = \frac{1}{Y_{\pi 1,2}}$$

$$\frac{Y_{c1}}{Y_{c2}} = \frac{Y_{\pi 1}}{Y_{\pi 2}} = -R_c R_\pi, \qquad \frac{Z_{c1}}{Z_{c2}} = \frac{Z_{\pi 1}}{Z_{\pi 2}} = -\frac{1}{R_c R_\pi} \tag{50f}$$

$Y_{c1,2}$ and $Y_{\pi 1,2}$ can be also expressed by network parameters as follows:

$$Y_{c1} = \frac{k_{zc}(1/n_l - k_l R_c)}{\omega L_0(1 - k_l^2)} = \frac{k_{zc}(L_2 - MR_c)}{\omega^2(L_1 L_2 - M^2)}$$

$$Y_{c2} = \frac{k_{zc}(n_l R_c - k_l)}{R_c \omega L_0(1 - k_l^2)} = \frac{k_{zc}(L_1 R_c - M)}{R_c \omega^2(L_1 L_2 - M^2)} \tag{50g}$$

$$Y_{\pi 1} = \frac{k_{z\pi}(1/n_l - k_l R_\pi)}{\omega L_0(1 - k_l^2)} = \frac{k_{z\pi}(L_2 - MR_\pi)}{\omega^2(L_1 L_2 - M^2)}$$

$$Y_{\pi 2} = \frac{k_{z\pi}(n_l R_\pi - k_l)}{R_\pi \omega L_0(1 - k_l^2)} = \frac{k_{z\pi}(L_1 R_\pi - M)}{R_\pi \omega^2(L_1 L_2 - M^2)} \tag{50h}$$

From Eq. (50f), we can understand the following relations, which was mentioned earlier:

$$Y_{c1} \neq Y_{c2}, \qquad Y_{\pi 1} \neq Y_{\pi 2}$$

$$Z_{c1} \neq Z_{c2}, \qquad Z_{\pi 1} \neq Z_{\pi 2} \tag{50i}$$

The equal sign is available only in the case satisfying Eq. (49a).

Relation Between Input Voltages and Currents, and Output Voltages and Currents Equations (50a)–(50d) are converted to Eq. (51a).

$$\boldsymbol{V} = T_1[\Lambda_{co}\boldsymbol{a} + \Lambda_{si}\boldsymbol{b}]$$

$$\boldsymbol{I} = -T_2[\Lambda_{si}\boldsymbol{a} + \Lambda_{co}\boldsymbol{b}]$$

$$\boldsymbol{a} = \begin{bmatrix} a_c \\ a_\pi \end{bmatrix}, \qquad \boldsymbol{b} = \begin{bmatrix} b_c \\ b_\pi \end{bmatrix}, \qquad \boldsymbol{V} = \begin{bmatrix} V_1 \\ V_2 \end{bmatrix}, \qquad \boldsymbol{I} = \begin{bmatrix} I_1 \\ I_2 \end{bmatrix}$$

$$\Lambda_{co} = \begin{bmatrix} \cos k_{zc} z & 0 \\ 0 & \cos k_{z\pi} z \end{bmatrix}$$

$$\Lambda_{si} = \begin{bmatrix} j\sin k_{zc}z & 0 \\ 0 & j\sin k_{z\pi}z \end{bmatrix}$$

$$T_1 = \begin{bmatrix} 1 & 1 \\ R_c & R_\pi \end{bmatrix}, \qquad T_2 = \begin{bmatrix} Y_{c1} & Y_{\pi 1} \\ Y_{c2}R_c & Y_{\pi 2}R_\pi \end{bmatrix}$$

$$a_{c,\pi} = v_+^{c,\pi} + v_-^{c,\pi}, \qquad b_{c,\pi} = v_-^{c,\pi} - v_+^{c,\pi} \tag{51a}$$

Denoting the voltages and the currents of lines 1 and 2 at $z = 0$ by V_{10} and V_{20} and I_{10} and I_{20}, respectively, Eq. (51a) becomes Eq. (51b) when $\Lambda_{co} = U$ (unit matrix) and $\Lambda_{si} = 0$ at $z = 0$:

$$\boldsymbol{V}_0 = \begin{bmatrix} V_{10} \\ V_{20} \end{bmatrix} = T_1\boldsymbol{a}, \qquad \boldsymbol{I}_0 = \begin{bmatrix} I_{10} \\ I_{20} \end{bmatrix} = -T_2\boldsymbol{b} \tag{51b}$$

Substituting Eq. (51b) into Eq. (51a), we obtain the following values when $z = l$:

$$\boldsymbol{V}_l = T_1(\Lambda_{co,l}T_1^{-1}\boldsymbol{V}_0 - \Lambda_{si,l}T_2^{-1}\boldsymbol{I}_0)$$
$$\boldsymbol{I}_l = -T_2(\Lambda_{si,l}T_1^{-1}\boldsymbol{V}_0 - \Lambda_{co,l}T_2^{-1}\boldsymbol{I}_0) \tag{52a}$$

where

$$\boldsymbol{V}_l = \begin{bmatrix} V_{1l} \\ V_{2l} \end{bmatrix}, \qquad \boldsymbol{I}_l = \begin{bmatrix} I_{1l} \\ I_{2l} \end{bmatrix}$$

$$\Lambda_{co,l} = \begin{bmatrix} \cos k_{zc}l & 0 \\ 0 & \cos k_{z\pi}l \end{bmatrix}$$

$$\Lambda_{si,l} = \begin{bmatrix} j\sin k_{zc}l & 0 \\ 0 & j\sin k_{z\pi}l \end{bmatrix}$$

From Eq. (52a), Eq. (52b) is obtained.

$$\boldsymbol{V}_0 = T_1(\Lambda_{co,l}T_1^{-1}\boldsymbol{V}_l + \Lambda_{si,l}T_2^{-1}\boldsymbol{I}_l)$$
$$\boldsymbol{I}_0 = T_2(\Lambda_{si,l}T_1^{-1}\boldsymbol{V}_l + \Lambda_{co,l}T_2^{-1}\boldsymbol{I}_l) \tag{52b}$$

For the case of the coupled line in a homogeneous medium, we have

$$\Lambda_{co,l} = (\cos k_z l)\begin{bmatrix} 1 & 0 \\ 0 & 1 \end{bmatrix}$$

$$\Lambda_{si,l} = (j\sin k_z l)\begin{bmatrix} 1 & 0 \\ 0 & 1 \end{bmatrix}$$

$$T_2 = T_1\Lambda(Y_c, Y_\pi) = T_1\begin{bmatrix} Y_c & 0 \\ 0 & Y_\pi \end{bmatrix}$$

$$R_cR_\pi = -1 \tag{52c}$$

Substituting Eq. (52c) into Eq. (52b), we get

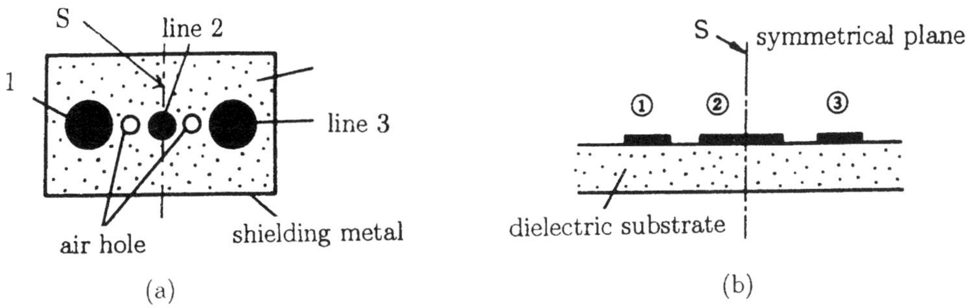

Figure 24. Examples of sectional views of symmetrical coupled three lines. (a) Symmetrical coupled three lines in a dielectric material with air holes; (b) symmetrical coupled three microstrip lines.

$$\boldsymbol{V}_0 = (\cos k_z l)\boldsymbol{V}_l + (j \sin k_z l)(T_1 \Lambda^{-1} T_1^{-1})\boldsymbol{I}_l$$
$$\boldsymbol{I}_0 = (j \sin k_z l)(T_i \Lambda T_1^{-1})\boldsymbol{V}_l + (\cos k_z l)\boldsymbol{I}_l$$

It is understood that Eq. (52d) coincides with Eq. (36a).

Symmetrical Coupled Three Lines

We explain the coupled three lines with the symmetrical sectional view about a single plane; examples are shown in Fig. 24.

Because the electromagnetic fields of a symmetrical structure about a symmetrical plane S consist of an even mode and an odd mode, one side of the electromagnetic field of the even and odd modes becomes the fields of Fig. 25a and 25b, respectively.

The mode of Fig. 25a consists of c and π modes described in Section 2.2, and the ratio of the voltages or the currents between lines 2 and 1 are shown in Eq. (53a) as provided by Eqs. (48a) and (48b):

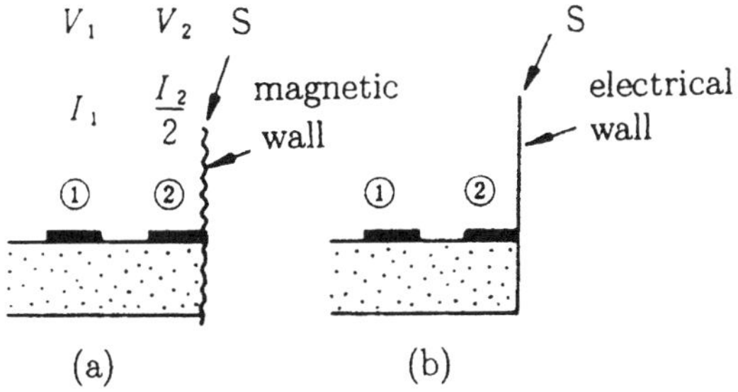

Figure 25. Equivalent construction for even (a) and odd mode (b) excitation on the lines of Fig. 24.

$$\frac{V^e_{2c,\pi}}{V^e_{1c,\pi}} = \frac{V^e_{2c,\pi}}{V^e_{3c,\pi}} = R^e_{c,\pi}$$

$$\frac{I^e_{2c,\pi}/2}{I^e_{1c,\pi}} = \frac{I^e_{2c,\pi}/2}{I^e_{3c,\pi}} = -\frac{1}{R_{\pi,c}} \tag{53a}$$

Next, in the case of the odd mode, the voltage of line 2 becomes zero and the currents of lines 1 and 2 of Fig. 25b take opposite signs, which results in the zero current of line 2 in Fig. 24. We, therefore, get the following:

$$\frac{V^0_3}{V^0_1} = \frac{I^0_3}{I^0_1} = -1, \qquad V^0_2 = I^0_2 = 0 \tag{53b}$$

By the relative relation of Eqs. (53a) and (53b), we get the following values of T_V and T_I directly.

$$T_V = \begin{bmatrix} 1 & 1 & 1 \\ 0 & R^e_c & R^e_\pi \\ -1 & 1 & 1 \end{bmatrix} \Lambda_V \tag{53c}$$

$$T_I = \begin{bmatrix} 1 & 1 & 1 \\ 0 & \dfrac{-2}{R^e_\pi} & \dfrac{-2}{R^e_c} \\ -1 & 1 & 1 \end{bmatrix} \Lambda_I \tag{53d}$$

where Λ_V and Λ_I are the arbitral real diagonal matrices satisfying the relation $T_I = (T_V^{-1})'$. If we temporarily call the first, second, and third rows of T_V and T_I as the α, β, and γ modes, the α mode corresponds to that of Fig. 26b and the β and γ modes correspond to the c and π modes, respectively, of Fig. 25a.

When the line parameters of Fig. 25a are expressed by those of Fig. 22, the characteristic admittance of the β and γ modes are obtained as shown in Table 2, where the current flowing in line 2 of Fig. 24 is twice the current flowing in line 2 of Fig. 25a. $Y_{c1,2}$ and $Y_{\pi 1,2}$ in Table 2 can be obtained by Eqs. (50g) and (50h).

The phase constant of the α mode obtained directly from Fig. 25b, and those of the β and γ modes are obtained from the c and π modes of Fig. 25a, the values of which can be obtained by Eq. (48c) together with the parameters of Figs. 22 and 25a.

When we show the distributed capabilities and inductances of Fig. 24 as shown in Fig. 26, the L and C matrices can be expressed by

Table 2. Characteristic Admittances of Each Line for Each Mode

Line	Admittance of each mode: Odd mode (α mode)	Even mode (c mode) (β mode)	Even mode (π mode) (γ mode)
1	Y_o	$Y_{\beta 1} = Y_{c1} = \dfrac{1}{Z_{c1}}$	$Y_{\gamma 1} = Y_{\pi 1} = \dfrac{1}{Z_{\pi 1}}$
2	0	$Y_{\beta 2} = 2Y_{c2} = \dfrac{2}{Z_{c2}}$	$Y_{\gamma 2} = 2Y_{\pi 2} = \dfrac{2}{Z_{\pi 2}}$
3	Y_o	$Y_{\beta 3} = Y_{c1} = \dfrac{1}{Z_{c1}}$	$Y_{\gamma 3} = Y_{\pi 1} = \dfrac{1}{Z_{\pi 1}}$

$$L = \begin{bmatrix} L_a & M_{ab} & M_{ac} \\ M_{ab} & L_b & M_{ab} \\ M_{ac} & M_{ab} & L_a \end{bmatrix}$$

$$C = \begin{bmatrix} C'_a & -C_{ab} & -C_{ac} \\ -C_{ab} & C'_b & -C_{ab} \\ -C_{ac} & -C_{ab} & C'_a \end{bmatrix} \tag{53e}$$

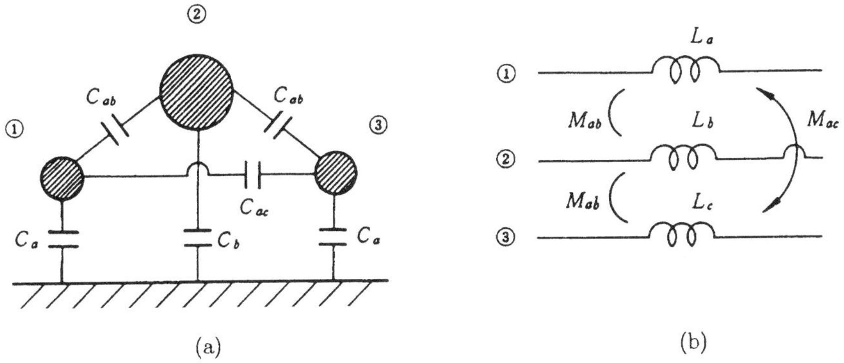

Figure 26. Distributed parameters (capacities and inductances) of symmetrical coupled three lines. (a) Sectional view and distributed capacities (F/m); (b) top view and distributed inductances (H/m).

where

$$C'_a = C_a + C_{ab} + C_{ac}$$
$$C'_b = C_b + 2C_{ab}$$

To make the even mode of Fig. 26 correspond to that in Fig. 24, we can use (Note 4)

$$L_1 = L_a + M_{ac}, \qquad L_2 = 2L_b, \qquad M = 2M_{ab}$$
$$C_1 = C'_a - C_{ac} = C_a + C_{ab}$$
$$C_2 = \frac{C'_b}{2} = \frac{C_b}{2} + C_{ab}, \qquad C_{12} - C_{ab} \tag{53f}$$

Using L_1, L_2, M_1, C_1, C_2, and C_{12} of Eq. (53f), we obtain L_0, C_0, n_l, n_c, k_l, and k_c from Eqs. (38a) and (38b), $k_{z,\beta}$ and $k_{z,\gamma}$ from Eq. (48c), $Y_{c1,2}$ and $Y_{\pi 1,2}$ from Eqs. (50g) and (50h), and $Y_{\beta 1,2}$ and $Y_{\gamma 1,2}$ from Table 2.

The parameters for the odd mode can be obtained directly by grounding line 2 of Fig. 26 as follows:

$$k_{z,\alpha} = \omega\sqrt{(L_a - M_{ac})(C_a + C_{ab} + 2C_{ac})} \tag{53g}$$

$$Y_0 = \sqrt{\frac{C_\alpha + C_{ab} + C_{ac}}{L_a - M_{ac}}} \tag{53h}$$

Substituting Eq. (50f) into the values of Table 2, we obtain

$$\frac{Y_{\beta 1}}{Y_{\beta 2}} = \frac{Y_{\gamma 1}}{Y_{\gamma 2}} = \frac{Y_{c1}}{2Y_{c2}} = \frac{Y_{\pi 1}}{2Y_{\pi 2}} = -\frac{R_c R_\pi}{2} = -\frac{R_c^e R_\pi^e}{2} \tag{53i}$$

The values of $R_c R_\pi$ in Eq. (53i) can be obtained by substituting Eq. (53f) into Eq. (48d).

Expressing the incident and reflected waves for the α and $\beta\gamma$ modes by $V_+^{\alpha,\beta,\gamma}$ and $v_-^{\alpha,\beta,\gamma}$, respectively, we obtain the voltages and currents of lines 1, 2, and 3 as the following

$$V_1 = \sum_{a=\alpha,\beta,\gamma} R_1^s(v_+^s e^{-jk_{z,s}z} + v_-^s e^{jk_{z,s}z}) \tag{53j}$$

$$V_2 = \sum_{a=\alpha,\beta,\gamma} R_2^s(v_+^s e^{-jk_{z,s}z} + v_-^s e^{jk_{z,s}z}) \tag{53k}$$

$$V_3 = \sum_{a=\alpha,\beta,\gamma} R_3^s(v_+^s e^{-jk_{z,s}z} + v_-^s e^{jk_{z,s}z}) \tag{53l}$$

$$I_1 = \sum_{a=\alpha,\beta,\gamma} Y_{s1} R_1^s(v_+^s e^{-jk_{z,s}z} - v_-^s e^{jk_{z,s}z}) \tag{53m}$$

$$I_2 = \sum_{a=\alpha,\beta,\gamma} Y_{s2}R_2^s(v_+^s e^{-jk_{z,s}z} - v_-^s e^{jk_{z,s}z}) \tag{53n}$$

$$I_3 = \sum_{a=\alpha,\beta,\gamma} Y_{s3}R_3^s(v_+^s e^{-jk_{z,s}z} - v_-^s e^{jk_{z,s}z}), \qquad s = \alpha, \beta, \gamma \tag{53o}$$

$$T_V = \begin{bmatrix} 1 & 1 & 1 \\ 0 & R_c^e & R_\pi^e \\ -1 & 1 & 1 \end{bmatrix} = \begin{bmatrix} R_1^\alpha & R_1^\beta & R_1^\gamma \\ R_2^\alpha & R_2^\beta & R_2^\gamma \\ R_3^\alpha & R_3^\beta & R_3^\gamma \end{bmatrix} \tag{53p}$$

$$Y_{\alpha 1} = Y_{\alpha 2} = Y_{\alpha 3} = Y_0 \tag{53q}$$

$$Y_{\beta 1} = Y_{\beta 3} \neq Y_{\beta 2}, \qquad Y_{\gamma 1} = Y_{\gamma 3} \neq Y_\gamma$$

Equations (53j)–(53o) can be transformed to Eq. (53r):

$$V_i = \sum_{s=\alpha,\beta,\gamma} R_i^s(a_s \cos k_{z,s}z + jb_s \sin k_{z,s}z)$$

$$I_i = -\sum_{s=\alpha,\beta,\gamma} Y_{si}R_i^s(ja_s \sin k_{z,s}z + b_s \cos k_{z,s}z)$$

$$i = 1, 2, 3, \qquad a_s = v_+^s + v_-^s, \qquad b_s = v_-^s - v_+^s \tag{53r}$$

Equation (53r) can be expressed in matrix form as follows:

$$V = T_1(\Lambda_{co}a + \Lambda_{si}b)$$

$$I = T_2(\Lambda_{si}a + \Lambda_{co}b) \tag{53s}$$

where

$$a = \begin{bmatrix} a_\alpha \\ a_\beta \\ a_\gamma \end{bmatrix}, \qquad b = \begin{bmatrix} b_\alpha \\ b_\beta \\ b_\gamma \end{bmatrix}, \qquad V = \begin{bmatrix} V_1 \\ V_2 \\ V_3 \end{bmatrix}, \qquad I = \begin{bmatrix} I_1 \\ I_2 \\ I_3 \end{bmatrix} \tag{53t}$$

$$\Lambda_{co} = \begin{bmatrix} \cos k_{z\alpha}z & 0 & 0 \\ 0 & \cos k_{z\beta}z & 0 \\ 0 & 0 & \cos k_{z\gamma}z \end{bmatrix}$$

$$\Lambda_{si} = \begin{bmatrix} j \sin k_{z\alpha}z & 0 & 0 \\ 0 & j \sin k_{z\beta}z & 0 \\ 0 & 0 & j \sin k_{z\gamma}z \end{bmatrix}$$

$$T_1 = \begin{bmatrix} 1 & 1 & 1 \\ 0 & R_c^e & R_\pi^e \\ -1 & 1 & 1 \end{bmatrix}, \qquad T_2 = \begin{bmatrix} Y_0 & Y_{\beta 1} & Y_{\gamma 1} \\ 0 & Y_{\beta 2}R_c^e & Y_{\gamma 2}R_\pi^e \\ -Y_0 & Y_{\beta 1} & Y_{\gamma 1} \end{bmatrix} \tag{53u}$$

Denoting the voltages and the currents at $z = 0$ by V_{10}, V_{20}, and V_{30}, and

I_{10}, I_{20}, and I_{30}, and then at $z = l$ by V_{1l}, V_{2l}, and V_{3l}, and I_{1l}, I_{2l}, and I_{3l}, we get

$$\begin{aligned} V_l &= T_1[\Lambda_{co,l}T_1^{-1}V_o - \Lambda_{si,l}T_2^{-1}I_o] \\ I_l &= -T_2[\Lambda_{si,l}T_1^{-1}V_o - \Lambda_{co,l}T_2^{-1}I_o] \end{aligned} \tag{54a}$$

$$\begin{aligned} V_0 &= T_1(\Lambda_{co,l}T_1^{-1}V_l + \Lambda_{si,l}T_2^{-1}I_l) \\ I_0 &= -T_2(\Lambda_{si,l}T_1^{-1}V_o - \Lambda_{co,l}T_2^{-1}I_l) \end{aligned} \tag{54b}$$

$$V_{0,l} = \begin{bmatrix} V_{10,l} \\ V_{20,l} \\ V_{30,l} \end{bmatrix}, \qquad I_{0,l} = \begin{bmatrix} I_{10,l} \\ I_{20,l} \\ I_{30,l} \end{bmatrix} \tag{55a}$$

$$\Lambda_{co,l} = \begin{bmatrix} \cos k_{z,\alpha}l & 0 & 0 \\ 0 & \cos k_{z,\beta}l & 0 \\ 0 & 0 & \cos k_{z,\gamma}l \end{bmatrix}$$

$$\Lambda_{si,l} = \begin{bmatrix} j\sin k_{z,\alpha}l & 0 & 0 \\ 0 & j\sin k_{z,\beta}l & 0 \\ 0 & 0 & j\sin k_{z,\gamma}l \end{bmatrix} \tag{55b}$$

2.3 Equivalent Networks of Distributed Coupled Lines with Finite Length

When we consider the coupled two lines as shown in Fig. 27, we get four ports. If we give the boundary condition such as open or shortened to the arbitrary two ports, we get the two-ports network consisting of the left two ports.

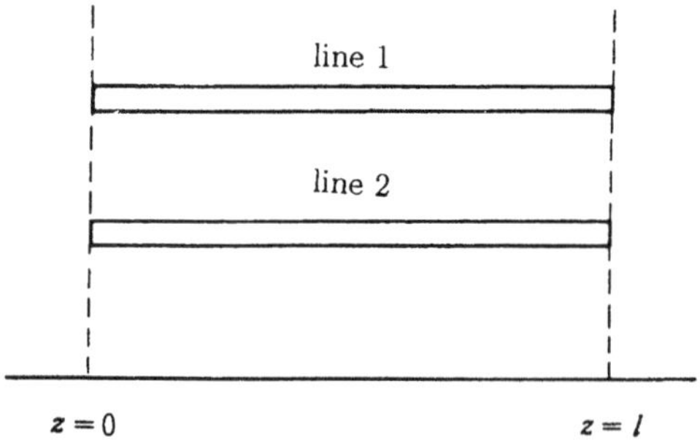

Figure 27. Coupled two lines with the length of l.

The equivalent network corresponding to the several kinds of boundary conditions are very useful in considering the characteristics. The boundary conditions can be transformed to that of each independent mode. In such a way, we can get the circuit matrices such as Z and Y or G matrix and the equivalent networks. The results are shown in Fig. 28. The structure of Fig.

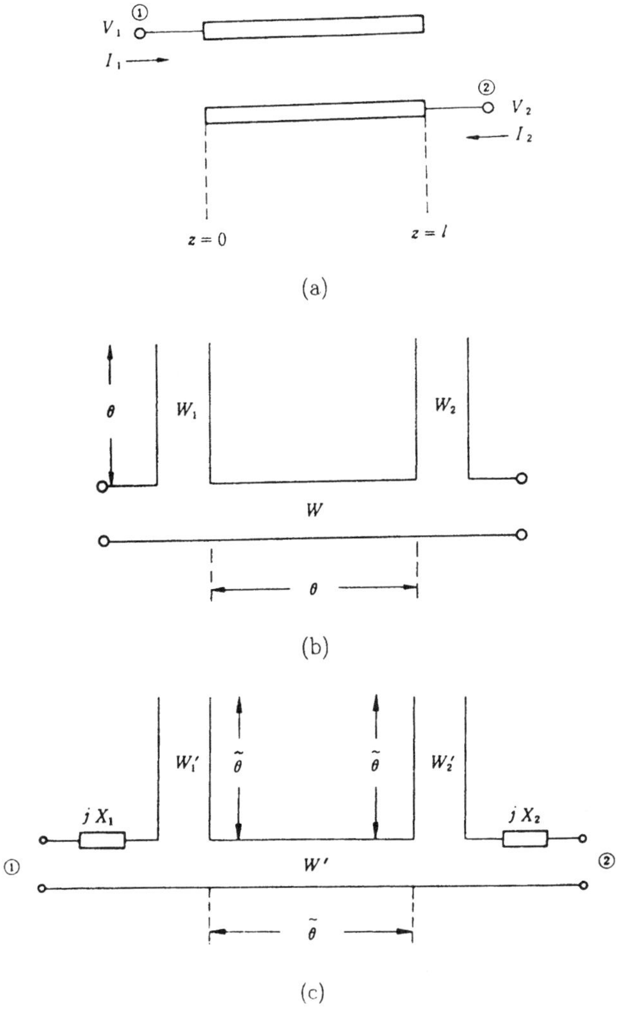

Figure 28. Example 1. (a) Structure; (b) case of homogeneous medium; (c) case of inhomogeneous medium.

28a takes the equivalent network of Figs. 28b and 28c in the case of homogeneous and inhomogeneous media, respectively. The network parameters are shown in Eqs. (56a) and (56b).

$$W = \frac{R_c}{1 + R_c^2}(W_c - W_\pi)$$

$$W_1 = \frac{(1 - R_c)W_c + R_c(1 + R_c)W_\pi}{1 + R_c^2}$$

$$W_2 = \frac{R_c(R_c - 1)W_c + (1 + R_c)W_\pi}{1 + R_c^2}$$

$$\theta = k_z l \tag{56a}$$

$$W_1' = \frac{-R_cR_\pi(1 - R_c)W_c + R_c(-1/R_\pi + 1)W_\pi}{R_c^2 - R_cR_\pi}$$

$$W_2' = \frac{-R_c^2R_\pi(R_c - 1)W_c + R_c(-R_\pi + 1)W_\pi}{R_c^2 - R_cR_\pi}$$

$$W' = \frac{R_c(-R_cR_\pi W_c - W_\pi)}{R_c^2 - R_cR_\pi}$$

$$X_1 = \frac{(-R_cR_\pi W_c + (R_c/R_\pi)W_\pi)(2/\pi)\Delta}{R_c^2 - R_cR_\pi}$$

$$X_2 = \frac{R_c^2(-R_cR_\pi W_c + (R_\pi/R_c)W_\pi)(\pi/2)\Delta}{R_c^2 - R_cR_\pi}$$

$$k_{c,\pi} \simeq \tilde{k}(1 \pm \Delta), \qquad \tilde{k}l = \frac{\pi}{2} \tag{56b}$$

In the symmetrical structure, R_c and R_π is obtained from

$$R_c = 1, \qquad R_\pi = -1 \tag{56c}$$

Therefore,

$$W = \frac{W_e - W_o}{2}, \qquad W_1 = W_0 \tag{56d}$$

where subscripts e and o denote the even and odd modes, respectively.

Considering $\tilde{k}l = \pi/2$, the network of Fig. 28b can be transformed to Fig. 29 which shows a bandpass filter.

Example 2 is presented in Fig. 30 and Eqs. (56e).

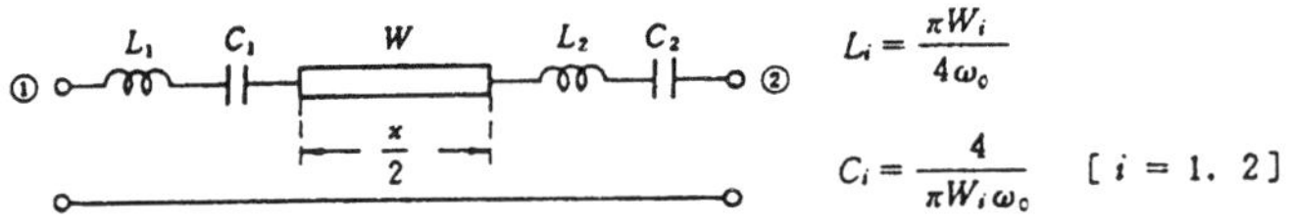

Figure 29. Equivalent network of Fig. 28b.

$$\frac{1}{W} = \frac{R_c}{1 + R_c^2}\left(\frac{1}{W} - \frac{1}{W_c}\right)$$

$$\frac{1}{W_1} = \frac{1}{1 + R_c^2}\left(\frac{1 + R_c}{W_c} + \frac{R_c^2 - 1}{W_\pi}\right)$$

$$\frac{1}{W_2} = \frac{1}{1 + R_c^2}\left(\frac{R_c(1 + R_c)}{W_c} + \frac{1 - R_c}{W_\pi}\right)$$

$$\frac{1}{W'} = \frac{-R_cR_\pi}{R_c^2 - R_cR_\pi}\left(\frac{1}{W_cR_\pi} + \frac{R_c}{W_\pi}\right)$$

$$\frac{1}{W_1'} = \frac{-R_cR_\pi}{R_c^2 - R_cR_\pi}\left(\frac{1 + 1/R_\pi}{W_c} + \frac{R_c(R_c - 1)}{W_\pi}\right)$$

$$\frac{1}{W_2'} = \frac{-R_cR_\pi}{R_c^2 - R_cR_\pi}\left(\frac{(1/R_\pi)(1/R_\pi - 1)}{W_c} + \frac{1 - R_c}{W_\pi}\right)$$

$$B_1' = \frac{-R_cR_\pi}{R_c^2 - R_cR_\pi}\left(\frac{1}{W_c} - \frac{1}{W_\pi}\right)\frac{\pi}{2}\Delta$$

$$B_2' = \frac{-R_cR_\pi}{R_c^2 - R_cR_\pi}\left(\frac{1}{W_cR_\pi^2} - \frac{1}{W_\pi}\right)\frac{\pi}{2}\Delta \qquad (56e)$$

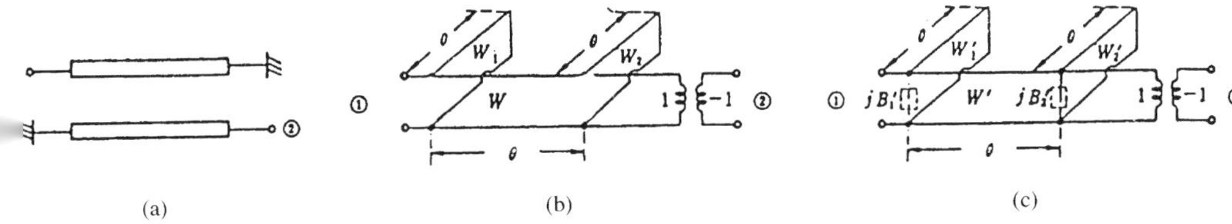

Figure 30. Example 2. (a) Structure; (b) the case of a homogeneous medium; (c) the case of an inhomogeneous medium.

In the case of a symmetry structure,

$$\frac{1}{W} = \frac{1}{2}\left(\frac{1}{W_\pi} - \frac{1}{W_c}\right) = \frac{1}{2}\left(\frac{1}{W_0} - \frac{1}{W_e}\right)$$

$$\frac{1}{W_1} = \frac{1}{W_2} = \frac{1}{W_c}$$

A third example is presented in Fig. 31 and Eqs. (56f).

$$W' = (1 + R_c^2)\frac{W_c W_\pi}{R_c^2 W_\pi + W_c}, \quad W = \frac{2W_c W_\pi}{W_c + W_\pi} = \frac{2W_e W_0}{W_e W_0} \quad \text{(symmetry)}$$

$$W' = (R_\pi - R_c)\frac{R_\pi W_c W_\pi}{W_c R_\pi^2 + W_\pi}$$

$$B = \frac{1}{R_\pi - R_c}\left(\frac{1}{W_c R_\pi} - \frac{R_\pi}{W_\pi}\right)\frac{\pi}{2}\Delta \tag{56f}$$

Example 4 is presented in Fig. 32 and Eqs. (56g).

$$Y_1 = \frac{1}{W_1} = \frac{1}{1 + R_c^2}\left(\frac{(1 + R_c)}{W_c} - \frac{R_c(1 - R_c)}{W_\pi}\right)$$

$$Y_2 = \frac{1}{W_2} = \frac{R_c}{1 + R_c^2}\left(\frac{1}{W_\pi} - \frac{1}{W_c}\right)$$

$$Y_3 = \frac{1}{W_3} = \frac{1}{1 + R_c^2}\left(\frac{R_c(1 + R_c)}{W_c} + \frac{(1 - R_c)}{W_\pi}\right)$$

In the case of symmetry,

$$Y_1 = Y_3 = \frac{1}{W_c} = \frac{1}{W_e} \quad (W_1 = W_e)$$

$$Y_2 = \frac{W_c - W_\pi}{2W_\pi W_c} = \frac{W_e - W_0}{2W_e W_0} \quad \left(W_2 = \frac{2W_e W_0}{W_e - W_0}\right)$$

In the case of an inhomogeneous medium,

$$Y_1' = \frac{R_\pi}{R_\pi - R_c}\left[\frac{1}{W_c}\left(1 - \frac{1}{R_\pi}\right) + \frac{R_c(R_c - 1)}{W_\pi}\right] = \frac{1}{W_1'}$$

$$B_1 = \frac{R_\pi}{R_\pi - R_c}\left[\frac{1}{W_c}\left(1 - \frac{1}{R_\pi}\right) + \frac{R_c(R_c + 1)}{W_\pi}\right]\frac{\pi}{2}\Delta$$

$$Y_2' = \frac{R_\pi}{R_\pi - R_c}\left(\frac{1}{W_c R_\pi} + \frac{R_c}{W_\pi}\right) = \frac{1}{W_2'}$$

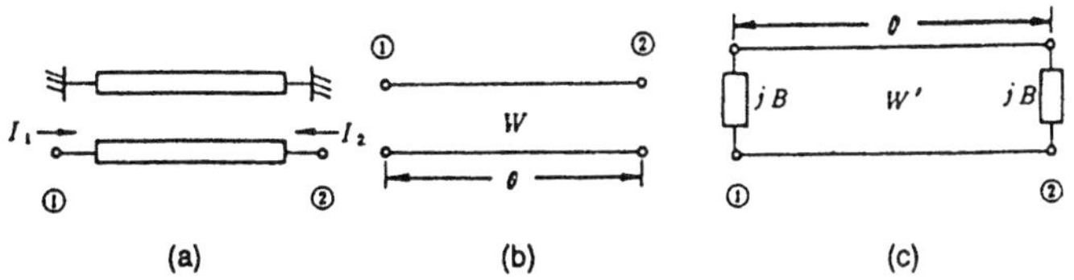

Figure 31. Example 3. (a) Structure; (b) homogeneous case; (c) inhomogeneous case.

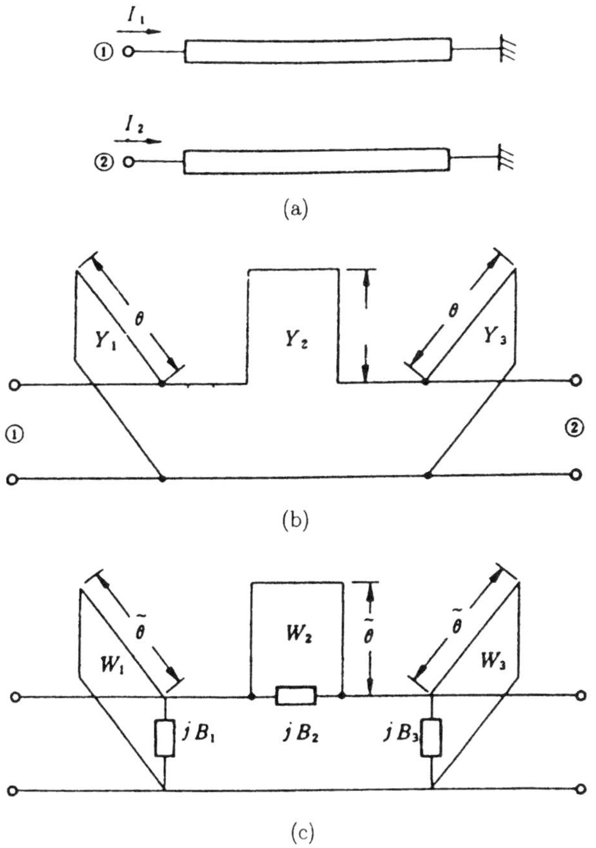

Figure 32. Example 4. (a) Structure; (b) homogeneous case; (c) inhomogeneous case.

$$B_2 = \frac{R_\pi}{R_\pi - R_c}\left(\frac{1}{W_c R_\pi} - \frac{R_c}{W_\pi}\right)\frac{\pi}{2}\Delta$$

$$Y_3' = \frac{R_\pi}{R_\pi - R_c}\left[\frac{1}{W_c}\left(\frac{1}{R_\pi^2} - \frac{1}{R_\pi}\right) + \frac{1}{W_\pi}(1 - R_c)\right]$$

$$B_3 = \frac{R_\pi}{R_\pi - R_c}\left[\frac{1}{W_c}\left(\frac{1}{R_\pi^2} - \frac{1}{R_\pi}\right) + \frac{1}{W_\pi}(1 + R_c)\right]\frac{\pi}{2}\Delta \qquad (56g)$$

Example 5 is presented in Fig. 33 and Eqs. (56h).

$$W_1 = \frac{1}{1 + R_c^2}[W_c(1 - R_c) + W_\pi R_c(R_c + 1)]$$

$$W_2 = \frac{1}{1 + R_c^2}(W_c - W_\pi)R_c$$

$$W_3 = \frac{1}{1 + R_c^2}[W_c R_c(R_c - 1) + W_\pi(1 + R_c)]$$

In the case of symmetry,

$$W_1 = W_3 = W_\pi = W_0$$

$$W_2 = \frac{W_c - W_\pi}{2} = \frac{W_e - W_0}{2}$$

$$W_1' = \frac{1}{R_\pi - R_c}\left[W_c(R_\pi - R_\pi R_c) + W_\pi\left(\frac{1}{R_\pi} - 1\right)\right]$$

$$X_1 = \frac{1}{R_\pi - R_c}\left[W_c(R_\pi - R_\pi R_c) + W_\pi\left(\frac{1}{R_\pi} + 1\right)\frac{\pi}{2}\Delta\right]$$

$$W_2' = \frac{1}{R_\pi - R_c}(W_c R_c R_\pi - W_\pi)$$

$$X_2 = \frac{1}{R_\pi - R_c}(W_c R_c R_\pi - W_\pi)\frac{\pi}{2}\Delta$$

$$W_3' = \frac{1}{R_\pi - R_c}[W_c(R_c^2 R_\pi - R_\pi R_c) + W_\pi(R_\pi - 1)]$$

$$X_3 = \frac{1}{R_\pi - R_c}[W_c(R_c^2 R_\pi - R_\pi R_c) + W_\pi(R_\pi + 1)]\frac{\pi}{2}\Delta \qquad (56h)$$

Examples 6 and 7 are presented in Fig. 34 and Eqs. (56i), and Fig. 35 and Eqs. (56j), respectively.

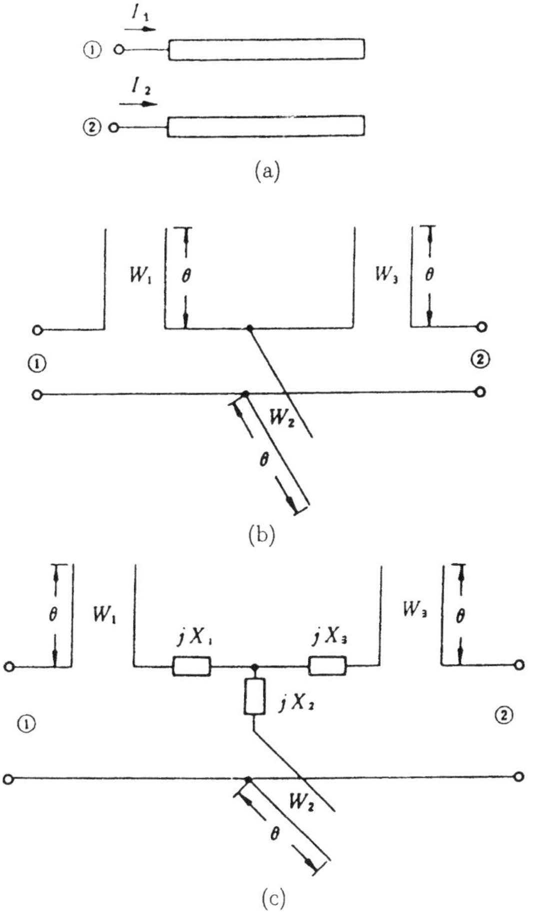

Figure 33. Example 5. (a) Structure; (b) homogeneous case; (c) inhomogeneous case.

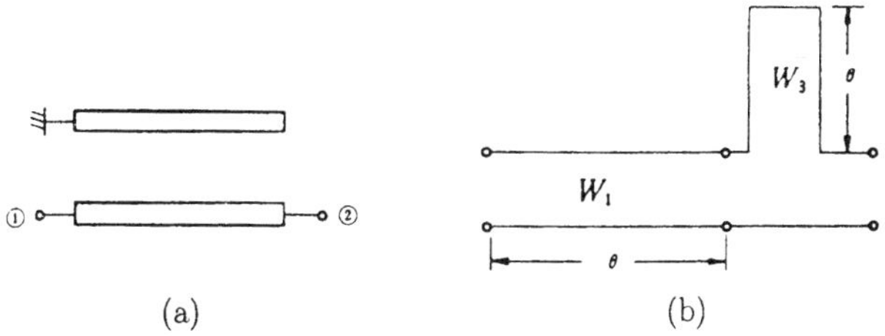

Figure 34. Example 6.

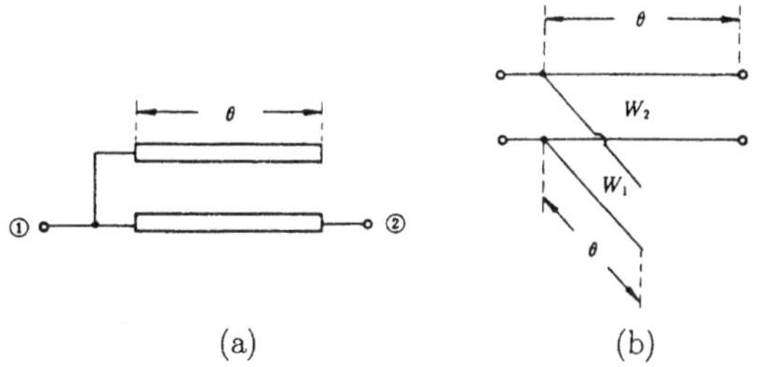

Figure 35. Example 7.

$$\frac{1}{W_1} = \frac{1}{1 + R_c^2}\left(\frac{R_c^2}{W_c} + \frac{1}{W_\pi}\right)$$

$$W_2 = \frac{R_c^2(W_c - W_\pi)^2}{(1 + R_c^2)(R_c^2 W_\pi + W_c)} \tag{56i}$$

$$\frac{1}{W_1} = \frac{R_c^2 W_c^2 + W_\pi^2}{W_c W_\pi (W_c R_c^2 + W_\pi)} + \frac{2R_c(W_\pi - W_c)}{(R_c^2 + 1)W_c W_\pi}$$

$$W_2 = \frac{W_c R_c^2 + W_\pi}{1 + R_c^2} \tag{56j}$$

Example 8 is represented in Fig. 36 and Eqs. (56k) and (56l).

$$W' = \frac{1 + R_c^2}{(R_c^2/W_\pi + 1/W_c)}$$

$$W = \frac{R_c^2}{1 + R_c^2}\frac{(W_c - W_\pi)^2}{(W_c R_c^2 + W_\pi)}$$

$$n = \frac{R_c(W_c - W_\pi)}{W_c R_c^2 + W_\pi} \tag{56k}$$

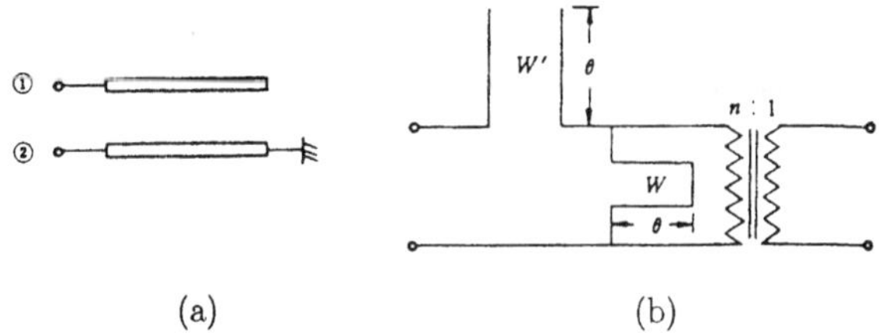

Figure 36. Example 8.

In the case of symmetry,

$$W' = \frac{2W_c W_\pi}{W_c + W_\pi} = \frac{2W_e W_0}{W_e + W_0}$$

$$W = \frac{(W_c - W_\pi)^2}{2(W_c + W_\pi)} = \frac{(W_e - W_0)^2}{2(W_e + W_0)}$$

$$n = \frac{W_c - W_\pi}{W_c + W_\pi} = \frac{W_e - W_0}{W_e + W_0} \tag{56l}$$

In Eqs. (56a)–(56f), the values of R_c, W_c, and W_π in the homogeneous case should be obtained from Eqs. (41c) and (41e). W_e and W_0 in the case of the symmetry should be obtained from Eq. (42b).

APPENDIX

Note 1

$T_I^{-1}YZT_I = \Lambda$, where Λ is the diagonal matrix. Taking the transpose matrix and substituting Eq. (17b), we get $T_I^t Z^t Y^t (T_I)^t = (T_V)^{-1}ZYT^* = (T_V^*)^{-1}ZYT^*$, because Z and Y are symmetric matrices. If $T_V^* = T_V$, we get Eq. (16e) equal to Eq. (16d).

Note 2

Expressing Z and Y by $Z = R + jX$ and $Y = G + jB$, R, G, $\partial B/\partial\omega$, and $\partial B/\partial\omega$ should be positive real symmetric matrices as shown in Chapter 2. X and B, however, are expressed by $X = \omega L$ and $B = \omega C$; therefore, $\partial X/\partial\omega = L$ and $\partial B/\partial\omega = C$ are positive real symmetric matrices.

Note 3

Substituting $\boldsymbol{V} = T_v \boldsymbol{v}$ and $\boldsymbol{I} = T_I \boldsymbol{i}$ into Eq. (33b),

$$T_v^{-1}LCT_v = T_v^{-1}LT_I T_I^{-1}CT_v = \Lambda_L' \Lambda_c' = \frac{k_z^2}{\omega^2} = \mu\varepsilon$$

Then,

$$\Lambda_L' = \mu\varepsilon(\Lambda_c')^{-1}$$

Note 4

When currents I_1, I_2, and I_1 are flowing in lines 1, 2, and 3, respectively, of Fig. 48b, the voltage induced in lines 1, 3, and 2, ΔV_1, ΔV_1, and ΔV_2, respectively, become

$$\Delta V_1 = [j\omega(L_a + M_{ac})I_1 + j\omega M_{ab}I_2]\Delta z \tag{1}$$

$$\Delta V_2 = [2j\omega M_{ab}I_1 + j\omega L_b I_2]\Delta z \tag{2}$$

Denoting the currents of line 2 of Fig. 47b by I_2' and $I_2 = I_2/2$. Substituting this relation into Eqs. (1) and (2), we get

$$\begin{bmatrix} \Delta V_1 \\ \Delta V_1 \end{bmatrix} = j\omega \begin{bmatrix} L_a + M_{ab} & 2M_{ab} \\ 2M_{ab} & 2L_b \end{bmatrix} \Delta z$$

Dividing half the construction of Fig. 26a by the magnetic wall at S, and comparing it with Fig. 22, we get the capacities in Eq. (53f).

REFERENCES

1. Tripathai, V. K., Asymmetric coupled transmission lines in an inhomogeneous medium, *IEEE Trans.*, *MTT-23*, 734–739 (1975).
2. Gupta, K. C., Garg, R., Bahl, I. J., *Microstrip Lines and Slot Lines*, Artech House, Inc., pp. 303–361.
3. Tripathi, V. K., Properties and applications of asymmetric coupled line structures in an inhomogeneous medium, in *Proc. 5th European Microwave Conference*, Hamburg, 1975, pp. 278–282.

4

Principle of Electromagnetic Resonators

1 THE BASIC CHARACTERISTICS OF A RESONATOR

1.1 Lossless Resonator

First, we consider the simple LC circuit as shown in Fig. 1a and connect the switch S in series. We supply the voltage v^{-0} to C by opening and closing switch S. The current flowing in the LC circuit, i, and the voltage of C, v, takes the values

$$i = \frac{v^{-0}}{\omega L} \sin \omega_0 t = I \sin \omega_0 t, \qquad I = \frac{v^{-0}}{\omega_0 L} \tag{1a}$$

$$v = -\frac{I}{\omega_0 C} \cos \omega_0 t \tag{1b}$$

Therefore, the magnetic energy W_m in inductor L and the electric energy W_e in capacitor C are calculated from Eqs. (2a) and (2b), respectively and change with time t, as shown in Fig. 1b.

$$W_m = \frac{L}{2} I^2 \sin^2 \omega_0 t \tag{2a}$$

$$W_e = \frac{C}{2} \frac{I^2}{(\omega_0 C)^2} \cos^2 \omega_0 t \tag{2b}$$

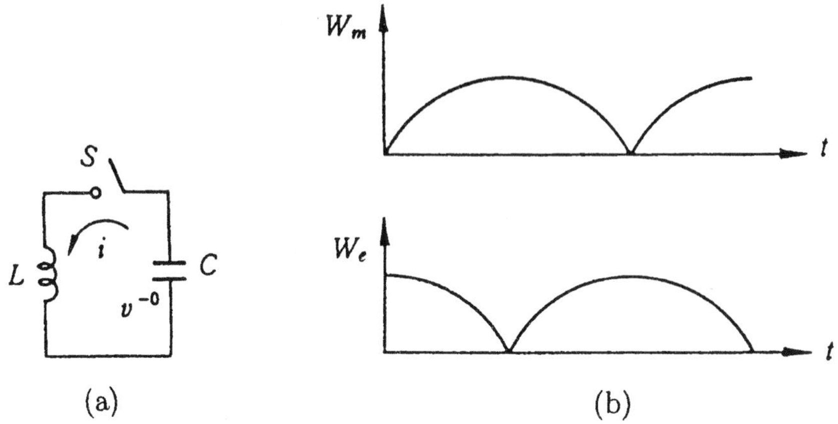

Figure 1. (a) The current flowing in a series LC resonant circuit; (b) the change in W_m and W_e with time.

As understood from Fig. 1b, W_m and W_e are replaced alternatively as time passes.

Because the maximum values of W_m and W_e should be equal, we get

$$\frac{L}{2}I^2 = \frac{I^2}{2\omega_0^2 L} \tag{3a}$$

then,

$$\omega_0 = \frac{1}{\sqrt{LC}} \tag{3b}$$

Next, we consider the distributed constant line. When we consider the case of

$$V_0^+ = -V_0^- \tag{4a}$$

we obtain

$$\dot{V} = -2jV_0^+ \sin\frac{2\pi}{\lambda_g}z \tag{4b}$$

$$\dot{I} = \frac{2V_0^+}{Z_0}\cos\frac{2\pi}{\lambda_y}z \tag{4c}$$

by substituting Eq. (4a) into Eq. (6b) in Chapter 1. Incidental values of voltage v and current i, therefore, can be obtained as

$$v = \sqrt{2}\ \mathrm{Re}(\dot{V}e^{j\omega t} = 2\sqrt{2}V_0^+ \sin \frac{2\pi}{\lambda_g} z \cos\left(\omega t - \frac{\pi}{2}\right)$$

$$i = \sqrt{2}\ \mathrm{Re}(\dot{I}e^{j\omega t}) = 2\sqrt{2}\,\frac{V_0^+}{Z_0} \cos \frac{2\pi}{\lambda_g} z \cos \omega t \qquad (4d)$$

As understood from Eq. (4d), the phase of the voltage or the current at the same given instant of time is the same phase or the reverse phase. The phase difference between the voltage and the current is 90°. This matter is also proved in the case of electric and magnetic fields inside of a resonant cavity.

The maximum values of the magnetic and electric energies $W_{m,\max}$ and $W_{e,\max}$ should also take the same values as the *LC* resonant circuit, which can be proved as follows.

In the lossless resonator, the pointing power toward the closed surface S surrounding the resonator should be zero. Therefore, (as described in Note 8 of Chapter 2),

$$\int\!\!\int_s \boldsymbol{E} \times \boldsymbol{H}^* \cdot \mathbf{n}\, dS = 2j\omega(\tilde{W}_m - \tilde{W}_e) = 0 \qquad (5a)$$

where $\mathbf{n}$ is the normal unit vector on S. Then,

$$\tilde{W}_m = \tilde{W}_e \qquad (5b)$$

where $\tilde{W}_m$ and $\tilde{W}_e$ are the time averages of the magnetic and electric energies. We can summarize the important characteristics of the lossless resonator as follows.

1. The maximum values or the time averages of the magnetic and electric energies are the same. The ratio of the values of the magnetic energy and the electric energy changes with time.
2. The magnetic field or the electric field at the same given instant of time is the same phase or the reverse phase. The phase difference between the magnetic and electric fields is 90°.

In the resonator with the distributed constant line, the maximum and the minimum points of the magnetic field (or the current) are the minimum and the maximum points of the electric field (or the voltage), respectively.

1.2 Lossy Resonator

After the voltage v^{-0} is supplied to the capacitor C of the *LC* resonant circuit of Fig. 2, the current flowing into the circuit in closing the switch S is calculated using

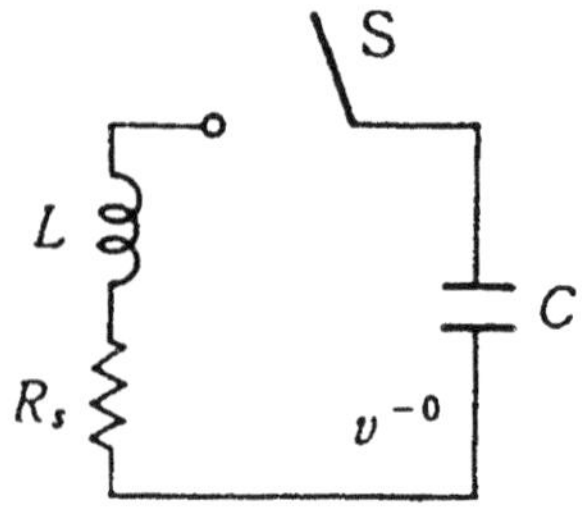

Figure 2. Lossy resonator.

$$i = -\frac{v^{-0}}{\omega L}\exp\left(-\frac{R_s}{2L}t\right)\sin\omega_0 t \tag{6a}$$

The maximum values of the magnetic energy included in L therefore is calculated using

$$W_{m,\max} = \frac{(v^{-0})^2}{2\omega^2 L}\exp\left(-\frac{R_s}{L}t\right) \simeq \frac{(v^{-0})^2}{2\omega^2 L}\left(1 - \frac{R_s}{L}t\right)$$

Then, the dispated energy W_0 in $t = 1/\omega_0$ sec is

$$\frac{P_0}{\omega} = W_0 = \frac{(v^{-0})^2}{2\omega_0^2 L}\frac{R_s}{\omega_0 L} \tag{6b}$$

where P_0 is the dissipated energy per second.

We denote the ratio of the initial maximum values of the magnetic energy W_m and P_0 as Q,

$$Q = \frac{\omega W_{m,\max}}{P_0} \tag{6c}$$

Since the time average of the magnetic energy $\tilde{W}_{m,\max}$ is half of $W_{m,\max}$, the time average of the total reactive energy $\tilde{W}_t$ is the same as $W_{m,\max}$.

Then,

$$Q = \frac{\omega\tilde{W}_t}{P_0} = \frac{\omega_0 L}{R_s} \tag{6d}$$

Therefore, we will define Q as

$$Q = \frac{\text{Time average of reactive energies}}{\text{Dissipated energy in the time for 1 radian}}$$

$$= \frac{\omega\tilde{W}_t}{P_0} \tag{6e}$$

In other words, we can say that the Q values in Fig. 2 take the values in Eq. (6d) based on the definition of Q of Eq. (6e). Because $\boldsymbol{E}$ and $\boldsymbol{H}$ in a resonant cavity can be expressed by

$$\begin{aligned} \boldsymbol{E} &= \mathscr{E} e^{-\omega' t} e^{j\omega t} \\ \boldsymbol{H} &= \mathscr{H} e^{-\omega' t} e^{j\omega t} \end{aligned} \tag{6f}$$

we have

$$e^{-\omega'/\omega'} = 1 - \frac{1}{2Q}$$

and

$$\omega' \simeq \frac{\omega}{2Q} \tag{6g}$$

by the definition of Q. Therefore, we can express the complex frequency as

$$\begin{aligned} \dot{\boldsymbol{E}} &= \varepsilon \exp\left[j\omega\left(1 + j\frac{1}{2Q}\right)t\right] \\ \dot{\boldsymbol{H}} &= h \exp\left[j\omega\left(1 + j\frac{1}{2Q}\right)t\right] \end{aligned} \tag{6h}$$

When the constant voltage E is supplied to the LCr_s series circuit in Fig. 3a, the current takes maximum values $I_{\max}$ at angular frequency ω_0 $(= 1/\sqrt{LC})$ and $I_{\max}/\sqrt{2}$ at $\omega_0 \pm \Delta\omega/2$, as shown in Fig. 3b.

By a simple calculation, we get

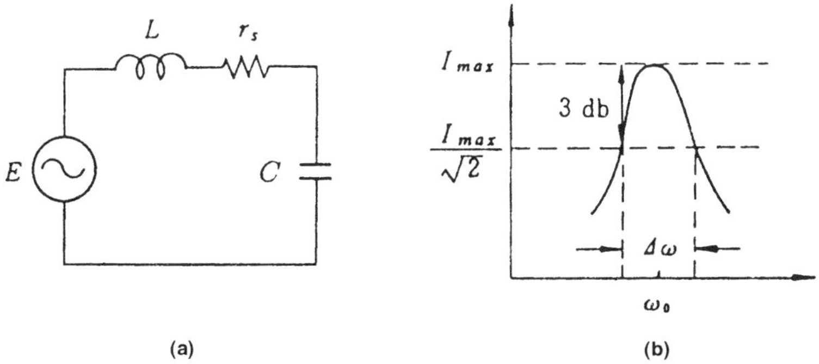

Figure 3. Constant voltage source was connected to LCr_s series circuit (a) and the frequency performance of the current is shown in (b).

$$\frac{\omega_0}{\Delta\omega} = \frac{\omega L}{r_s} = Q \tag{7a}$$

In the same way, when the constant current source is connected to C of the LC parallel resonant circuit as in Fig. 4a, the voltage of C takes maximum values V_{max} at ω_0 $(= 1/\sqrt{LC})$ and $V_{max}/\sqrt{2}$ at $\omega_0 \pm \Delta\omega/2$, as shown in Fig. 4b. By a simple calculation, we can also obtain the following relation:

$$\frac{\omega_0}{\Delta\omega} = \omega CR = Q \tag{7b}$$

Equations (7a) and (7b) can be introduced also in the case of a cavity.

When we excite a cavity with constant current or constant voltage, we have the time average of the reactive energies, $\tilde{W}_t$ or $\tilde{W}'_t$, together with the dissipated energies per second, P_0 or P'_0, respectively, as shown in Figs. 5a and 5b.

The impedance z of Fig. 3a and the admittance y of Fig. 3b at $\omega_0 + \Delta\omega/2$ take the values

$$z = P_0 + j\Delta\omega_0\tilde{W}_t \tag{8a}$$

$$y = P'_0 + j\Delta\omega_0\tilde{W}'_t \tag{8b}$$

These equations can be easily introduced from Eqs. (18f) and (19c) of Chapter 2 together with the equations of the frequency variation of Eqs. (22a) and (22b) of Chapter 2.

In Eqs. (8a) and (8b), we denote the angular frequency $\omega_0 \pm \Delta\omega_0/2$ when $P_0 = \Delta\omega_0\tilde{W}_t$ or $P'_0 = \Delta\omega_0\tilde{W}'_t$. In this case, we get

$$\frac{1}{\Delta\omega_0} = \frac{\tilde{W}_t}{P_0} = \frac{\tilde{W}'_t}{P'_0} \tag{8c}$$

Substituting Eq. (8c) into Eq. (6d), we get

$$Q = \frac{\omega_0}{\Delta\omega_0} \tag{8d}$$

which coincides with Eqs. (7a) and (7b). This results show that the Q values can be obtained from a 3-dB ($I_{max}/\sqrt{2}$ or $V_{max}/\sqrt{2}$) bandwidth ratio in Fig. 3b or Fig. 4b as Eq. (8d) in the case of the general resonant circuit including the cavities.

1.3 *Q* Values of a Cavity with Several Ports Terminated by an Absorber

For a cavity with several ports terminated by absorbers, the total reactive energy decreases with the time, because a part of the reactive energy is

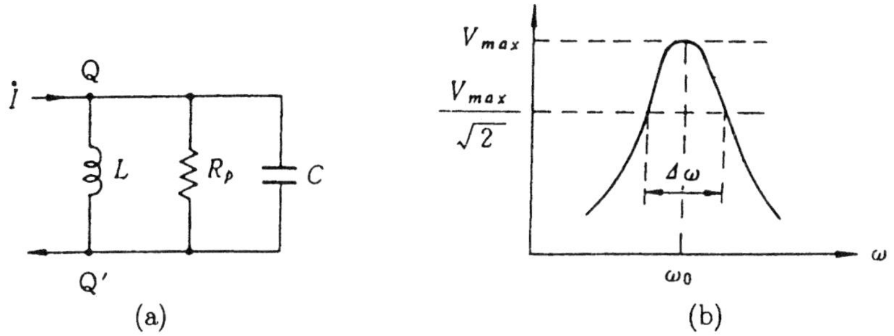

Figure 4. (a) Constant current source was connected between C; (b) the frequency performance of the voltage between C is shown in (b).

dissipated in the absorbers. If a port is excited by the source with internal resistance as in Fig. 6, we have to consider the dissipative energy at the internal resistance of the source. When we denote the dissipated energies per second at ports 1, 2, . . . , n by $P_1, P_2, \ldots, P_n$, the Q values defined by Eq. (9a) are called the external Q values, and we denote it by Q_e.

$$Q_e = \frac{\omega \tilde{W}_t}{P_e} = \frac{\omega \tilde{W}_t}{P_1 + P_2 + \cdots + P_n} \tag{9a}$$

On the other hand, the Q values described in Eq. (8d) is called the no-load Q, because we are considering the dissipated power inside the cavity only. When it is denoted by Q_0, we already have the relation in Eq. (9b) from Eq. (6e):

$$Q_0 = \frac{\omega \tilde{W}_t}{P_0} \tag{9b}$$

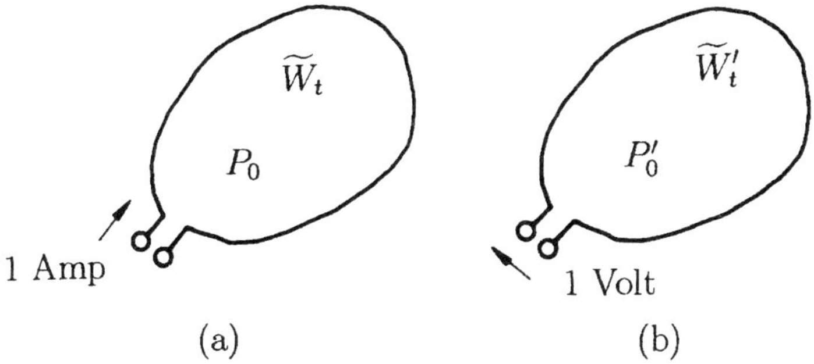

Figure 5. Constant current (a) or voltage (b) excitation of a cavity.

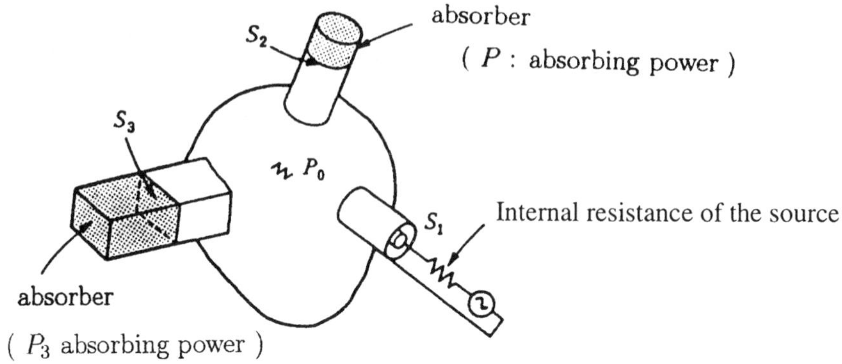

Figure 6. Loaded Q of the cavity with ports 1, 2, and 3, where the port 1 is excited by a source and ports 2 and 3 are loaded by absorbers.

The Q values considering the total dissipative energies inside and outside (connected ports) of a cavity is called the loaded Q, which is denoted by Q_L. Therefore, we have

$$\frac{1}{Q_L} = \frac{1}{Q_0} + \frac{1}{Q_e} \tag{9c}$$

As an example, we will consider the cavity with two ports, where one port is excited by the source and another port is the load, as shown in Fig. 7a. The circuit works as a bandpass filter (described later), and the frequency performance is shown in Fig. 7b.

The insertion loss of the bandpass filter, L (dB), takes the values in Eq. (10a) at the center frequency f_0, and the values of Eq. (10b) at the frequency $f_0 + \Delta f$.

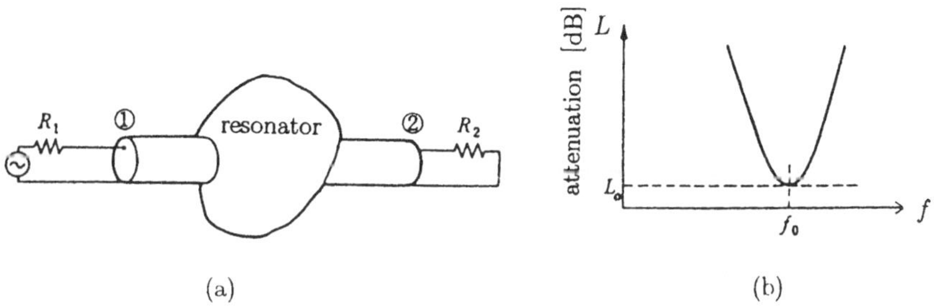

Figure 7. (a) Resonator with two ports, where one port is excited by the source and another port is loaded; (b) the frequency performance of (a).

$$L_0\ (\text{dB}) = 10 \log_{10} \left(\frac{1}{4} \frac{(\beta_1 + \beta_2 + 1)^2}{\beta_1 \beta_2} \right)$$

$$\beta_1 = \frac{Q_0}{Q_{e1}}, \qquad \beta_2 = \frac{Q_0}{Q_{e2}} \tag{10a}$$

where Q_0 is the no-load Q of the resonator, and Q_{ei} ($i = 1, 2$) is the external Q of the ith port

$$L\ (\text{dB}) = 10 \log_{10} \left\{ \frac{1}{4} \frac{(\beta_1 + \beta_2 + 1)^2}{\beta_1 \beta_2} \left[1 + \left(2Q_L \frac{\Delta f}{f_0} \right)^2 \right] \right\} \tag{10b}$$

where Q_L is the loaded Q in the case of Fig. 7a.

2 CONSTRUCTION OF SEVERAL KINDS OF RESONATORS AND THEIR CHARACTERISTICS

A transmission line of length $s\lambda_g/2$ ($s = 1, 2, \ldots$, λ_g is the guided wavelength) can be a resonator by shortening or opening both ends. A transmission line of length $(2s + 1)\lambda_g/4$ also can be a resonator by shortening one end and opening the other end. As transmission lines, a coaxial line, a waveguide, a microstrop line, a fine line, and others are used. As other examples of resonators, the cavity made by a surrounding metal wall and the special two-dimensional metal patterns made on a dielectric substrate are used as resonators. We will show the construction and the characteristics such as Q values of several kinds of resonators mentioned above.

2.1 Coaxial Resonators

Figure 8 shows the side view of a coaxial resonator with shortened ends. The current flows on the surface of the inner conductor, the outer conductor, and the both shortened ends of the conductors, of which the ohmic losses deteriorate the Q values caused by conductive loss, Q_c.

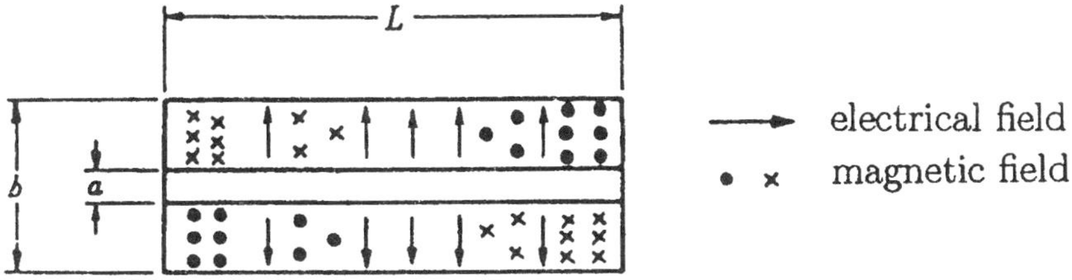

Figure 8. A coaxial resonator with shortened ends.

In the case of the lossless material, the resonant frequency $f_{r,\varepsilon\mu}$ takes the values of Eq. (11), where f_r is the resonant frequency of the $\lambda_g/2$ coaxial resonator with vacuum on half of the material:

$$f_r = \frac{v_0}{2L}$$

$$f_{r,\varepsilon\mu} = \frac{f_r}{\sqrt{\varepsilon_r \mu_r}} \tag{11}$$

where ε_r is the relative dielectric constant and μ is the relative permeability.

When the material filled inside the coaxial line has the loss components, it deteriorates the Q values of the resonator, the Q values of which will be denoted by $Q_{\varepsilon\mu}$. In such a lossy material, however, the dielectric constant and the permeability can be shown by the complex numbers $\dot{\varepsilon}$ and $\dot{\mu}$, respectively as follows [also shown in Eq. (62a) of Chapter 1]:

$$\dot{\varepsilon} = \varepsilon' - j\varepsilon'', \qquad \dot{\mu} = \mu' - j\mu'' \tag{12}$$

The real parts of Eq. (12) (i.e., ε' and μ') show the lossless components, and the imaginary parts, ε'' and μ'', show the loss components of the material. By using Eq. (12), we can show the total Q values of the cavity,

$$\frac{1}{Q} = \frac{1}{Q_c} + \frac{1}{Q_{\varepsilon\mu}}, \qquad Q_c = \frac{\lambda_g}{\delta_s}\left(4 + \frac{\lambda_g}{2b}\frac{1 + b/a}{\ln(b/a)}\right)^{-1}$$

$$\frac{1}{Q_{\varepsilon\mu}} = \frac{1}{Q_\varepsilon} + \frac{1}{Q_\mu}, \qquad Q_\varepsilon = \frac{\varepsilon'}{\varepsilon''}, \qquad Q_\mu = \frac{\mu'}{\mu''} \tag{13a}$$

where δ_s is the skin depth. Q_ε and Q_μ are sometimes expressed by

$$\frac{1}{Q_\varepsilon} = \tan\delta_\varepsilon, \qquad \frac{1}{Q_\mu} = \tan\delta_\mu \tag{13b}$$

Q_c of Eq. (13a) takes the maximum values when

$$\frac{b}{a} = 3.6 \tag{14a}$$

The characteristics impedance Z_c is calculated from

$$Z_c = \frac{75}{\sqrt{\varepsilon_r}} \quad (\Omega) \tag{14b}$$

The practical construction of the coaxial resonator of the length of $\lambda_g/4$ is shown in Fig. 9. As shown in Fig. 9a, the open end of the coaxial line, S, faces the shielding case. In this region, the electric field takes the place of the z component, which is the evanescent E wave. This electrical field

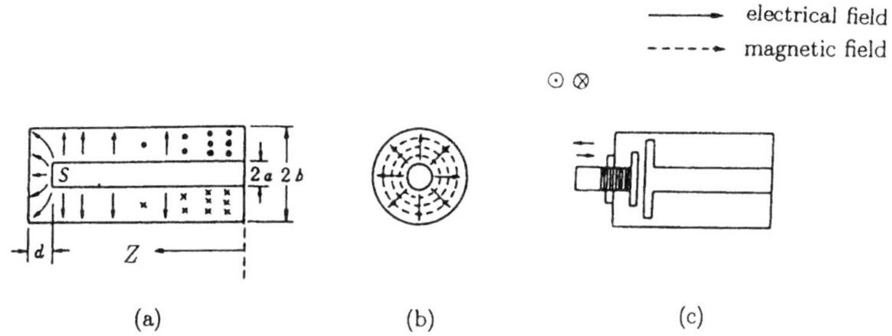

Figure 9. The $\lambda_g/4$ coaxial resonator and the electromagnetic field. (a) Side view; (b) sectional view; (c) construction to adjust the resonant frequency.

makes an equivalent capacitive load at the open end of the coaxial line. Denoting this capacity by C, it is calculated from

$$C = C_0 + C_f$$

$$C_0 - \frac{0.0885\pi a^2}{d} \quad \text{(pF)}$$

$$C_f \simeq 0.524a + \frac{0.0094b}{\ln^2(b/a)} \quad \text{(pF)} \tag{15}$$

a, b, and d are measured in centimeters.

When the length of the axial length becomes shorter than the radius direction, as shown in Fig. 10c, the electromagnetic fields change to those of the radial transmission line rather than the coaxial line; its equivalent network is shown in Fig. 10d.

2.2 Waveguide Resonator

As mentioned earlier, a waveguide of length $s\lambda_g/2$ with shortened ends can be a resonator. Typical constructions are the rectangular cavity and the circular cavity as shown in Figs. 11a and 11b, respectively, and they have an infinite number of resonant frequencies corresponding to their modes.

In the case of the empty cavities, the resonant frequencies and the corresponding wavelengths in free space are shown in the following equations:

$$f_0 = \frac{v_0}{\lambda_0} \tag{16a}$$

where v_0 is the light velocity = 3×10^8 m/sec, and

$$\lambda_0 = \frac{2}{[(m/a)^2 + (n/b)^2 + (s/c)^2]^{1/2}} \quad \text{(Fig. 11a)} \tag{16b}$$

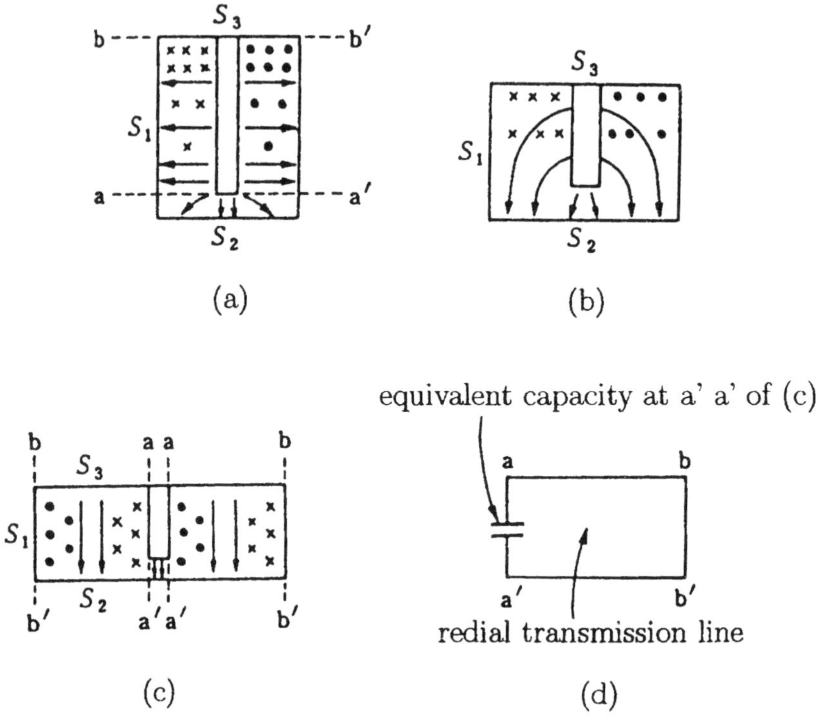

Figure 10. The construction and the electromagnetic field in the case of several relative lengths between the axial and radius directions. (a) Coaxial line; (b) construction between (a) and (c); (c) radius larger than axial length; (d) equivalent network of (c).

$$\lambda_0 = \frac{2}{[(s/l)^2 + (x'_{mn}/\pi a)^2]^{1/2}}, \quad \text{TE}^{\square}_{mns} \text{ (Fig. 11b)} \tag{16c}$$

$$\lambda_0 = \frac{2}{[(s/l)^2 + (x_{mn}/\pi a)^2]^{1/2}}, \quad \text{TM}^{\square}_{mns} \text{ (Fig. 11b)} \tag{16d}$$

In Eqs. (16c) and (16d), the values of x_{mn} and x'_{mn} are the nth root of $J_m(x)$ and $J'_m(x)$ and they take the values given in Table 1.

When the material with relative dielectric constant ε_r and relative permeability μ_r are filled in the cavities, the resonant frequencies are reduced to the values given by

$$f_{\mu,\varepsilon} = \frac{f_0}{\sqrt{\mu_r \varepsilon_r}} \tag{17}$$

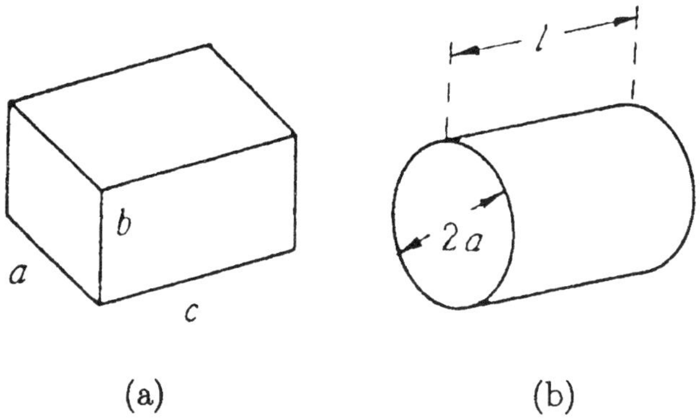

Figure 11. Typical waveguide resonators. (a) The rectangular cavity; (b) the circulator cavity.

Next, the values of Q_c caused by the surface resistance of the metal wall surrounding the cavities can be calculated as shown in Fig. 12, which presents values for the case of the copper for the metal. When the cavities are filled by the material with a relative dielectric constant of ε_r and relative permeability μ_r, the values of Q_c are reduced to the values given by

Table 1. The Values of x_{mn} and x'_{mn}

	m					
n	0	1	2	3	4	5
			(a) Values of x_{mn}			
1	2.405	3.832	5.136	6.380	7.588	8.771
2	5.520	7.106	8.417	9.761	11.095	12.339
3	8.654	10.173	11.620	13.015	14.372	—
4	11.792	13.324	14.796	—	—	—
			(b) Values of x'_{mn}			
1	3.832	1.841	3.054	4.201	5.317	6.416
2	7.016	5.331	6.706	8.015	9.282	10.520
3	10.173	8.536	9.969	11.346	12.682	13.987
4	13.324	11.706	13.170	—	—	—

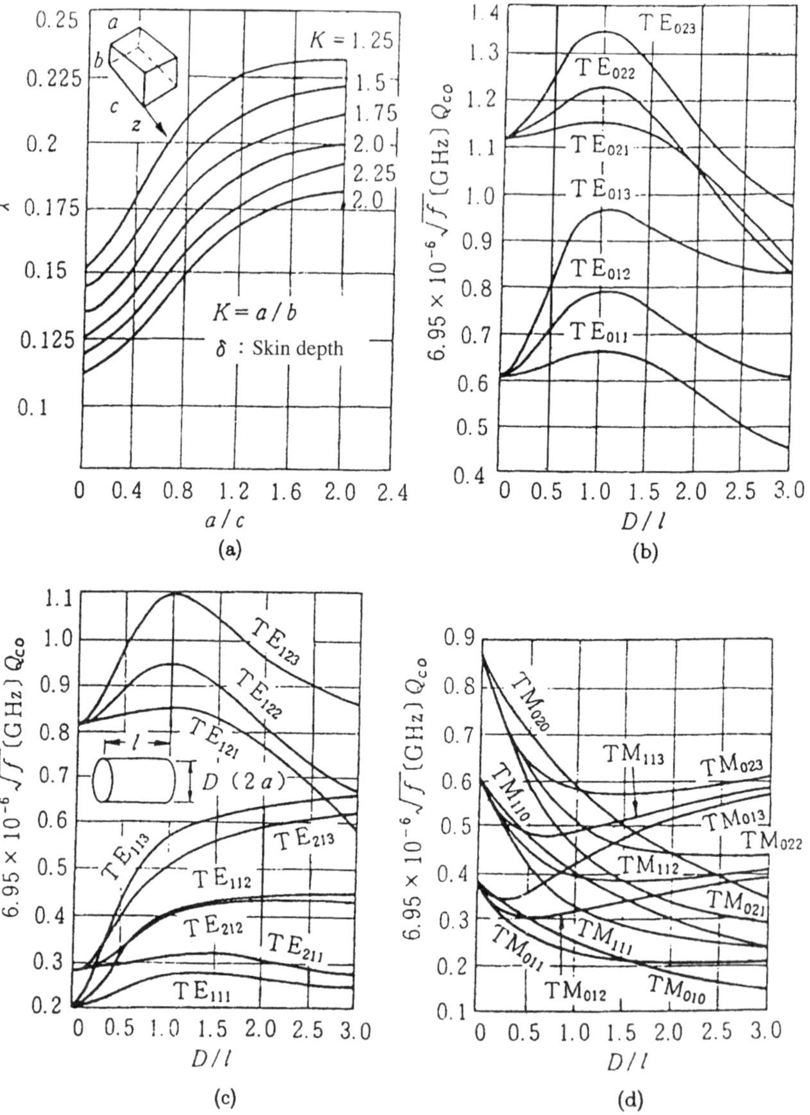

Figure 12. Q_c of several modes of waveguide resonators. (a) $TE^{\square}_{101}$ rectangular cavity; (b) TE°_{0} circular cavity (in the case of copper); (c) TE° circular cavity (copper); (d) TM° circular cavity (copper).

$$Q_{c,\varepsilon,\mu} = \frac{Q_{co}}{\sqrt{\varepsilon_r \mu_r}} \tag{18}$$

This is because Q_c is proportional to l/δ and l is proportional to $1/\sqrt{\varepsilon_r\mu_r}$, where l is a, b, and c in Figs. 11a and 11b; see Note 1 in the Appendix to this chapter.

If the material is lossy and takes the values Q_ε and Q_μ corresponding to the losses of the dielectric constant and permeability, respectively, the total Q values, Q, is calculated from

$$\frac{1}{Q} = \frac{1}{Q_{c,\varepsilon\mu}} + \frac{1}{Q_\varepsilon} + \frac{1}{Q_\mu} \tag{19}$$

2.3 Microstrip Resonator

As mentioned in Chapter 1, Section 1, the microstrip line is the sectional view in Fig. 13a, the electromagnetic fields take place as shown in Fig. 13b, and the current density becomes large at the edges of the microstrip, as shown in Fig. 13c.

The electric fields exist not only in the dielectric substrate but also in air, as shown in Fig. 13b, which provides the effective capacity between the microstrip line and the ground. The magnetic fields surrounding the mi-

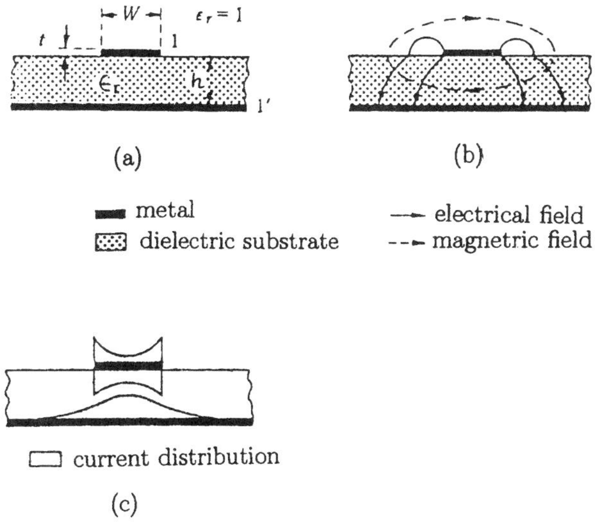

Figure 13. (a) Sectional view of a microstrip line; (b) electromagnetic field distribution; (c) current distribution.

crostrip exist not only in the substrate, sometimes made by magnetic material, but also in air, which provides the effective inductance. Denoting the effective capacity per meter by C_{eff} (F/m) and the effective inductance per meter by L_{eff} (H/m), we can get the characteristic impedance Z_c, the phase velocity v_p, and the guided wave length λ_g as follows:

$$Z_c = \sqrt{\frac{L_{\mathrm{eff}}}{C_{\mathrm{eff}}}} \tag{20a}$$

$$v_p = \frac{1}{\sqrt{C_{\mathrm{eff}} L_{\mathrm{eff}}}} \tag{20b}$$

$$\lambda_g = \frac{v_p}{f} \tag{20c}$$

Considering the capacitance C_0 (F/m) and the inductance L_0 (H/m) in the case when the substrate is replaced by air ($\varepsilon_r = \mu_r = 1$), C_{eff} and L_{eff} can be shown by

$$C_{\mathrm{eff}} = \varepsilon_{r,\mathrm{eff}} C_0 \tag{21a}$$

$$L_{\mathrm{eff}} = \mu_{r,\mathrm{eff}} L_0 \tag{21b}$$

The equations show that the C_{eff} and L_{eff} are the values of the microstrip line placed in the infinite medium with dielectric constants $\varepsilon_{\mathrm{eff}}$ and permeability of μ_{eff}.

In the case of a conventional dielectric substrate, as we have

$$\mu_{r,\mathrm{eff}} = 1 \quad \text{because } \mu_r = 1$$

$$L_{\mathrm{eff}} = L_0 \tag{22}$$

Eqs. (20a), (20b), and (20c) become Eqs. (23a), (23b), and (23c), respectively:

$$Z_c = \frac{Z_0}{\sqrt{\varepsilon_{r,\mathrm{eff}}}} \quad (Z_0 = \text{the values of } Z_c \text{ for } \varepsilon_{r,\mathrm{eff}} = 1) \tag{23a}$$

$$v_p = \frac{v_0}{\sqrt{\varepsilon_{r,\mathrm{eff}}}} \quad (v_0 = 3 \times 10^8 \text{ m/sec}) \tag{23b}$$

$$\lambda_g = \frac{\lambda_0}{\sqrt{\varepsilon_{r,\mathrm{eff}}}} \quad \left(\lambda_0 = \text{free-space wavelength} = \frac{3 \times 10^8}{f}\ \text{m}\right) \tag{23c}$$

Z_0 and $\varepsilon_{r,\mathrm{eff}}$ is calculated using the values of Eq. (24a) [1] and Eq. (24b) [2]:

$$Z_0 = 30 \ln \left\{ 1 + \frac{4h}{W_0} \left[\frac{8h}{W_0} + \sqrt{\left(\frac{8h}{W_0}\right)^2 + \pi^2} \right] \right\}$$

$$W_0 = W + \Delta W$$

$$\Delta W = \frac{t}{\pi} \ln \frac{4e}{\{(t/h)^2 + [\pi^2(W/t + 1.1)^2]^{-1}\}^{1/2}}$$

$$Z_c = \frac{Z_0}{\sqrt{\varepsilon_{r,\text{eff}}}} \tag{24a}$$

$$\varepsilon_{r,\text{eff}} = \frac{\varepsilon_r + 1}{2} + \frac{\varepsilon_r - 1}{2} \left(1 + \frac{10h}{W}\right)^{1/2} \tag{24b}$$

Although Eq. (24b) is obtained based on a TEM (transverse electromagnetic) wave, the values deviate corresponding with the frequency when the frequency becomes higher and the wave is not TEM. The values of $\varepsilon_{r,\text{eff}}$ is obtained as

$$\varepsilon_{\text{eff}} = \varepsilon_r + \frac{\varepsilon_W - \varepsilon_r}{1 + P(f)}$$

$$P(f) = P_1 P_2[(0.1844 + P_3 P_4)10fh]^{1.5763}$$

$$P_1 = 0.2749 + (0.6315 + (1 + 0.157fh)^{20} 0.525) \left(\frac{W}{h}\right) - 0.06568 \exp\left(-8.751 \frac{W}{h}\right)$$

$$P_2 = 0.3362 \; [1 - \exp(-0.034426)] \tag{24c}$$

The microstrip resonator, however, can be realized by using the length $\lambda_g/2$ with both ends shortened, as shown in Fig. 14a, or with both ends open, as shown in Fig. 14b, or using the length $\lambda_g/4$ with one shortened end together with opening the other end, as shown in Fig. 14c.

The Q values of the microstrip resonator, however, consists of Q_c caused by the conductive loss, Q_ε caused by the dielectric loss of the substrate, Q_μ caused by the magnetic loss in the case of the magnetic substrate, and Q_r caused by the radiation loss. Denoting the total Q values by Q_0, Q_0 can be obtained from

$$\frac{1}{Q_0} = \frac{1}{Q_c} + \frac{1}{Q_\varepsilon} + \frac{1}{Q_\mu} + \frac{1}{Q_r} \tag{25}$$

Such Q values are described below.

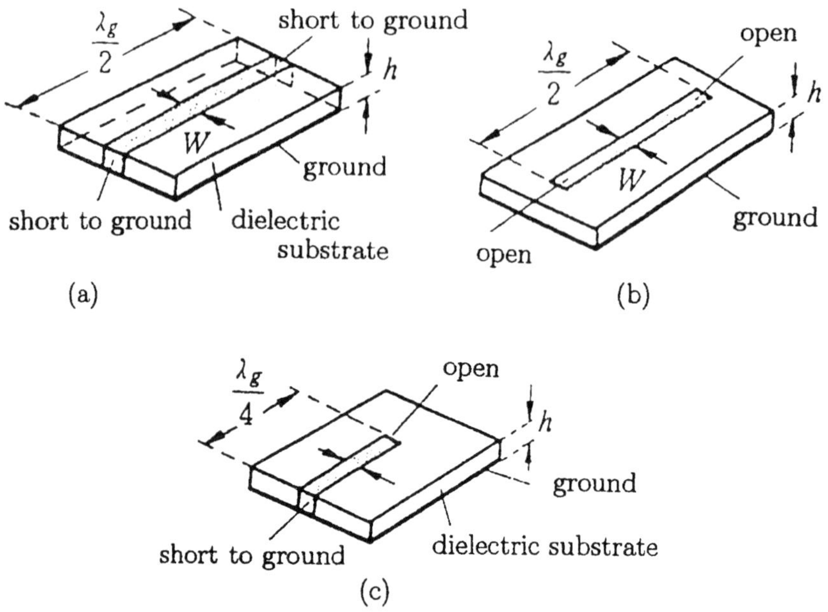

Figure 14. Several constructions of microstrip resonators. (a) $\lambda_g/2$ resonator; (b) both ends open $\lambda_g/2$ resonator; (c) $\lambda_g/4$ resonator.

Q_c

When $h \ll \lambda_g/2$ and $\lambda_g/4$, the conductive loss at the portions of the shortened ends are negligible. In such a case, Q_c can be related to the attenuation constant α_c as follows:

$$Q_c = \frac{\pi}{\lambda_g \alpha_c} \tag{26}$$

For a simple case such as

$$W \gg h \tag{27a}$$

the current distribution becomes almost uniform, which results in the Z_c of Eq. (27b):

$$\begin{aligned} Z_c &= 120\pi \sqrt{\frac{\mu_r}{\varepsilon_r}} \frac{h}{W} \\ &= \frac{120\pi}{\sqrt{\varepsilon_r}} \frac{h}{W} \quad (\text{for } \mu_r = 1) \end{aligned} \tag{27b}$$

In this case, α_c is calculated from

$$\alpha_c = \frac{R_s}{WZ_c} \left(\frac{\text{Np}}{\text{m}}\right)$$
$$= \frac{8.68R_s}{WZ_c} \left(\frac{\text{dB}}{\text{m}}\right) \tag{27c}$$

From Eqs. (27b) and (27c) we get

$$dQ_c = \frac{W\pi Z_c}{\lambda_g R_s} = \frac{W\pi Z_0}{\lambda_0 R_s} = \frac{120\pi^2 h}{\lambda_0 R_s} \tag{27d}$$

where Z_0 is the characteristic impedance at $\varepsilon_r = 1$.

In the case of a 1-GHz microstrip resonator with copper, Q_c can be obtained from Eq. (27d) by using $R_s = 8.29 \times 10^{-3}\ \Omega$:

$$Q_c = \frac{120\pi^2}{300 \times 8.29 \times 10^{-3}} = 476 \tag{27e}$$

when $h = 1$ mm.

When W/h decreases, the current concentrates at the edges of the microstrip line, as shown in Fig. 13c, which results in an increase in the conductive loss and α_c. The values of α_c is shown in Fig. 15 [4]. We will describe the process to obtain Q_c using Fig. 15.

Step 1 When w, h, t, and ε_r are given, Z_c and $\varepsilon_{r,\text{eff}}$ are obtained from Eqs. (24a) and (24b).

Step 2 From the values of W/h and t/h, $\alpha_c Z_c h/R_s = A$ (dB) is obtained from Fig. 15.

Step 3 α_c (Np) is obtained from the values of A as

$$\alpha_c = \frac{A}{8.68} \frac{R_s}{Z_c h} \tag{28a}$$

Therefore, Q_c can be obtained from Eq. (26) as follows:

$$Q_c = \frac{8.68\pi}{A} \frac{Z_c}{R_s} \frac{h}{\lambda_g}$$
$$= \frac{8.68\pi}{A} \frac{Z_0}{R_s} \frac{h}{\lambda_0} \tag{28b}$$

because

$$Z_c = \frac{1}{\sqrt{\varepsilon_{r,\text{eff}}}}, \qquad \lambda_g = \frac{\lambda_0}{\sqrt{\varepsilon_{r,\text{eff}}}}$$

It is understood that Q_c is not affected by the values of $\sqrt{\varepsilon_{r,\text{eff}}}$.

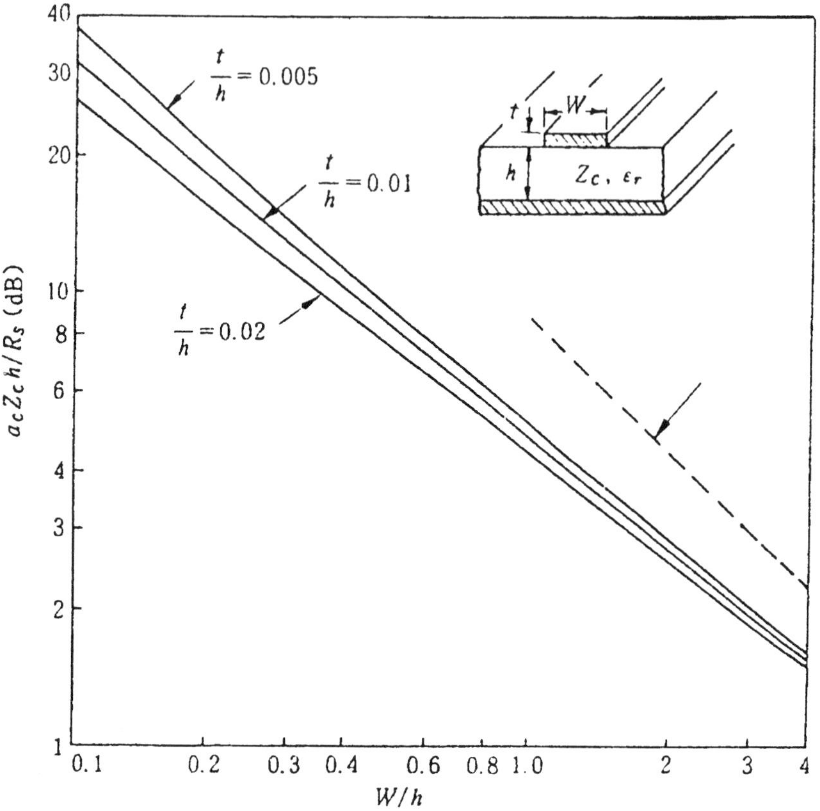

Figure 15. Relationship between the attenuation constant α_c and the sectional construction of the microstrip line.

The process mentioned above is the method to obtain Q_c from Fig. 15. Substituting the values of Eq. (59b) of Chapter 2 (i.e.,

$$R_s = 8.29 \times 10^{-3} R_{s,r} \sqrt{f}$$

where $R_{s,r}$ is shown in Table 4 of Chapter 1) for R_s of Eq. (28b), we get

$$Q_c = 10.965 \frac{Z_0 h}{R_{s,r} A} \sqrt{f} \tag{28c}$$

If the values of Q_c for $h = 1$ mm, $f = 1$ GHz, $R_{s,r} = 1$, and several W/h ($Q_{c,\mathrm{Cu}}$) are obtained, Q_c can be calculated as

$$Q_c = Q_{c,\mathrm{Cu}} \frac{h\sqrt{f}}{R_{s,r}} \tag{28d}$$

The values of $Q_{c,\mathrm{Cu}}$, however, are calculated from 10.965 Z_0/A, where A is obtained by Fig. 15 and Z_0 is obtained by Eq. (24a), the results of which are shown in Fig. 16. For example, we will obtain the values of Q_c for h = 1.6 mm, W = 3.2 mm, t = 0.032 mm, λ_0 = 750 mm (f = 400 MHz), and ε_r = 2.5. In this case, because we get W/h = 2 and t/h = 0.02, we can obtain $Q_{c,\mathrm{Cu}}$ = 394 from Fig. 16. These results are also obtained from Steps 1–3.

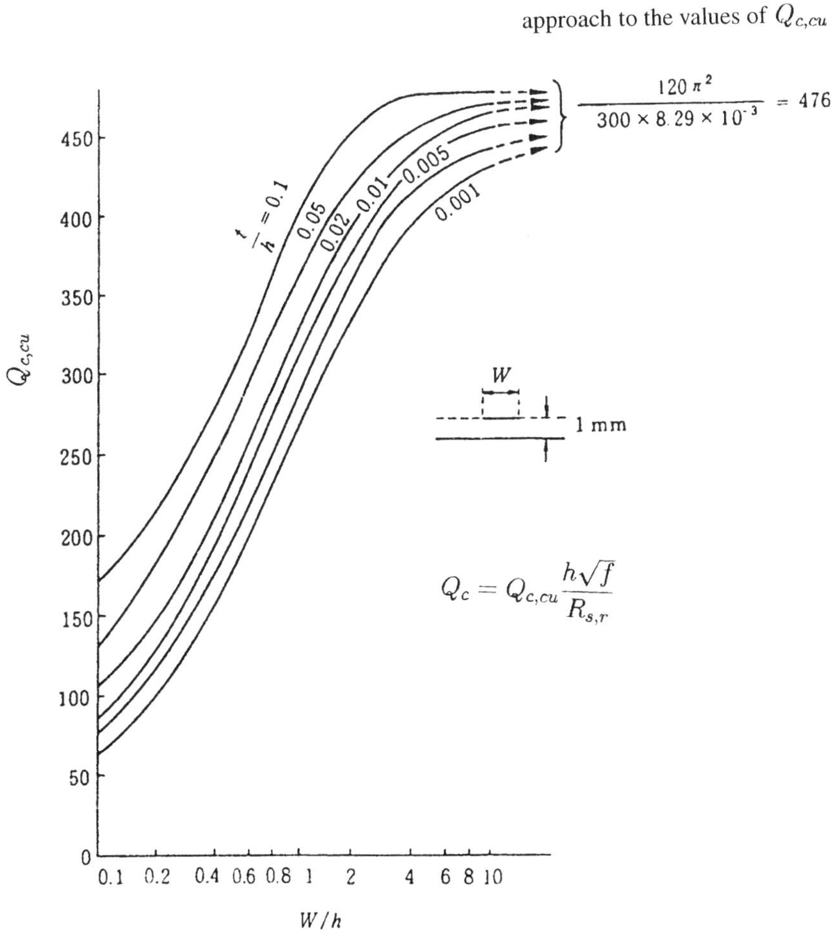

Figure 16. Q_c of microstrip with h = 1 mm at 1 GHz.

We can summarize the main characteristics of Q_c as follows:

1. Q_c increases for thicker microstrips.
2. An increase Q_c corresponding to an increase W/h, which slows in the region $W/h > 2$.
3. Q_c increases in proportion to h.
4. Q_c increases in proportion to $\sqrt{f}$.

$Q_{\varepsilon,\text{micro}}$

Substituting the complex dielectric constant of Eq. (12) into Eq. (24b), we get

$$\dot{\varepsilon}_{r,\text{eff}} = \frac{\varepsilon_r' + 1}{2} + \frac{\varepsilon_r' - 1}{2}\left(1 + \frac{10h}{W}\right)^{-1/2} - j\frac{\varepsilon_r''}{22}\left[1 + \left(1 + \frac{10h}{W}\right)^{-1/2}\right]$$

$$= \varepsilon_{r,\text{eff}}' - j\varepsilon_{r,\text{eff}}'' \tag{29a}$$

$$Q_\varepsilon = \frac{\varepsilon_{r,\text{eff}}'}{\varepsilon_{r,\text{eff}}''} \tag{29b}$$

By using the definition

$$\frac{\varepsilon_r'}{\varepsilon_r''} = \frac{1}{\tan\delta_\varepsilon}$$

we obtain values of Q_ε in Eq. (29b) as follows:

$$Q_\varepsilon = \frac{\varepsilon_r' + 1 + (\varepsilon_r' - 1)(1 + 10h/W)^{1/2}}{\varepsilon_r' \tan\delta_\varepsilon[1 + (1 + 10h/W)^{-1/2}]} \tag{29c}$$

For example, when we consider the case

$$\varepsilon_r' = 2.5, \qquad \frac{1}{\tan\delta_\varepsilon} = 400, \qquad \frac{W}{h} = 2$$

we get

$$Q_\varepsilon = \frac{3 + 1.5(1 + 5)^{-1/2}}{2.5 \times (1/400)[1 + (1 + 5)^{-1/2}]} = \frac{3.61 \times 400}{3.52} = 410$$

It is understood that Q_ε is higher than the Q values of the substrate. The reason is that the electric fields exist in the lossless air region together with the substrate including the loss components.

Q_r

Waves are radiated from the open end and shortened end in Fig. 14, which result in the deterioration of the microstrip resonators. Denoting the radiation power by P_{rad} and the time average of the total reactive power by $\tilde{W}_t$, we can show the values caused by radiation, Q_r, as

$$Q_r = \frac{\omega \tilde{W}_t}{P_{\text{rad}}} \tag{30}$$

P_{rad} is obtained for the incident power, P_{in}, to portion to the radiation by Lewin [5] as follows:

$$\frac{P_{\text{rad}}}{P_{\text{in}}} = 2\pi\eta_0 \left(\frac{h}{\lambda_0}\right)^2 \frac{F_i}{Z_c}, \quad \eta_0 = 120\pi \tag{31}$$

Open end:

$$F_1 \simeq \frac{8}{3\varepsilon_{r,\text{eff}}} \tag{32a}$$

Shortened end:

$$F_2 \simeq \frac{16}{15\varepsilon_{r,\text{eff}}^2} \tag{32b}$$

To obtain the values of Q_r, we must obtain the relationship between $\tilde{W}_t$ and P_{in}. The result is shown in Eq. (33a) (see Note 2):

$$\tilde{W}_t = \frac{P_{\text{in}}}{v_p} \lambda \tag{33a}$$

Substituting Eq. (31) into Eq. (30), we get

$$Q_r = \frac{\omega \tilde{W}_t}{P_{\text{rad},i} + P_{\text{rad},j}} = \frac{\omega \tilde{W}_t}{P_{\text{in}}} \frac{P_{\text{in}}}{P_{\text{rad},i} + P_{\text{rad},j}} = 2\pi \frac{P_{\text{in}}}{P_{\text{rad},i} + P_{\text{rad},j}} \tag{33b}$$

[1: Fig. 14(b)   2: Fig. 14(a)]

In the case of Figs. 14a and 14b, therefore, $i = j = 2$ and $i = j = 1$, respectively. From Eqs. (31) and (33b), we get

$$Q_r = \frac{\pi P_{\text{in}}}{P_{\text{rad},1,2}} = \frac{Z_c}{\eta_0 F_{1,2}(h/\lambda_0)^2} \qquad \left(\frac{\lambda}{2} \text{ resonator}\right) \tag{33c}$$

[1: Fig. 14(b)   2: Fig. 14(a)]

In the case of Fig. 14c, because $\tilde{W}_t$ becomes halved and $P_{\text{rad}} = P_{\text{rad},1} + P_{\text{rad},2}$, we get

$$Q_r = \frac{\pi P_{\text{in}}}{P_{\text{rad},1} + P_{\text{rad},2}} = \frac{Z_c}{\eta_0 (F_1 + F_2)(h/\lambda_0)^2} \qquad \left(\frac{\lambda}{4} \text{ resonator}\right) \tag{33d}$$

Practical Example 1

In the case of $\varepsilon_{r,\text{eff}} = 7$, $Z_c = 50\ \Omega$, $h = 0.6$ mm, and a $\lambda/2$ resonator of 10 GHz with open both ends, we get

$$F_1 = \frac{8}{3 \times 7} = \frac{8}{21}, \qquad \lambda_0 = 30 \text{ mm}$$

$$Q_r = \frac{50}{2 \times 120\pi \times 8/21 \times (0.6/30)^2} = 435.2$$

Practical Example 2

In the case of $\varepsilon_{r,\text{eff}} = 2$, $Z_c = 30\ \Omega$, $h = 1.6$ mm, and a $\lambda/4$ resonator of 0.4 GHz, we get

$$F_1 = \frac{8}{3 \times 2} = \frac{4}{3}, \qquad F_2 = \frac{16}{15 \times 2^2} = \frac{4}{15}, \qquad \lambda_c = 750 \text{ mm}$$

$$Q_r = \frac{30}{2 \times 120\pi(4/3 + 4/15)(1.6/750)^2} = 5464$$

2.4 Dielectric Resonator

Low-Loss High-Dielectric Ceramics

Recently, the low-loss high-dielectric ceramics with a small temperature coefficient were developed. They are used for the small, high-Q resonators. Several materials are shown in Tables 2 and 3.

The complex dielectric constant $\dot{\varepsilon}(\omega)$, where ω is the angular frequency can be expressed by Eq. (34) [18]:

$$\dot{\varepsilon}(\omega) = \varepsilon(\infty) = \frac{4\pi\rho}{\omega_T^2 - \omega^2 + jr\omega} \tag{34}$$

where $4\pi\rho$ is the strength of the lattice vibration, ω_T is the resonant angular frequency of the lattice vibration, and γ is the attenuation constant of the lattice vibration.

Because $\omega_r/2\pi$, generally, exists from 3000 to 30,000 GHz, $\omega_r \gg \omega$ in the microwave frequency range. Substituting this condition into Eq. (34), we get

$$\varepsilon'(\omega) = \frac{4\pi\rho}{\omega_T^2}$$

$$\frac{1}{Q} = \tan\delta = \frac{\varepsilon''(\omega)}{\varepsilon'(\omega)} = \frac{r}{\omega_T^2}\,\omega$$

$$\dot{\varepsilon} = \varepsilon'(\omega) = j\varepsilon''(\omega) \tag{35}$$

As understood from Eq. (35), the relative dielectric constant ε_r' is constant and $\tan\delta$ is inversely proportional to the frequency f; that is, the values of Qf is constant. The values of Qf, therefore, are the important constant to evaluate the microwave ceramics.

Table 2. Performance of Several High-Dielectric Ceramics

Materials	ε_r	$Q_{\varepsilon m}$	τ_f (ppm/°C)	f_m (GHz)	Ref.
$MgTiO_3-CaTiO_3$	21	8000	0	7	6
$Ba(Mg, Ta)O_3$	25	16000	3	10	7
$Ba(Mg, Ta)O_3$	25	35000	4	10	8
$Ba(Sn, Mg, Ta)O_3$	25	20000	0	10	9
$Ba(Mg, Ta)O_3-Ba(Zn, Ta)O_3$	27	15000	0	10	10
$Ba(Zn, Nb)O_3-Ba(Zn, Ta)O_3$	30	14000	0	12	11
$Ba(Zr, Zn, Ta)O_3$	30	10000	0	10	12
$(Ca, Sr, Ba)ZrO_3$	30	4000	5	11	13
$BaO-TiO_2-WO_3$	37	8800	2	6	14
$(Zr, Sn)TiO_4$	38	7000	0	7	15
$Ba_2Ti_gO_{20}$	40	8000	2	4	16
$Sr(Zn, Nb)O_3-SrTiO_3$	43	5000	−5 to +5	5	10
$BaO-Sm_2O_3-5TiO_2$	77	4000	15	2	17
$BaO-PbO-Nb_2O_3-TiO_2$	90	5000	0	1	15

Note: $Q_{\varepsilon m}$ = 1/tan δ = measured values of Q; f_0 = frequency to measure the values of Q; ε_r = relative dielectric constant.

Q values at a constant frequency of 10 GHz was plotted corresponding to the relative dielectric constant in Fig. 17.

Typical Resonators with High-Dielectric Ceramics

There are the coaxial resonator and several dielectric resonators corresponding to the several modes, which do not have any metal wall except the shielding metal plate to protect the radiation power. Typical constructions and modes are shown in Fig. 18. Because the dielectric constant is very large, the dielectric resonators (DR) are small and the Q values are quite high. As a reference, the resonators with the high-dielectric ceramics were compared with conventional empty cavities, as shown in Table 4. We can

Table 3. Higher-Dielectric Ceramics

Materials	ε_r	$Q_{\varepsilon m} f_m$	τ_f
TiO_2	104	40000	460
$CaTiO_3$	180	7000	910
$SrTiO_3$	304	3300	1700

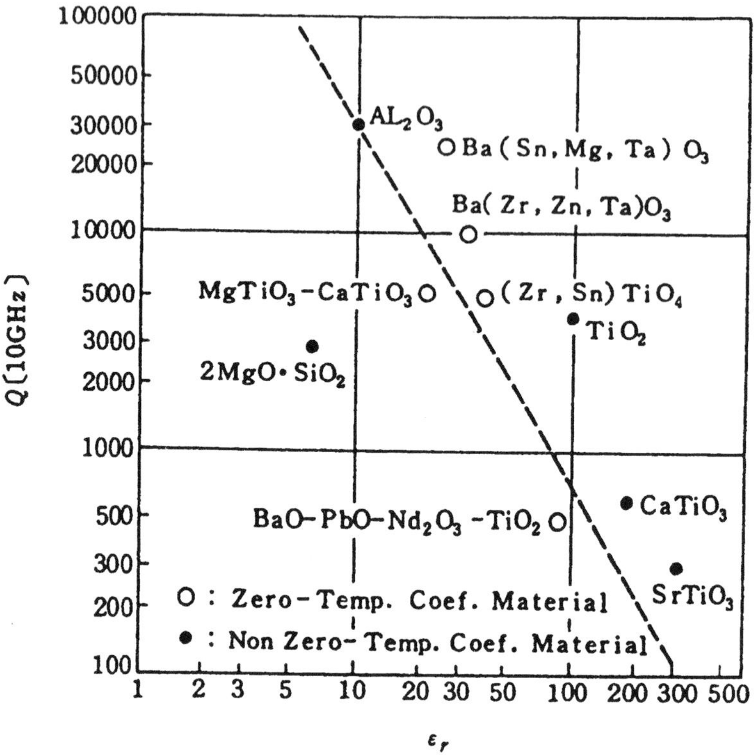

Figure 17. Relationship between Q values and ε_r for several materials.

understand how the small and high Q's are obtained from the DR and conventional cavities.

Coaxial Resonator with High-Dielectric Ceramics

As shown in Fig. 19, the coaxial line is filled by the high-dielectric ceramics, where S_1 is open without metal and S_2 is shortened with a metal membrane. This resonator is made by the ceramic with a center hole with the diameter

---

Figure 18. Typical resonators with high-dielectric ceramics. (a) Coaxial resonator; (b) $TE^{\circ}_{01\delta}$ DR in shielding case; (c) TE°_{010} DR in shielding case; (d) $TM^{\square}_{110}$ in shielding case; (e) $TM_{01\delta}$ in shielding case; (f) $EH_{11\delta}$ in shielding case; (g) $HE_{11\delta}$ in shielding case.

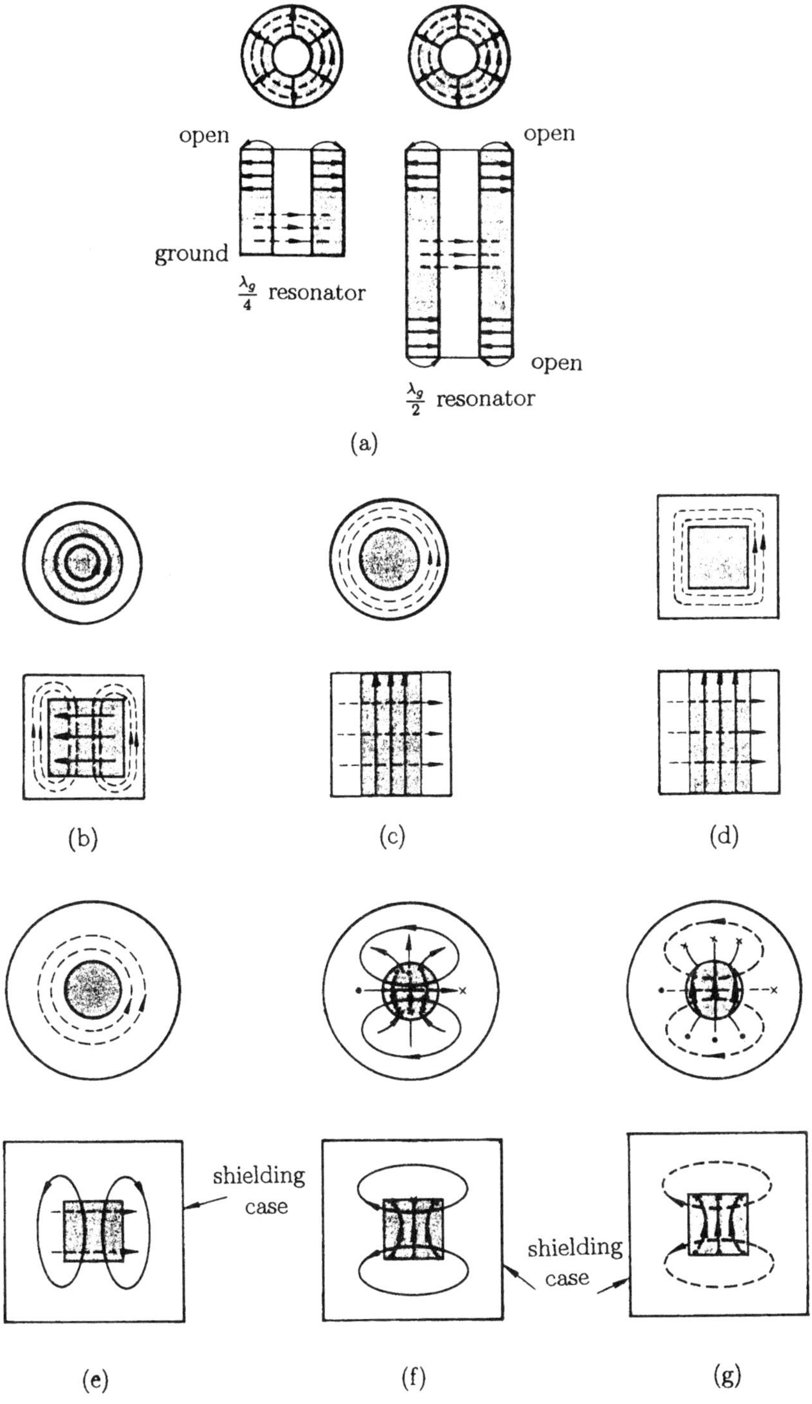
open
open
ground
$\frac{\lambda_g}{4}$ resonator
open
$\frac{\lambda_g}{2}$ resonator
(a)
(b)
(c)
(d)
shielding case
shielding case
(e)
(f)
(g)

Table 4. Comparison Between the Dielectric Resonators and Conventional Cavities

	Resonant mode	Outside size (mm)	Volume cm^3	Q_0	Q_C	Q_D
Empty cavities	TE°_{01}	$524\phi \times 209H$	45,000	70,000		
	$(1/4)\lambda_g$ TEM	$10\phi \times 4\phi \times 75.0$	5.0	1,210		
Dielectric resonator	$TE^{\circ}_{01\delta}$	$163\phi \times 65t^{a}$	1,400	40,000	180,000	51,000
	$TM^{\square}_{110}$	$63.5 \times 63.5 \times 52.5^{b}$	210	11,900	15,600	50,000
	$(1/4)\lambda_g$ TEM	$10\phi \times 4\phi \times 12$	0.8	1,010	1,030	50,000

Note: H = height of the empty cavity; h = height of the DR. Resonant frequency = 1 GHz; relative dielectric constant = 38; conductivity of copper = 5.8×10^7 ($1/\Omega$ m).
[a] $TE^{\circ}_{01\delta}$ size = 54 mm $\phi \times$ 22 mm h.
[b] $TE^{\square}_{110}$ size = 14 m $\times$ 14 mm $\times$ 52.5 mm h.

of $2a$ and it is coated by a metal membrane. The resonant angular frequency ω_r is

$$\omega_r = \frac{c\pi}{2l\sqrt{\varepsilon_r}} \quad (c = \text{light velocity})$$

$$= \frac{\pi\lambda_0}{2\sqrt{\varepsilon_r}} \quad (\lambda_0 = \text{free-space wavelength}) \tag{36}$$

The Q values, Q_0, are calculated from

$$\frac{1}{Q_0} = \frac{1}{Q_c} + \frac{1}{Q_\varepsilon} + \frac{1}{Q_r}$$

Q_c are the Q values caused by the metal membrane's surface resistance; Q_ε is that caused by the dielectric ceramics' tan δ_ε; Q_r is that caused by the radiation from S_1 of Fig. 19a.

As indicated in Eq. (18), Q_c decreases because ε_r increases. Q_ε is very large, as shown in Table 1. Q_r is also very large in the coaxial line because

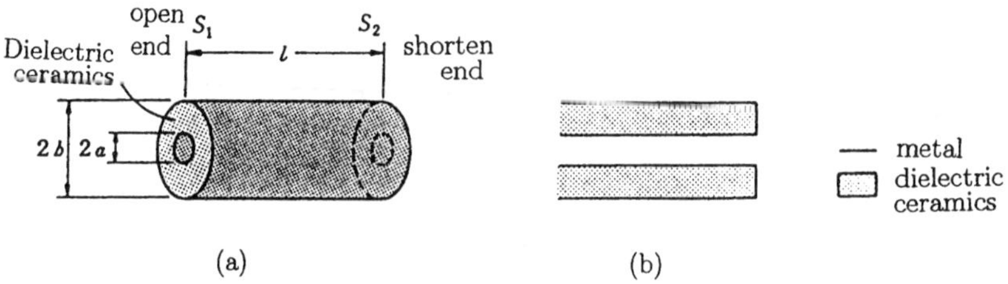

Figure 19. $\lambda_g/4$ coaxial dielectric resonator. (a) Construction; (b) side view.

the electrical fields always have the pair with opposite directions, which cancel out the radiation toward the axial direction of the coaxial line and the radiation toward the other direction is also small in the case of $2b \ll \lambda_0$. Therefore, Q_0 can be expressed by

$$\frac{1}{Q_0} = \frac{1}{Q_c} + \frac{1}{Q_\varepsilon} \tag{37a}$$

Q_c is calculated from Eq. (37b), which is introduced by substituting $4l$ for λ_g of Eq. (13a):

$$Q_c = \sqrt{2\sigma\omega\mu_0}\,\frac{\ln(b/a)}{1/b + 1/a + (2/l)\ln(b/a)} \tag{37b}$$

Q_ε can be obtained from the values of Table 1, as

$$Q_\varepsilon = \frac{Q_{\varepsilon m} f_m}{f_0} \tag{37c}$$

where f_0 is the resonant frequency.

Q_c, for the case when $b/a = 3.6$ is calculated from Eq. (14a). Under such a condition, the Q_c can be calculated by using conductivity of 70% of $6.17 \times 10^7/\Omega$ m, which is that of pure silver as shown in Table 4 of Chapter 2. The values of Q_c calculated by the above conditions are shown in Fig. 20. As understood from Fig. 20, Q_c increases corresponding to the higher frequency, the larger radius, and the lower ε_r.

Dielectric Resonators

Several TE and TM Modes In the coaxial resonator, metal walls such as the inner conductor, the outer conductor, and the short conductor at the end were required. However, considering the dielectric rod as shown in Figs. 21a–21c, the electromagnetic wave can propagate. This rod is called the dielectric rod guide. Therefore, we can suppose that a rod with the length of $s\lambda_g/2$ will become a resonator, as shown in Figs. 22a–22c. The electromagnetic wave, however, cannot disappear suddenly at the end of the rod, but decreases exponentially on the outside of the rod along the axis in the region of air. Generally, the length of the rod L becomes shorter than $s\lambda_g/2$. This means $s < 1$ for the dominant mode and it is shown by $\delta = 1$. Therefore, the rod resonates in the length $(\delta + s)\lambda_g/2$ ($s = 0, 1, \ldots$).

In the case of TE and TM modes, they are denoted by $TE_{m,n,\delta+s}$ and $TM_{m,n,\delta+s}$. For example, $TE_{m,n,s}$ and $TM_{m,n,s}$ correspond to $TE_{m,n,s-1+\delta}$ and $TM_{m,n,s-1+\delta}$ modes of the waveguide resonator.

Examples of TE and TM modes are shown in Fig. 23. We can understand that the magnetic fields are almost perpendicular to the surface S, which includes the circumference of the dielectric rod, and are parallel to the axis

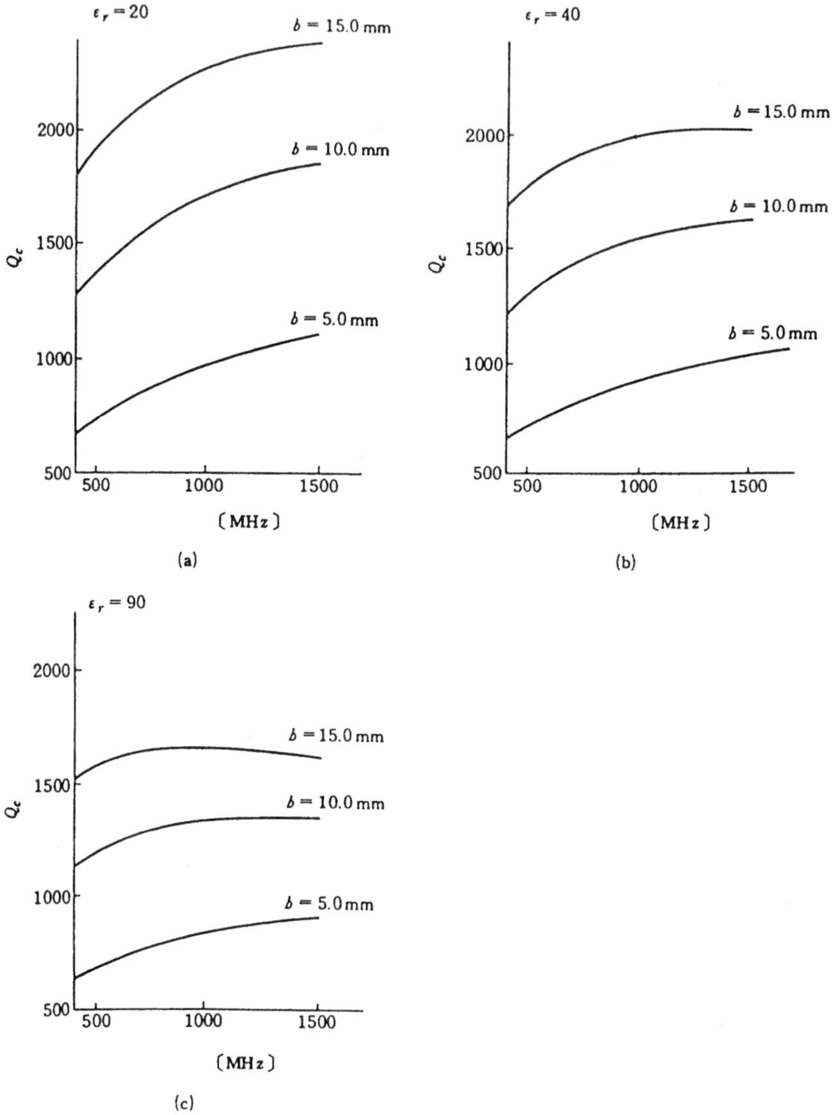

Figure 20. Q_c for several ε_r and sizes corresponding to the frequencies.

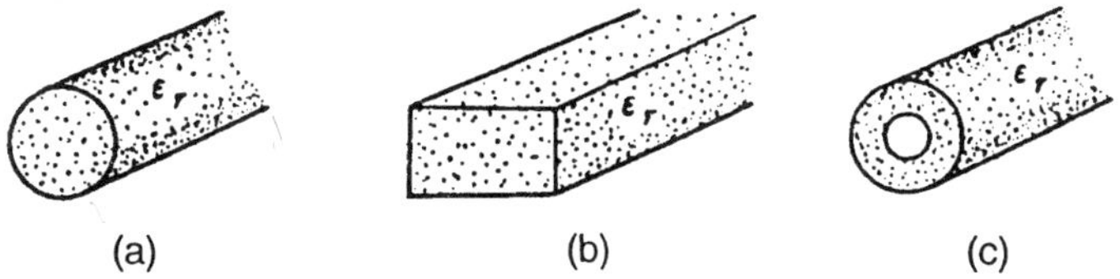

Figure 21. Dielectric rod guides.

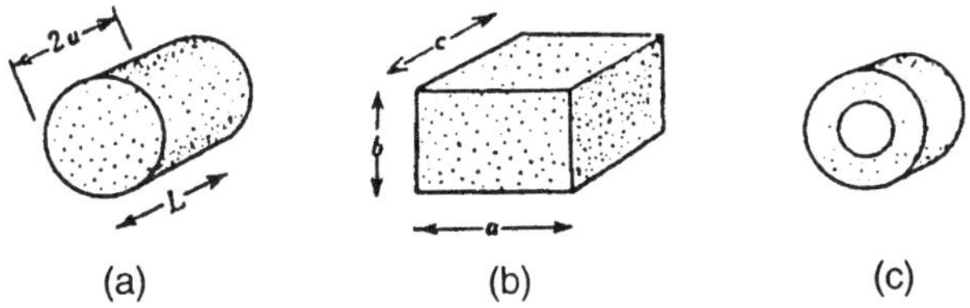

Figure 22. Dielectric resonator with a finite length of dielectric rod.

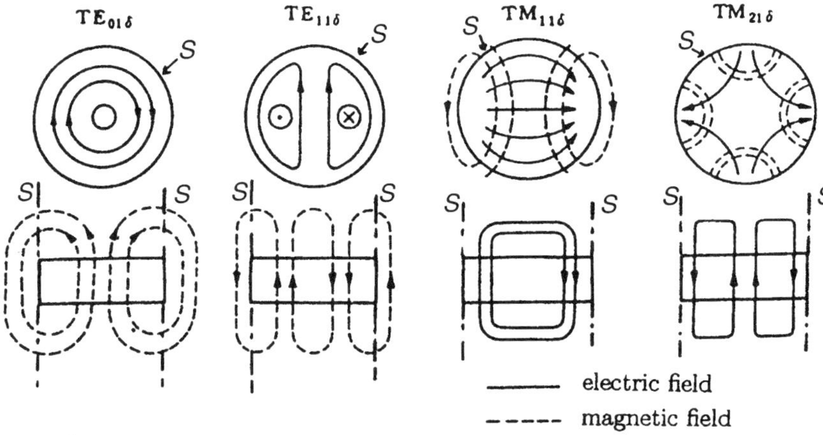

Figure 23. Typical TE and TM modes of a dielectric resonator.

of the rod. In other words, the S plane shown in Fig. 23 could be the magnetic wall.

Cohn obtained the resonant frequencies of the dielectric resonator by assuming the S plane to be the magnetic wall [19]. Therefore, we can consider the waveguide surrounded by the magnetic wall as the metal wall of a conventional waveguide. The right part of the waveguide is filled with the high relative dielectric constant ε_r and the left part is filled by the air or the low relative dielectric constant.

The waveguide with ε_r is the transmission waveguide and the other part is usually the cutoff waveguide. Such a waveguide can be expressed by the equivalent distributed line explained in Chapter 4, Section 1.

When we use the equivalent distributed line, we get Fig. 24a.

In Fig. 24, the region $-L/2 < z < L/2$ is part of the DR and the transmission line. Therefore, k_z and Z_d are real positive numbers. On the other hand, the region $z < -L/2$ and $z > L/2$ is the evanescent (cutoff) region. Therefore, α_1 and α_2 are real numbers and Z_{a1} and Z_{a2} imaginary numbers. These values are shown in the following equations:

$$Z_d = \frac{\omega\mu_0}{k_z}, \qquad Z_{ai} = j\frac{\omega\mu_0}{\alpha_i} \quad (i = 1, 2) \quad \text{in the case of the TE mode} \tag{38a}$$

$$Z_d = \frac{k_z}{\omega\varepsilon_0\varepsilon_r}, \qquad Z_{ai} = \frac{\alpha_i}{j\omega\varepsilon_0} \quad (i = 1, 2) \quad \text{in the case of the TM mode} \tag{38b}$$

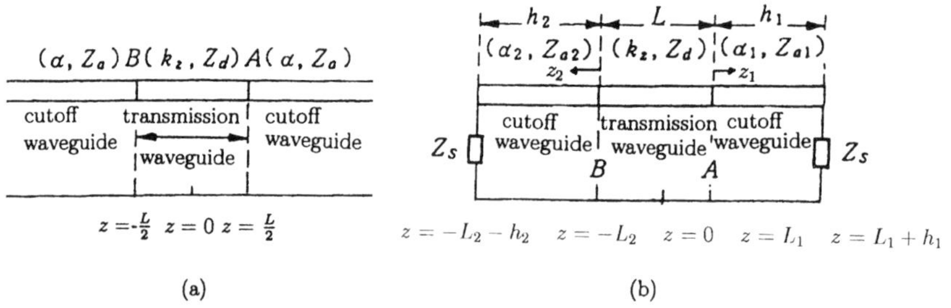

Figure 24. The equivalent distributed lines of the waveguide made by the surrounding magnetic wall, which contains the circumference and is parallel to the axis. (a) The length of the cutoff waveguide is infinite. (DR is placed in free space.); (b) the cutoff waveguides are terminated by the metal walls. (DR is placed between two parallel shielded metal plates.)

$$k_z = \sqrt{\varepsilon_r \left(\frac{2\pi}{\lambda_0}\right)^2 - k_t^2} = \sqrt{\varepsilon_r}\,\frac{2\pi}{\lambda_0}\sqrt{1 - \left(\frac{\lambda_0}{\lambda_c}\right)^2} \tag{38c}$$

$$\alpha_i = \sqrt{k_t^2 - \varepsilon_{r,i}k_0^2} = \sqrt{\varepsilon_r}\,\frac{2\pi}{\lambda_0}\sqrt{\left(\frac{\lambda_0}{\lambda_c}\right)^2 - \frac{\varepsilon_{r,i}}{\varepsilon_r}} \tag{38d}$$

where $\varepsilon_{r,i}$ is the relative dielectric constant of region i ($i = 1, 2$) and ε_r is the relative dielectric constant of the DR. Equations (38c) and (38d) are introduced from the relationship between the phase constants:

$$k_z^2 + k_t^2 = \varepsilon_r k_0^2 \tag{38e}$$

$$-\alpha_i^2 + k_t^2 = \varepsilon_{r,i}k_0^2 \quad (i = 1, 2) \tag{38f}$$

where k_t is the constant determined by the sectional shape of the waveguide surrounding the magnetic wall, that is, the sectional view of the dielectric resonator. For example, in the case of the circular dielectric rod with radius a,

$$\begin{aligned} &k_t a = x_\rho \\ &J_m(x_\rho) = 0 \quad \text{in the case of the TE}^\circ \text{ mode} \\ &J'_m(x_\rho) = 0 \quad \text{in the case of the TM}^\circ \text{ mode} \end{aligned} \tag{38g}$$

In a simple case of $\varepsilon_{r1,2} = 1$, we get the equations to obtain the resonant frequency from the equivalent network of Fig. 24a together with Eqs. (38a)–(38g) as follows:

$$\left.\begin{aligned} &x_z \tan\left(x_z\xi - s\,\frac{\pi}{2}\right) = x_\alpha \\ &J_m(x_\rho) = 0 \end{aligned}\right\} \text{TE}^\circ_{m,n,\delta+s} \text{ mode}$$

$$\left.\begin{aligned} &x_z \tan\left(x_z\xi = s\,\frac{\pi}{2}\right) = \varepsilon_r x_\alpha \\ &J'_m(x_\rho) = 0 \end{aligned}\right\} \text{TM}^\circ_{m,n,\delta+s} \text{ mode}$$

$$x_\rho^2 + x_z^2 = \varepsilon_r x_0^2$$

$$x_\rho^2 - x_\alpha^2 = x_0^2$$

$$x_z = k_z a, \qquad x_\rho = k_t a, \qquad x_\alpha = \alpha a$$

$$x_0 = k_0 a = \frac{2\pi}{\lambda_0}\,a, \qquad \lambda_0 = \frac{v}{f}, \qquad \xi = \frac{L}{2a} \tag{39}$$

where f is the frequency, v is the velocity of light, λ_0 is the wavelength in free space, a is the radius, and L is the length of the rod. The calculated results are shown in Fig. 25. In the same way, they can be obtained for the dielectric resonator of the hexahedron; the results are shown in Fig. 26.

$TE_{01\delta}$ Mode and $TE^{\square}_{11\delta}$ Mode The $TE_{01\delta}$ mode is used most often because it has a small size. This is understood from Fig. 25, where the curve of $TE_{01\delta}$ is situated in the lowest side for $\xi < 0.5$. Values of ξ should be chosen for the values where the next higher mode is most separated at the higher frequency. For example, when we use the ceramics of $\varepsilon_r = 35$, ξ should be 0.35, where the next higher mode is $TE_{11\delta}$. In this meaning, ξ is usually chosen at the values between 0.25 and 0.4. The electromagnetic field exists as shown in Fig. 27a. The mode of the hexahedron corresponding to $TE^{\square}_{01\delta}$ is the $TE^{\square}_{11\delta}$ mode, as shown in Fig. 27b.

As there are no metal walls in the dielectric resonators, basically there exists no conductive loss. However, the dielectric resonator of $TE^{\circ}_{01\delta}$ and $TE^{\square}_{01\delta}$ modes radiate the power, which makes the resonator's Q values decrease. For example, $TE^{\circ}_{01\delta}$ DR (dielectric resonator) in a free space has Q

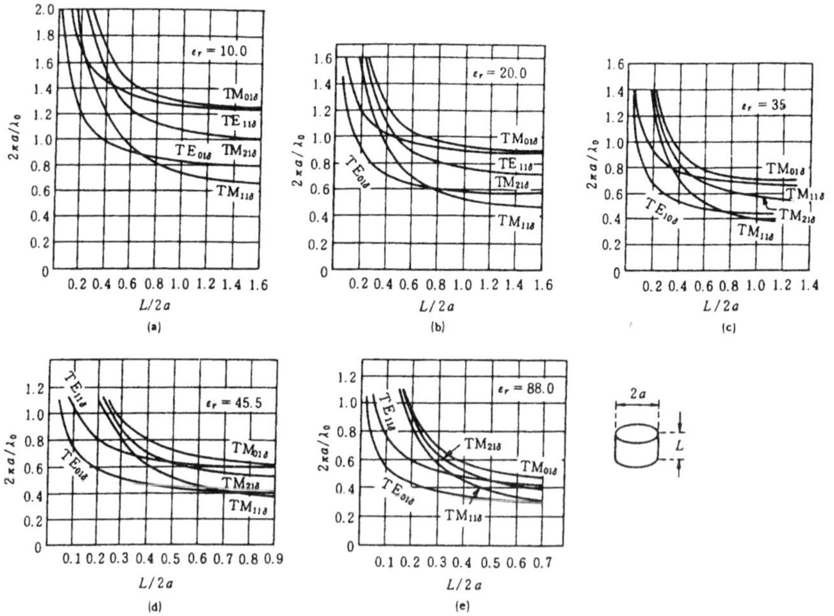

Figure 25. First approximated values of resonant frequencies of the dielectric resonator in the shape of a circular rod ($\lambda_0 = v/f$).

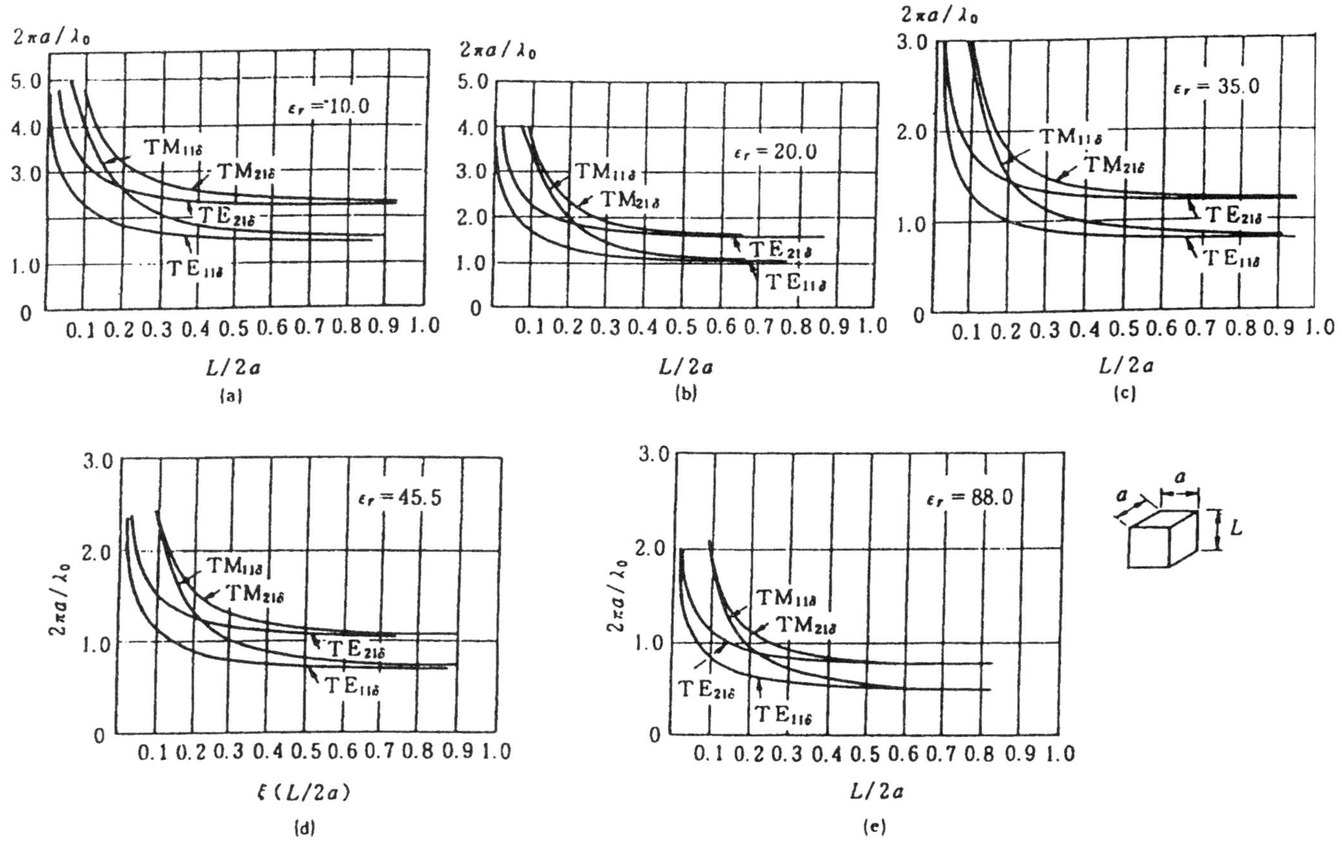

Figure 26. First approximated values of resonant frequencies of the dielectric resonator in the shape of a hexahedron ($\lambda_0 = v/f$).

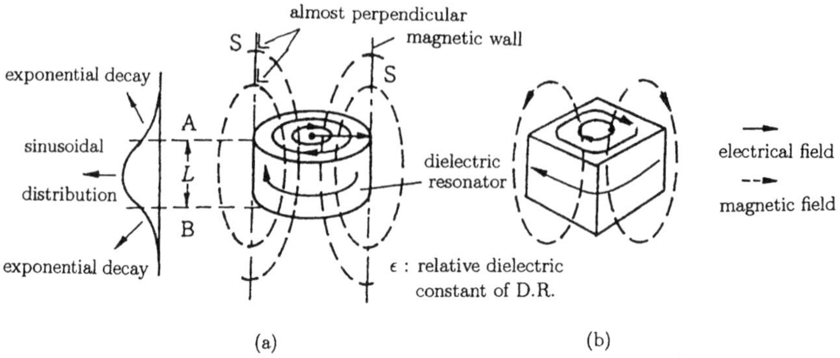

Figure 27. The electromagnetic field distribution of the $TE^{\circ}_{01\delta}$ mode (a) and the $TE^{\square}_{11\delta}$ mode (b).

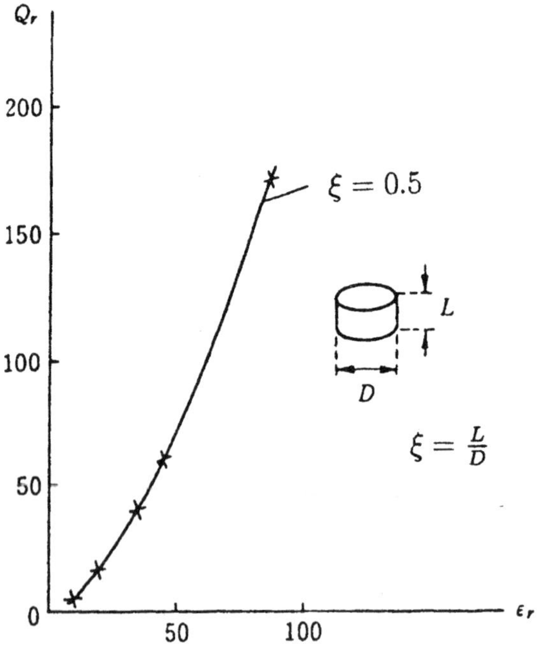

Figure 28. Q values caused by the radiation of $TE^{\circ}_{01\delta}$ mode DR.

values caused by the radiation, which can be calculated as shown in Fig. 28 [20].

We find that Q_r is much lower than $Q_\varepsilon = 1/\tan\delta$. We, therefore, need the shielding metal plates to protect the radiation, as shown in Fig. 29. In Fig. 29, A and B planes are the shielding plates, and S_{m1} is the magnetic wall which makes the waveguide terminated by A and B plates; S_{m2} is the magnetic wall, where the circular electrical fields become maximum. The distances L_1 and L_2 in Fig. 24b are different in the case of $h_1 \neq h_2$, and $L_1 = L_2 = L/2$ in the case of $h_1 = h_2$ or h_1 and h_2 equal ∞.

Because the metal walls have a surface impedance Z_s of $(1 + j)R_s$, as described in Eqs. (56b) and Eq. (58) of Chapter 1, the equivalent distributed line is as shown in Fig. 24b. The values of R_s are the conductive loss and make Q_c finite.

When we show the place of S_{m2} in Fig. 29 by $z = 0$, the position of the A and B planes are shown by $z = L_1 + h_1$ and $z = L_2 - h_2$, respectively. Because S_{m2} is the open surface, we can obtain the resonant condition under the assumption $R_s \ll Z_{ai}$, as follows:

$$\frac{j\omega\mu_0}{\alpha_i}\tanh\alpha_i h_i = j\frac{\omega\mu_0}{k_z}\cot k_z L_i \quad (i = 1, 2)$$

Therefore,

$$k_z L_i = \cot^{-1}\left(\frac{k_z}{\alpha_i}\tanh\alpha_1 h_1\right) \; (i = 1, 2) \tag{40}$$

Calculating the electrical field $\boldsymbol{e}_{to}$ at $z = 0$ as

$$\boldsymbol{e}_{to} = \boldsymbol{i}_\theta E_0 \frac{J_1(k_t r)}{J_1(k_t a)} \tag{41}$$

We can calculate the time average of the total reactive energy $\tilde{W}_t$ and the dissipated power P_{c1} and P_{c2} on S_1 and S_2, respectively, in Fig. 29. From

$$Q_c = \frac{\omega\tilde{W}_t}{P_{c1} + P_{c2}}$$

we obtain the Q values caused by shielding plates as follows, where we also summarized the necessary equations discussed earlier:

$$Q_c = \frac{\omega\mu_0(L_1 + L_2)\varepsilon_r\left(1 + \dfrac{\sin 2k_z L_1 + \sin 2k_z L_2}{2k_z(L_1 + L_2)}\right)\left(\dfrac{2\pi}{\lambda_0}\right)^2}{\left[2\left(\dfrac{\alpha_1^2\cos^2 k_z L_1}{\sinh^2\alpha_1 h_1} + \dfrac{\alpha_2^2\cos^2 k_z L_2}{\sinh^2\alpha_2 h_2}\right)R_s\right]^{-1}} \tag{42a}$$

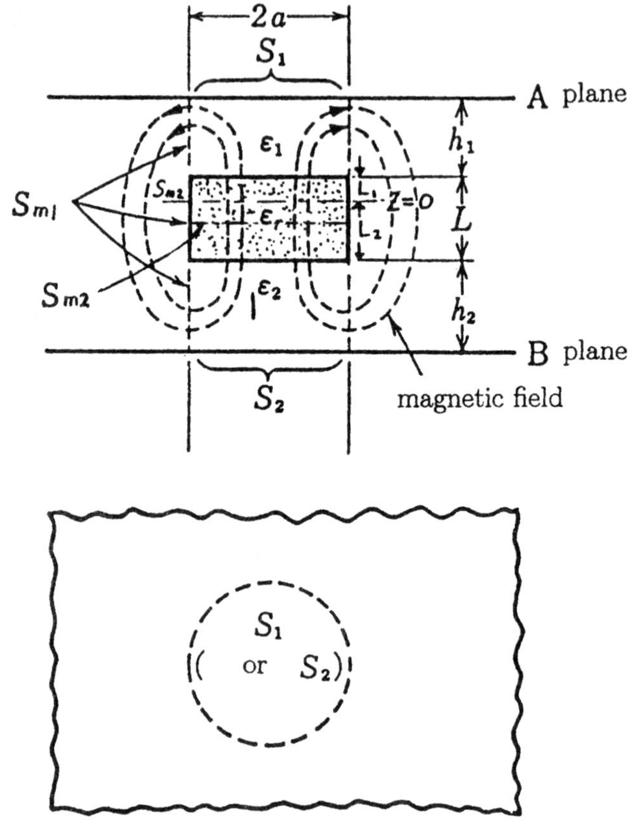

Figure 29. $TE^{\circ}_{01\delta}$ mode DR with shielding metal planes A and B.

$$k_z L_{1,2} = \cot^{-1}\left(\frac{k_z}{\alpha_{1,2}} \tanh \alpha_{1,2} h_{1,2}\right) \tag{42b}$$

$$k_z = \sqrt{\varepsilon_r}\,\frac{2\pi}{\lambda_0}\sqrt{1 - \left(\frac{\lambda_0}{\lambda_c}\right)^2} \tag{42c}$$

$$\alpha_i = \sqrt{\varepsilon_r}\,\frac{2\pi}{\lambda_0}\sqrt{\left(\frac{\lambda_0}{\lambda_c}\right)^2 - \frac{\varepsilon_{r,i}}{\varepsilon_r}} \tag{42d}$$

$$\frac{2\pi a}{\lambda_c}\sqrt{\varepsilon_r} = 2.405 \tag{42e}$$

(See Note 3.)

When ε_r, ε_1, ε_2, a, h_1, h_2, λ_0, and R_s are given, we get λ_c, k_z, α_i, L_1, L_2, and Q_c. For example, in the case of $\varepsilon_r = 35$, $\varepsilon_1 = \varepsilon_2 = 1$, $a = 0.003$ m, $h_1 = h_2 = 0.0006$ m, $\lambda_0 = 0.03$ m ($f_0 = 10$ GHz), and $R_s = 0.0253$ Ω (silver at 10 GHz), we obtain $\lambda_c = 0.0464$ [Eq. (42e)], $k_z = 942.2$ [Eq. (42c)], $\alpha_1 = \alpha_2 = 771.52$ [Eq. (42d)], $L_1 = L_2 = 0.00115$ [Eq. (42b)], and $Q_c = 3810.8$ [Eq. (42a)].

Even if we use the dielectric resonator with Q_ε values of 7000–10,000, as shown in Table 1, the total Q values deteriorates to 2467–2559. If $h_1 = h_2 = 0.0016$ m, Q_c values exponentially increased to 41,000, which results in total Q values of 5980–8040.

As another example, in the case of $\varepsilon_{r1} = 9.8$, $h_1 = h_2 = 0.0016$ m, $\varepsilon_r = 35$, $\varepsilon_{r2} = 1$, $\lambda_0 = 0.03$ m, and $R_s = 0.0253$, we get $\alpha_1 = 460.3$, $\alpha_2 = 771.52$, $k_z = 942.2$, $L_1 = 0.0007$ m, $L_2 = 0.00115$ m, $L = 0.0022$, and $Q_c = 14101$, the values of which are decreased from 41,000 in the case of $\varepsilon_r = 1$. This means that the evanescent mode of region 1 of ε_{r1} decreases slowly in the case of the lower ε_{r1} and the induced currents on the metal plate A increases, which deteriorates Q_c caused by conductive loss.

Another important effect caused by the shielding metal plates is increase of the resonant frequency rather than the free-space resonant frequency. This is explained by the perturbation theory and is also the equivalent network of Fig. 24b.

In the resonator bounded by a closed surface S as shown in Fig. 30a, the resonant frequency f_0 is deviated to $f_0 + \Delta f$ by deforming the surface S to that of the solid line of Fig. 30b.

When the time average of the magnetic energy and the electric energy in the portion surrounded by S'' and S are denoted by $\Delta\tilde{W}_m$ and $\Delta\tilde{W}_e$, Δf can

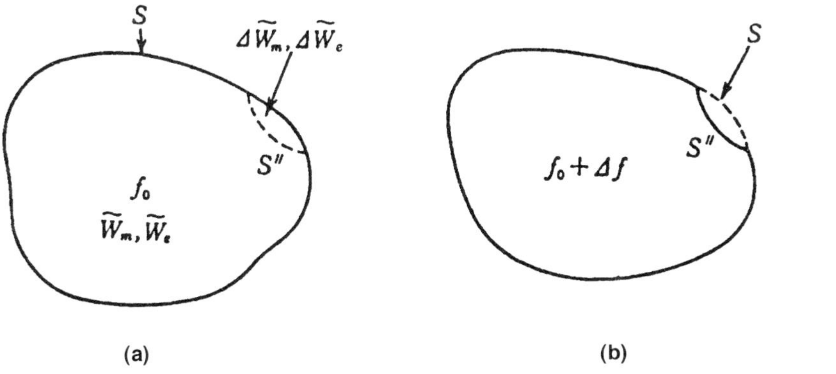

Figure 30. The deviation of the resonant frequency by the perturbation of the form.

be obtained by the perturbation theory as follows (see Appendix 2 of the volume):

$$\frac{\Delta f}{f_0} = \frac{\Delta \tilde{W}_m - \Delta \tilde{W}_e}{\tilde{W}_t} \tag{43}$$

From Eq. (43), we see that the resonant frequency f_0 becomes higher by pushing the portion concentrated with magnetic energy, and lower by pushing the portion concentrated with electrical energy. This can be imagined from the simple example of an *LC* resonator; that is, the resonant frequency becomes higher by bringing the metal plate closer to the coil, and lower by making the distance between electrodes of the condenser shorter, which means pushing the electrode.

In the $TE^{\square}_{01\delta}$ dielectric resonator, regions of ε_1 and ε_2 in Fig. 29, the magnetic energies are more dominant than the electric energies, because the regions are the cutoff waveguides of TE mode. This was explained in Eq. (47a) of Chapter 1. As an example, the changing of the resonant frequency was calculated by Eq. (40) together with Eqs. (38a)–(38d), under the condition $h_1 = h_2$; the results are shown in Fig. 31. As shown in Fig. 31, the resonant frequency increases when the shielding plates approach the DR.

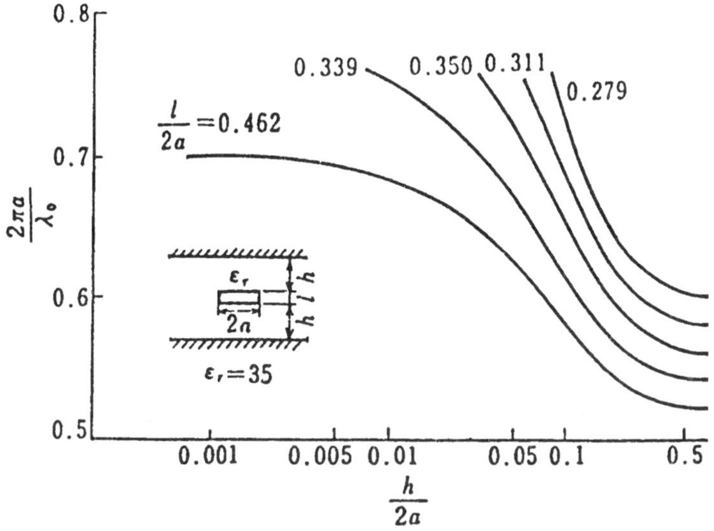

Figure 31. The resonant frequency of the $TE^{\circ}_{01\delta}$ mode of the DR corresponding to the distance between the shielding metal plates and the DR.

2.5 Helical Resonator

The spiral coil is installed inside the shielding case, as shown in Fig. 32, where Figs. 32b and 32c show the cylindrical and square cases, respectively. One of the ends of the coil is grounded to the bottom of the case. Assuming the coil to be stretched out, the line can be considered a distributed constant line grounded at one end. If the length of the line is a quarter wavelength, it becomes a resonator. To make the size of the distributed constant line smaller, the line is wound as a spiral shape. Although this is quite a bold concept, the inductance is increased by the coupling between the neighboring coils, which results in the total length of the coil being smaller than a quarter wave.

The distributed inductance and the capacitance per unit length, for example, per inch, are obtained as [21,22].

$$L = 0.025n^2d^2\left[1 - \left(\frac{d}{D}\right)^2\right]\left(\frac{\mu\text{H}}{\text{in.}}\right) \tag{44a}$$

$$C = \frac{0.75}{\log_{10}(D/d)}\left(\frac{\text{pF}}{\text{in.}}\right) \tag{44b}$$

where n is the number of turns of the coil and inch is show the length for the direction of the center axis of the coil.

Equations (44a) and (44b) are available under the conditions of Eq. (44c).

$$\begin{aligned}
&1.0 < \frac{b}{d} < 4.0\\
&0.45 < \frac{d}{D} < 0.6\\
&0.4 < \frac{d_0}{\tau} < 0.6 \quad \left(\text{in the case of } \frac{b}{d} = 1.5\right),\ \tau = \frac{1}{n}\\
&0.5 < \frac{d_0}{\tau} < 0.7 \quad \left(\text{in the case of } \frac{b}{d} = 4.0\right)
\end{aligned} \tag{44c}$$

Therefore, the characteristic impedance Z_0 and the phase constant β of the distributed constant line with L and C as Eqs. (44a) and (44b) are calculated from Eqs. (45a) and (45b), respectively.

$$Z_0 = \sqrt{\frac{L}{C}} = 182.6nd\left\{\left[1 - \left(\frac{d}{D}\right)^2\right]\log_{10}\left(\frac{D}{d}\right)\right\}^{1/2} \quad (\Omega) \tag{45a}$$

$$\begin{aligned}
\beta &= \frac{2\pi}{\lambda_g} = (2\pi f_0 \times 10^6)(\sqrt{LC} \times 10^{-9})\\
&= 2\pi f\sqrt{LC} \times 10^{-3} \quad (f_0 \text{ in MHz})
\end{aligned} \tag{45b}$$

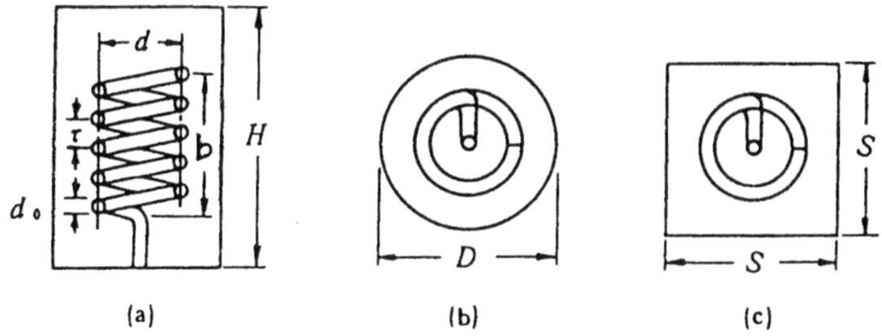

Figure 32. Construction of a helical resonator.

Because the length, b, of Fig. 32b should be $\lambda_g/4$, b in the ideal case, b', must be

$$b' = \frac{\lambda_g}{4} = \frac{250}{f_0\sqrt{LC}} \tag{45c}$$

from Eq. (45b).

We, however, have the fringe effect at the open end of the coil and the capacities existing at the coil itself, which make the values of b' smaller than that of Eqs. (45c):

$$b = 0.94b' = \frac{235}{f_0\sqrt{LC}} \tag{45d}$$

The values of Eq. (45d) were obtained empirically. Because the total number of turns of the coil, N, is

$$N = bn \tag{45e}$$

we get the results in Eq. (45f) by substituting Eq. (45e) into Eqs. (44a), (44b), and (45d):

$$N = \frac{1720}{f_0 D(d/D)}\left(\frac{\log_{10}(D/d)}{1 - (d/D)^2}\right)^{1/2} \tag{45f}$$

Using Eq. (45e), we obtain the values of n from

$$n = \frac{N}{b} = \frac{1720}{f_0 D^2(b/d)(d/D)^2}\left(\frac{\log_{10}(D/d)}{1 - (d/D)^2}\right)^{1/2} \tag{45g}$$

Next, the values of the no-load Q, Q_u, is determined by the ohmic resistance of the helical resonator itself, R_c, and the shielding case, R_s. The values of Q_u are obtained from Eq. (45h) [21,22]:

$$Q_u = \frac{2\pi f_0 L}{R_s + R_c}$$

$$R_c = (0.083 \times 10^{-3}) \left(\frac{\phi}{nd_0}\right) n^2 \pi d f_0^{1/2} \quad \left(\frac{\Omega}{\text{in.}}\right)$$

$$R_s = \frac{9.37 n^2 b^2 (d/2)^4 (1.724 f_0)^{1/2}}{b[D^2(b + d)/8]^{4/3}} \sqrt{\frac{\rho_s}{\rho_{Cu}}} \times 10^{-4} \quad \left(\frac{\Omega}{\text{in.}}\right) \tag{45h}$$

where ρ_s and ρ_{Cu} are the relative resistance of the metal of the shielding case and copper.

ϕ/nd_0 of Eq. (45h) is the coefficient according to the fringe effect and it is 3.7 in the range of Eq. (44c). In fact, the values of nd_0 take the minimum values in the range of Eq. (44c) and R_c takes the minimum values [23].

For the case when the coil and the shielding case are made of copper, we have Q_u from Eq. (45i) by using the relation $\rho_s = \rho_{Cu}$ and Eqs. (45h) and (44a):

$$Q_u = 600 \frac{d/D - (d/D)^3}{(\phi/nd_0)(1 + R_s/R_c)} Df^{1/2} \tag{45i}$$

In Eq. (45i), $Q_u/Df^{1/2}$ takes the maximum values at $d/D \simeq 0.555$ [23]. By using the relation, we get [22]

$$Q_u \simeq 50 D f_0^{1/2} \tag{46a}$$

For example, to get the maximum values of Q_u, we use

$$d/D = 0.55 \tag{46b}$$

and

$$\frac{b}{d} = 1.5 \tag{46c}$$

$$nd_0 = 0.5 \tag{46d}$$

in the range of Eq. (44c).

We obtain Eq. (47) by substituting Eqs. (46b) and (46c) into Eqs. (45f) and (45a)

$$N = \frac{1900}{f_0 D}$$

$$Z_0 = \frac{98000}{f_0 D} \quad (\Omega) \tag{47}$$

Obtaining the relationship between Q_u and D (cm) from Eq. (46a), we get the results of Fig. 33 [23], where we used the relation of 1 in. = 2.54 cm. In the same way, when we show the Eq. (47) by D (cm), we obtain the values of N and Z_0 from

$$N = \frac{4800}{f_0 D}$$

$$Z_0 = \frac{249000}{f_0 D} \tag{48}$$

In the case of Fig. (32c), several relationships can be obtained by using [22]

$$S = 1.2D \tag{49}$$

The diameter of the wire, d_0, should be larger than the skin depth δ. Usually,

$$d_0 > 5\delta \tag{50}$$

is used. Finally, the height of the shielding case, H, as shown in Eq. (51), is recommended to give the maximum Q of the resonator under the condition

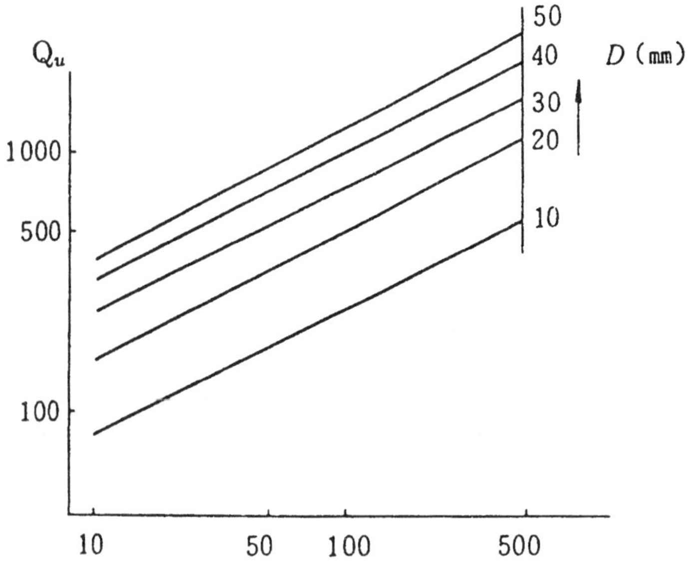

Figure 33. The relationship of the no-load Q, Q_u, and frequencies.

of the same volume as the case [21,22]:

$$H \simeq b + \frac{D}{2} \tag{51}$$

2.6 E-Plane Resonator

When we insert a metal plane inside the $TE_{10}^{\square}$ waveguide in parallel to the E plane, the region becomes a cutoff waveguide. If we take out a part of the metal on the plane as shown in Figs. 34a–34c, we can make a resonator.

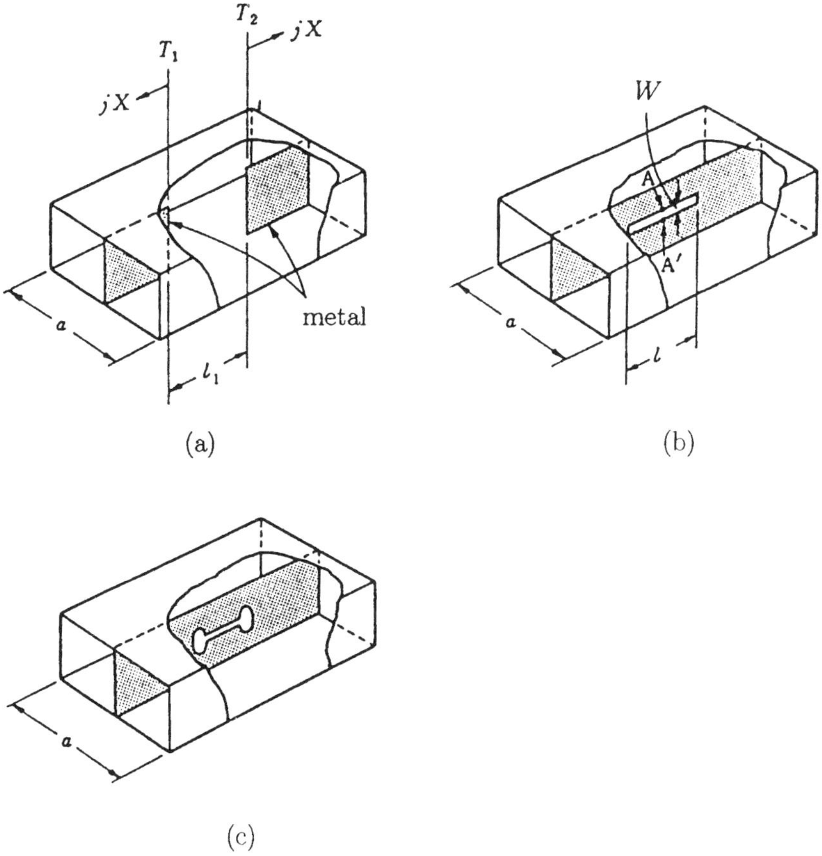

Figure 34. Examples of the construction of an E-plane resonator. (a) Two-metal-insert resonator; (b) slot resonator; (c) dumbbell slot resonator.

The resonant electromagnetic fields take place at the taken out pattern, and the resonant reactive energies cannot move to the region where there is no metal plane, because the reactive power generated at the resonator decreases enough at the cutoff region. This fact can make the resonator. Figure 34b is the slot resonator made on a metal plate. If the width of the slot, W, is much smaller compared to the free-space wavelength (i.e., $W \ll \lambda$), the slot is resonant at $1 \simeq \lambda/2$, because the cutoff frequency of the waveguide in the region of the slot becomes extremely low. This means that the guided wavelength approaches the free-space wavelength, which results in making a TEM half-wavelength resonator.

However, when W increases, the cutoff frequency becomes higher, and increasing the guided wavelength, λ_g, results in increasing the values of l/λ_0. These results are shown in the curves $a/\lambda_0 = 0.6$ and 0.68 of Fig. 35 [24]. On the other hand, when a is large, the increase in the values of λ_g are not great because λ_g approaches λ_0, and the inductance at the short end increases [25]. This inductance serves to decrease the resonant frequency, which requires the shorter length of l. The situation is shown in the curve $\alpha/\lambda_0 = 0.84$ of Fig. 35.

Therefore, the width of the slot, W, contributes not only to the guided wavelength but also to the equivalent inductances at both short ends of the slot. Such effects contribute to the length l in the opposite way; the results are shown in Fig. 35. In the case of $a/\lambda_0 = 0.76$, the effect is balanced; that is, the resonant frequency does not greatly effect the width. However, in the

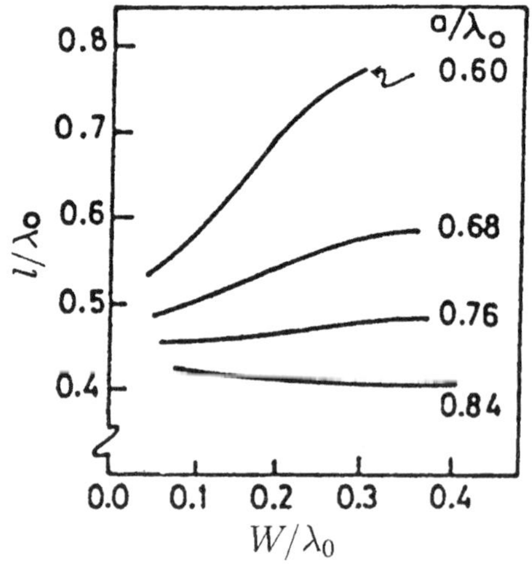

Figure 35. The relationship among l, W, and a for the resonant wavelength.

case of $W \simeq 0$, all curves should gather at the point $l/\lambda_0 = 0.5$, as mentioned earlier.

As a reference, the electromagnetic distributions of the slot resonator are shown in Fig. 36, which shows not only the resonant fields but also the decays in the cutoff waveguide regions.

When the width of the slot is expanded to the values of b of the waveguide, we get the construction in Fig. 34a. By using the proper length of the metal plate, that is, metal inductive strips in both sides, we get the one-stage BPF, and for the cascade connection of the construction, we get the several stages of BPF [26].

The slot line resonator of Fig. 34b can be deformed to the pattern shown in Fig. 34c, which is called the dumbbell slot resonator. By being narrow at the center of the slot, the equivalent capacity increases and the resonant frequency decreases. On the other hand, by being wide at the edge portion of the slot, the equivalent inductances increase, which results in decreasing the frequency. Many kinds of pattern deformed from the slot resonator can be considered.

Generally, unloaded Q values of around 1600 are obtained at the X band [27].

The metal plate with the pattern can be made on the dielectric substrate. Such a substrate can be also inserted in a waveguide parallel to the E plane of the $TE_{10}^{\square}$ waveguide. By such a technology, several kinds of resonator can be made in the same way. Because the metal plate is inserted parallel to the E plane, the circuits made by making the proper pattern in the metal plate

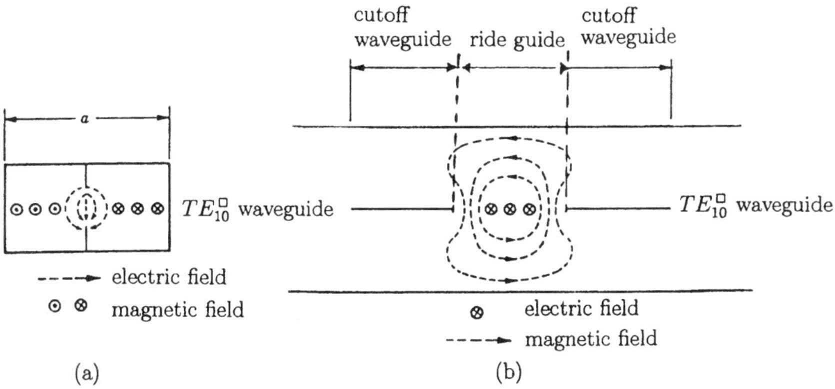

Figure 36. Electromagnetic fields of the E-plane slot resonator. (a) Sectional view; (b) top view.

are called the E plane of the fin line. A detailed explanation is provided in Ref. 28.

2.7 Planar Resonators

Planar resonators are used for microwave integrated circuits (MIC) and microstrip antennas. The approximate resonant frequencies can be obtained under the assumption of a magnetic wall in the circumference of the metal patterns, which is used for resonators.

As typical examples, there are circular and rectangular resonators, as shown in Figs. 37a and 37b, respectively, and their patterns are made on the substrate, as shown in Fig. 37c, in the case of the circular disk resonator.

The approximate resonant frequencies of the circular and resonators can be obtained using Eqs. (52) [29] and (53) [30]:

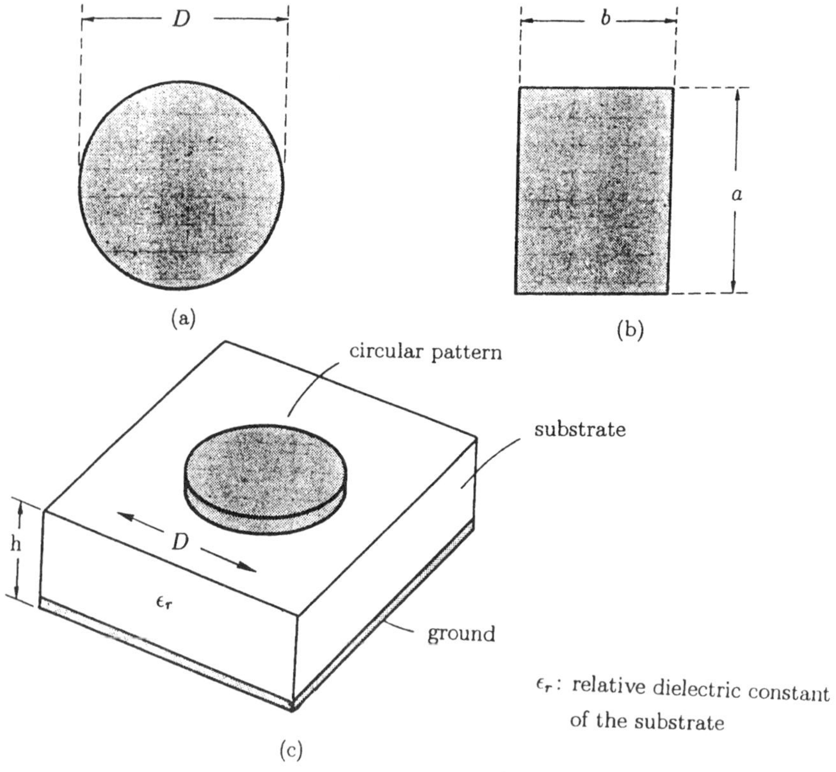

Figure 37. Typical planar resonators. (a) Pattern of circular resonator; (b) pattern of rectangular resonator; (c) construction of circular disk resonators.

$$f_0 = \frac{c}{\sqrt{\varepsilon_r}} \frac{j'_{nm}}{\pi D}, \qquad J_n(j'_{nm}) = 0 \quad \text{(circular resonator)} \tag{52}$$

$$f_0 = \frac{c}{\sqrt{\varepsilon_r}} \left[\left(\frac{n}{2a} \right)^2 + \left(\frac{m}{2b} \right)^2 \right]^{1/2} \quad \text{(rectangular resonator)} \tag{53}$$

In Eq. (52), the subscripts n and m indicate the number of circumferential full-wave field variations and radical half-wave field variations, and C is the light velocity. In Eq. (53), subscripts n and m indicate the numbers of half-wave field variations of the directions of the lengths a and b, respectively. For example, the fields distribution of the circular disk resonators are shown in Fig. 38 together with the corresponding resonant frequencies.

Because there are equivalent capacities in the periphery of the patterns, the equivalent size of the patterns becomes a little bit larger, which results in decreasing the resonant frequencies a little. For example, the effective diameter, $D_e = D + \Delta D$ are obtained as [31]

$$D_e = 2R_e$$

$$R_e = R \left[1 + \frac{2d}{\pi \varepsilon_r R} \left(\ln \frac{\pi R}{2d} + 1.7726 \right) \right]^{1/2} \tag{54}$$

Equation (54) is used for $d \ll R$, low values of ε_r, and the TM_{110} mode [32].

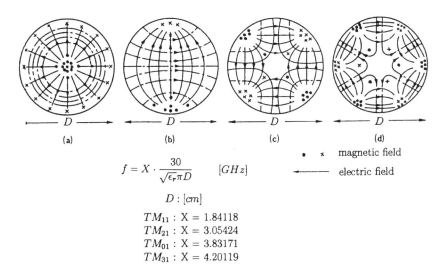

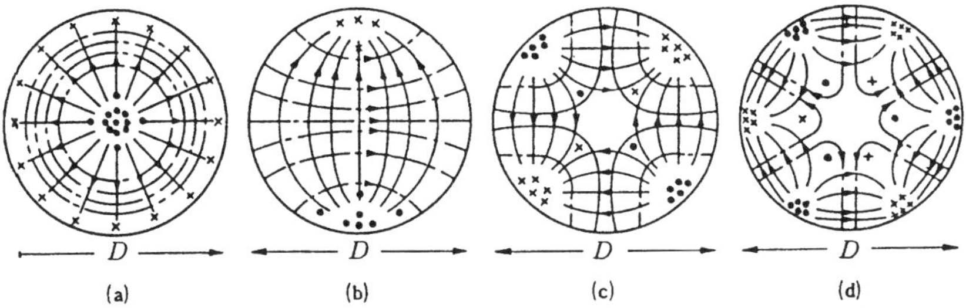

• × magnetic field

⟵ electric field

$$f = X \cdot \frac{30}{\sqrt{\epsilon_r} \pi D} \quad [GHz]$$

D : [cm]

TM_{11} : X = 1.84118
TM_{21} : X = 3.05424
TM_{01} : X = 3.83171
TM_{31} : X = 4.20119

Figure 38. Fields distributions and resonant frequencies of typical circular disk resonators. (a) TM_{01} mode; (b) TM_{11} mode; (c) TM_{21} mode; (d) $TM_{31}x - \lambda$.

APPENDIX

Note 1

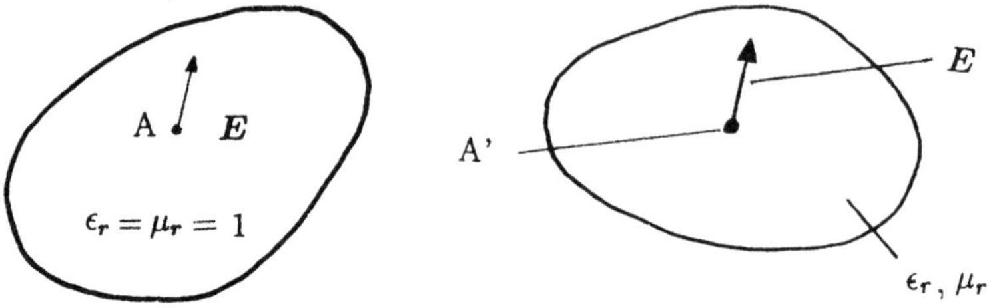

(a) (b)

We consider the two similar cavities with the same resonant frequency, where one of them is empty and the other includes the material of the relative dielectric constant ε_r and the relative permeability μ_r. The former cavity is larger in the size with $\sqrt{\varepsilon_r \mu_r}$ times. When we consider the same electric field $\boldsymbol{E}$ at the corresponding points A and A′, the magnetic fields $\boldsymbol{H}$ should be $\sqrt{\varepsilon_r/\mu_r}$ times, by Maxwell's equation

$$\nabla \times \boldsymbol{E} = -j\omega\mu\boldsymbol{H} \tag{1}$$

Denoting the time average of the reactive energy of part a of the figure by $\tilde{W}_{to}$, that of part b should be

$$\omega\tilde{W}_t = \mu\tilde{W}_{to}\frac{\varepsilon_r}{(\varepsilon_r\mu_r)^{3/2}} \tag{2}$$

Denoting consumed power of part a by P_{do}, that of part b should be

$$P_d = P_{do}R_s\frac{\varepsilon_r}{\mu_r}\frac{1}{\varepsilon_r\mu_r} \tag{3}$$

Considering $R_s = \sqrt{\omega\mu/2\rho}$, $\delta = \sqrt{2/\omega\mu\sigma}$ than $R_s = (\omega\mu/2)\sigma$, we obtain Eq. (4) from Eq. (3):

$$P_d = P_{do}\frac{\sigma}{\mu_r} \tag{4}$$

From Eqs. (2) and (4), the Q values caused by the conductive loss on the metal surface, Q_c, can be obtained as follows:

$$Q_c = Q_{co}\frac{\varepsilon_r\mu_r}{(\varepsilon_r\mu_r)^{3/2}} = \frac{Q_{co}}{\sqrt{\varepsilon_r\mu_r}} \tag{5}$$

where $Q_{co} = \omega\tilde{W}_{to}/P_{do}$. Equation (5) is Eq. (18) of the text.

Note 2

In the resonator, there is the standing wave. Denoting the effective values of the voltage at the maximum point of the voltage by $V_{\max}$, the incident voltage is $V_{\max}/2$. Then, the incident power P_{in} is

$$P_{\text{in}} = \left(\frac{V_{\max}}{2}\right)^2 \frac{1}{Z_c} \tag{1}$$

On the other hand, we have the relationship of Eq. (2):

$$W_e = \frac{C}{2}\int_0^{\lambda/2} V_{\max}^2 \cos^2\left(\frac{2\pi}{\lambda}\right) z\, dz = \frac{CV_{\max}^2}{8}\lambda \tag{2}$$

where C is the capacity of the microstrip line per meter. Therefore, we get

$$\tilde{W}_t = 2\tilde{W}_e = Z_c P_{\text{in}} C\lambda = \frac{P_{\text{in}}}{v_p}\lambda \tag{3}$$

Note 3

In the case of $TE_{01\delta}^{\circ}$, $J_0(k_t a) = 0$. Then $k_z a = 2.405$. On the other hand, $k_t = \sqrt{\varepsilon_r} k_0 = \sqrt{\varepsilon_r} 2\pi/\lambda_0 = \sqrt{\varepsilon_r} 2\pi/\lambda_c$ from Eq. (38e), where we substitute $k_z = 0$. Then we get the equation in the text.

REFERENCES

1. Wheeler, H. A., Transmission-line properties of a strip on a dielectric sheet on a plane, *IEEE Trans. Microwave Theory Tech.*, *MTT-25*, 631–647 (1977).
2. Schneider, M. V., Microstrip line for microwave integrated circuits, *B.S.T.J.*, 1421–1444 (May–June, 1969).
3. Kirshning, M. and Jansen, R. H., Accurate model for effective dielectric constant of microstrip with validity up to millimeter-wave frequencies, *Electron. Lett.*, *18*, 272–273 (1982).
4. Pucel, R. A., Mosse, D. J., and Hartwig, C. P., Losses in microstrip, *IEEE Trans. Microwave Theory Tech.*, *MTT-16*, 342–349 (1968).
5. Lewin, L., Radiation from discontinuities in strip line, *Proc. IEE*, *107C*, 163–170 (1960).
6. Wakino, K., et al., Dielectric materials for dielectric resonator, in *1976 Joint Convention Record of Four Institutes of Electrical Engineers*, 1976.
7. Nomura, S., et al., *Jpn. J. Appl. Phys.*, *21*, LG24 (1982).
8. Hiuga, T., Matsumoto, K., and Ichimura, H., Dielectric properties of BMT series ceramics at microwave frequencies, IECE Tech. Rep. CPM 86-31, 1986, p. 41.
9. Tamura, H., et al., High-*Q* Dielectric resonator material for millimeter-wave frequencies, in *Proc. of the 3rd U.S.–Japan Seminar on Dielectric and Piezoelectric Ceramics.*

10. Takasugi, A. and Kitoh, R., National Convention Record of IECE Japan, SC-9-6, 1988.
11. Kawashima, S., et al., Ba(Zn, Ta)O_3 ceramics with low dielectric loss at microwave frequencies, *J. Am. Ceram. Soc.*, *66*, 421 (1983).
12. Tamuri, H., et al., Improved high-*Q* dielectric resonator with complex perovskite structure, *J. Am. Ceram. Soc.*, *67*, C-59 (1984).
13. Yamaguchi, T., et al., (Ca, Sr, Ba)zirconate ceramics for microwave dielectric resonator, Annual Report of Study Group on Applied Ferroelectrics in Japan, 29, XXIX-159-1017, 1980.
14. Nishigaki, S., et al., Dielectric properties of $BaO-TiO_2-WO_3$ system at microwave frequency, Abstract of the 3rd U.S.–Japan Seminar on Dielectric and Piezoelectric Ceramics, 1986, p. 55.
15. Wakino, K., et al., Microwave characteristics of (Zr, Sn)TiO_4 and $BaO-PbO-Nd_2\ O_3-TiO_2$ dielectric resonators, *J. Am. Ceram. Soc.*, *67*, 421 (1983).
16. O'Bryan, H. M., Jr., et al., A new $BaO-TiO_2$ compound with temperature-stable high permittivity and low microwave loss, *J. Am. Ceramic. Soc.*, *57*, 450 (1974).
17. Kawashima, S., et al., Microwave dielectric materials and their applications, 1980 Annual Report of Study Group on Applied Ferroelectrics in Japan, 30, XXX-164-1036, 1980.
18. Cochran, W., *The Dynamics of Atoms in Crystals*, Edward Arnold Ltd., London, 1973.
19. Cohn, S. B., Microwave bandpass filters containing high-*Q* dielectric resonators, *IEEE Trans. Microwave Theory Tech.*, *MTT-16*, 218–227 (1968).
20. Konishi, Y., New theoretical concept of thin electromagnetic wave absorber using resonant and scattering effect of dielectric and magnetic materials, *IEEE Trans. Broadcasting*, *B41*(9), 94–100 (1995).
21. Macalpine, W. W. and Schildknecht, R. O., Coaxial resonator with helical inner conductor, *Proc. IRE*, 2099–2105 (December 1959).
22. Zverev, A. I. and Blinchikoff, H. J., Realization of a filter with helical components, *IRE Trans. Compon. Parts*, 99–109 (September 1961).
23. Okamoto, Takahashi, and Kato, Helical resonator filter, *Fujitsu*, *19*(2) (1968).
24. Konishi, Y., Uenakada, K., and Hosino, N., The design of planar circuit mounted in waveguide and application to low noise 12 GHz converter, *IEEE MTT-S Int. Microwave Symp. Digest*, 168–170 (1974).
25. Konishi, Y. and Matsumura, H., Short end effect of ridge guide with planar circuit mounted in a waveguide, *IEEE Trans. Microwave Theory Tech.*, *MTT-27*, 168–170 (1979).
26. Konishi, Y. and Uenakada, K., The design of a bandpass filter with inductive strip-planar circuit mounted in waveguide, *IEEE Trans. Microwave Theory Tech.*, *MTT-22*, 86–873 (1974).
27. Konishi, Y., Planar circuit mounted in waveguide technique as a downconverter, *IEEE Trans. Microwave Theory Tech.*, *MTT-26*, 716–719 (1978).
28. Bharathi Bhat,shiban K.koul, *Analysis, Design and Applications of Fin Lines*, ArTech House.

29. Watkins, J., Circular resonant structures in microstrip, *Electron. Lett.*, *5*(21), 524–525 (1969).
30. Napoli, L. S. and Hughes, J. J., A simple technique for the accurate determination of the microwave dielectric constant for microwave integrated circuit substrates, *IEEE Trans. Microwave Theory Tech.*, *MTT-19*(7), 664–665 (1971).
31. Shen, L. C., Long, S. A., Alleding, M. R., and Walton, M. D., Resonant frequency of a circular disk, printed-circuit antenna, *IEEE Trans. Antennas Propag.*, *AP-25*(7), 595–596 (1977).
32. Araki, K. and Itoh, T., Hankel transform domain analysis of open circular microstrip radiating structures, *IEEE Trans. Antennas Propag.*, *AP-29*(1), 84–89 (1981).

5

Principle of Filters

In this chapter, the physical meaning to constructing filters and important concepts such as the coupling between neighboring resonators and the external Q caused by the coupling between the port and the resonator are described.

1 THE KINDS AND THE APPLICATION OF THE FILTER

Filters are used to pass or eliminate specific frequency bands. The frequency performances of the attenuations are shown in Fig. 1. Filters are classified as follows: the low-pass filter (LPF), the high-pass filter (HPF), the bandpass filter (BPF), and the band-elimination filter (BEF) or the band-rejection filter (BRF).

To express the attenuation, we have the following several units. The supplied power to the load with R_0, which is the same power to the internal impedance of the power source as shown in Fig. 2a, takes the maximum values, P_{max}. When the filter is connected between the power source and the load with impedance R_L, as shown in Fig. 2b, the power supplied to the load is denoted by P_L. In this case, the attenuation A is defined as

$$A = \frac{1}{2} \log_e \frac{P_{max}}{P_L} \quad \text{(Np)} \tag{1a}$$

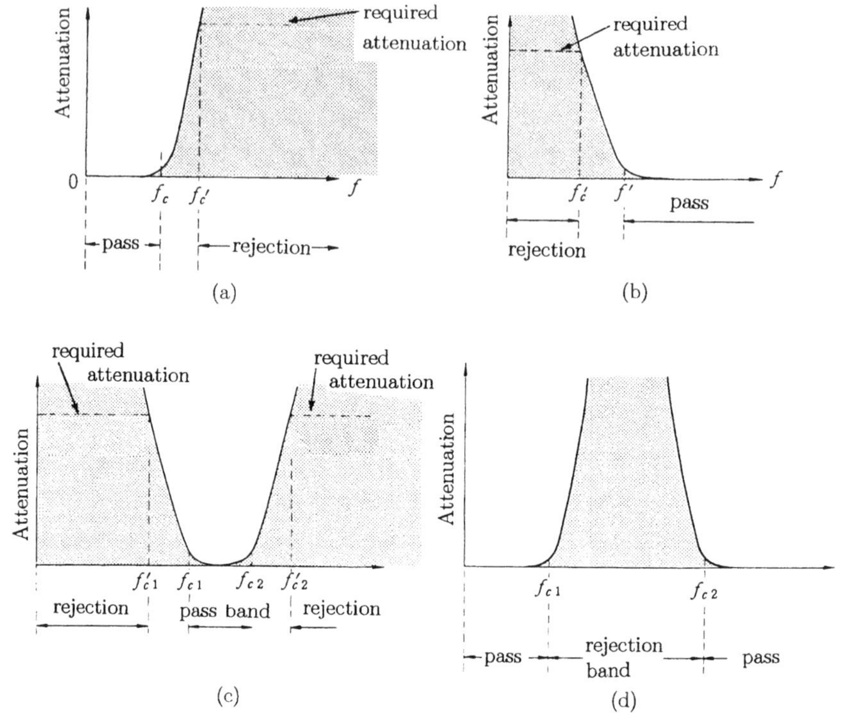

Figure 1. Frequency performance of several kinds of filters. (a) LPF; (b) HPF; (c) BRF; (d) BEF.

$$A = 10 \log_{10} \frac{P_{max}}{P_L} \quad (\text{dB}) \tag{1b}$$

When we use voltages as shown in Fig. 2, we get

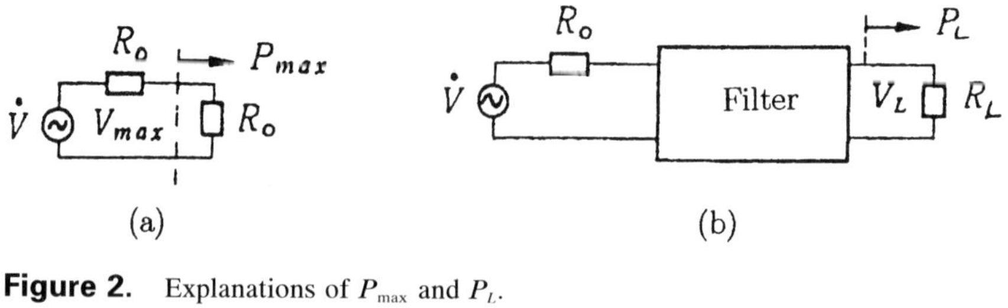

Figure 2. Explanations of P_{max} and P_L.

$$A = \log_e \frac{V_{max}}{V_L} \quad \text{(Np)} \tag{1c}$$

$$A = 20 \log_{10} \frac{V_{max}}{V_L} \quad \text{(dB)} \tag{1d}$$

We obtain the following relationship from Eqs. (1a) and (1b):

$$A \text{ (dB)} = 8.68A \text{ (Np)} \tag{2}$$

Some applications of the filters are as follows:

LPF

(a) Output of the oscillator to eliminate the harmonics generated by the nonlinearity of the oscillator
(b) Output of the mixer to pass only the intermediate frequency
(c) Input of the receiver to reject the unwanted higher frequencies
(d) Realize the wideband bandpass filter by combinating with HPF

HPF

(a) Output of the multiplier to eliminate the lower frequencies
(b) Input of the receiver to reject the unwanted lower frequencies

BPF

(a) Output of the oscillator to take out the required frequency only
(b) Input of the receiver and the amplifier to pass the required frequency only

BEF

(a) Output of the oscillator to eliminate the harmonic frequencies together with LPF
(b) Output of TV transmitter as the vestigial sideband filter (VSBF) to eliminate the sharp slope at the edge of the pass band together with BPF

2 PRINCIPLE OF CONSTRUCTION OF FILTER

2.1 Lumped Element Filter (1)

LPF

The lumped element LPF is constructed as shown in Fig. 3 when the source impedance is 1 Ω and the cutoff angular frequency, ω_c, is 1. Therefore, when the source impedance is R_0 Ω and the cutoff angular frequency is ω_c, the values of L_k and C_k in Fig. 5 must take the values of Eq. (3).

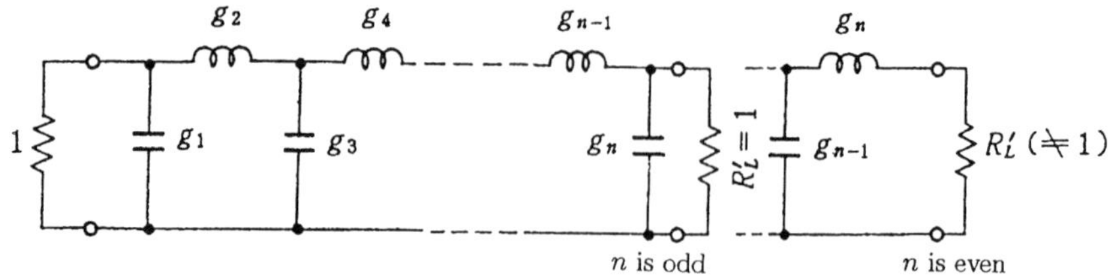

Figure 3. Construction of lumped element LPF with a source impedance of 1 Ω, which is normalized from the values of Fig. 33 and with $\omega_c = 1$.

$$L_k = \frac{R_0}{\omega_c} g_k \quad \text{(H)}$$

$$C_k = \frac{1}{\omega_c R_0} g_k \quad \text{(F)} \tag{3}$$

Generally, the LPFs are designed with frequency performances of the so-called maximum flat and the equal ripple characteristics, as shown in Figs. 4a and 4b, respectively.

The maximum flat characteristics is expressed by Eq. (4a) and g_k is calculated from Eq. (4b):

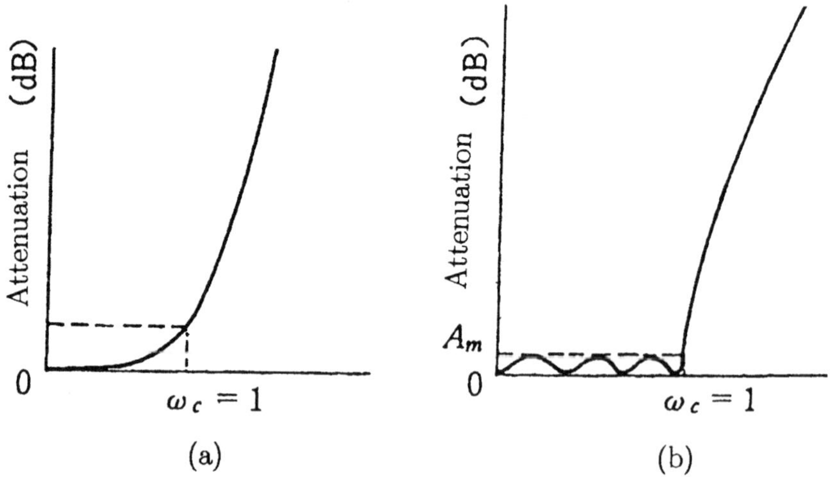

Figure 4. Typical performances of LPF. (a) Maximum flat characteristics; (b) equal ripple characteristics.

$$A = 10 \log_{10}\left[1 + \left(\frac{\omega}{\omega_c}\right)^{2n}\right] \tag{4a}$$

$$g_k = \sin\left(\frac{(2k-1)\pi}{2n}\right), \quad k = 1, 2, \ldots, n \tag{4b}$$

R'_L of Fig. 3 and R_L of Fig. 2 are calculated from Eqs. (4c) and (4d), respectively.

$$R'_L = 1 \tag{4c}$$

$$R_L = R_0 \tag{4d}$$

The values calculated from Eq. (4b) are shown in Table 1, and R'_L of Fig. 3 always equals 1.

Next, the equal ripple characteristics are expressed by Eq. (5a), and g_k is calculated from Eq. (5b). R'_L of Fig. 3 and R_L of Fig. 2 are calculated from Eqs. (5c) and (5d), respectively.

$$A = 10 \log_{10}\left[1 + (10^{A_m/10} - 1)\cos^2\left(n \cos^{-1}\frac{\omega}{\omega_c}\right)\right], \quad \omega \le \omega_c$$

$$A = 10 \log_{10}\left[1 + (10^{A_m/10} - 1)\cosh^2\left(n \cosh^{-1}\frac{\omega}{\omega_c}\right)\right], \quad \omega \ge \omega_c \tag{5a}$$

$$g_1 = \frac{2\sin(\pi/2n)}{\sinh(\beta/2n)}$$

$$g_{k+1} = \frac{4\sin[(2k-1)\pi/2n]\sin[(2k+1)\pi/2n]}{\sinh^2(\beta/2n) + \sin^2(k\pi/2n)g_k}, \quad k = 1, 2, \ldots, n-1$$

$$g_{n+1} = 1 \ (n \text{ is odd})$$

$$g_{n+1} = \coth^2\left(\frac{\beta}{4}\right) \ (n \text{ is even}), \qquad \beta = \ln\left(\coth\frac{A_m}{17.37}\right) \tag{5b}$$

$$R'_L = 1 \quad (n \text{ is odd})$$

$$R'_L = \tanh^2\left(\frac{\beta}{4}\right), \qquad \beta = \ln\left(\coth\frac{A_m}{17.34}\right) \quad (n \text{ is even}) \tag{5c}$$

The calculated values of Eq. (5b) are shown in Table 2.

We, therefore, construct the LPF with L and C as shown in Fig. 5, where the L_k and C_k values are obtained from Eq. (3), with g_k values as given in Table 2.

HPF

Lumped element HPF is constructed as shown in Fig. 6. The values of C'_k and L'_k can be obtained by replacing ω/ω_c in LPF by ω_c/ω in HPF. The values of the elements are obtained from Eq. (6):

Table 1. g_k Values of LPF with the Maximum Flat Characteristics

n	g_1	g_2	g_3	g_4	g_5	g_6	g_7	g_8	g_9	g_{10}	g_{11}
1	2.0000	1.0000									
2	1.4142	1.4142	1.0000								
3	1.0000	2.0000	1.0000	1.0000							
4	0.7654	1.8478	1.8478	0.7654	1.0000						
5	0.6180	1.6180	2.0000	1.6180	0.6180	1.0000					
6	0.5176	1.4142	1.9319	1.9319	1.4142	0.5176	1.0000				
7	0.4450	1.2470	1.8019	2.0000	1.8019	1.2470	0.4450	1.0000			
8	0.3902	1.1111	1.6629	1.9616	1.9616	1.6629	1.1111	0.3902	1.0000		
9	0.3473	1.0000	1.5321	1.8794	2.0000	1.8794	1.5321	1.0000	0.3473	1.0000	
10	0.3129	0.9080	1.4142	1.7820	1.9754	1.9754	1.7820	1.4142	0.9080	0.3129	1.0000

$$C'_k = \frac{1}{R_0\omega_c g_k} \quad \text{(F)}$$

$$L'_k = \frac{R_0}{\omega_c g_k} \quad \text{(H)} \tag{6}$$

BPF

The BPF can be realized by the replacement of L and C in Fig. 5 with the series resonant circuit and the parallel resonant circuit, respectively. We also have the typical frequency performance of the maximum flat characteristic, as shown indicated by the solid line of Fig. 7a and the equal ripple characteristics, as indicated by the dotted line of Fig. 7a. We have the bandwidth ratio defined as the ratio $\Delta f/f_0$ as shown in Fig. 7a, where Δf is defined as the bandwidth less than 3 dB attenuation for the maximum flat and less than A_m dB attenuation for the equal ripple. A_m is the value of the ripple of the attenuation in the pass band.

If we represent the kth series inductance and series capacitance by L_{sk} and C_{sk}, respectively, and the kth parallel inductance and capacitance by L_{pk} and C_{pk}, respectively, we get their values using

$$L_{sk} = \frac{R_0 g_k}{w\omega_0}, \qquad L_{pk} = \frac{wR_0}{\omega_0 g_k}$$

$$C_{sk} = \frac{w}{\omega_0 R_0 g_k}, \qquad C_{pk} = \frac{g_k}{w\omega_0 R_0} \tag{7}$$

In Fig. 7, the subscripts of s and p were omitted.

Table 2. Values of LPF with the Equal Ripple Characteristics

n	g_1	g_2	g_3	g_4	g_5	g_6	g_7	g_8	g_9	g_{10}	g_{11}
					0.1 dB-Ripple						
1	0.3052	1.000									
2	0.8430	0.6220	1.3554								
3	1.0315	1.1474	1.0315	1.0000							
4	1.1088	1.3061	1.7703	0.8180	1.3554						
5	1.1468	1.3712	1.9750	1.3712	1.1468	1.0000					
6	1.1681	1.4039	2.0562	1.5170	1.9029	0.8618	1.3554				
7	1.1811	1.4228	2.0966	1.5733	2.0966	1.4228	1.1811	1.0000			
8	1.1897	1.4346	2.1199	1.6010	2.1699	1.5640	1.9444	0.8778	1.3554		
9	1.1956	1.4425	2.1345	1.6167	2.2053	1.6167	2.1345	1.4425	1.1956	1.0000	
10	1.1999	1.4481	2.1444	1.6265	2.2253	1.6418	2.2046	1.5821	1.9628	0.8853	1.3554
					0.2-dB Ripple						
1	0.4342	1.0000									
2	1.0378	0.6745	1.5386								
3	1.2275	1.1525	1.2275	1.0000							
4	1.3028	1.2844	1.9761	0.8468	1.5386						
5	1.3394	1.3370	2.1660	1.3370	1.3394	1.0000					
6	1.3598	1.3632	2.2394	1.4555	2.0974	0.8838	1.5386				
7	1.3722	1.3781	2.2756	1.5001	2.2756	1.3781	1.3722	1.0000			
8	1.3804	1.3875	2.2963	1.5217	2.3413	1.4925	2.1349	0.8972	1.5386		
9	1.3860	1.3938	2.3093	1.5340	2.3728	1.5340	2.3093	1.3938	1.3860	1.0000	
10	1.3901	1.3983	2.3181	1.5417	2.3904	1.5536	2.3720	1.5066	2.1514	0.9034	1.5386
					0.5-dB Ripple						
1	0.6986	1.0000									
2	1.4029	0.7071	1.9841								
3	1.5963	1.0967	1.5963	1.0000							
4	1.6703	1.1926	2.3661	0.8419	1.9841						
5	1.7058	1.2296	2.5408	1.2296	1.7058	1.0000					
6	1.7254	1.2479	2.6064	1.3137	2.4758	0.8696	1.9841				
7	1.7372	1.2583	2.6381	1.3444	2.6381	1.2583	1.7372	1.0000			
8	1.7451	1.2647	2.6564	1.3590	2.6964	1.3389	2.5093	0.8796	1.9841		
9	1.7504	1.2690	2.6678	1.3673	2.7239	1.3673	2.6678	1.2690	1.7504	1.0000	
10	1.7543	1.2721	2.6754	1.3725	2.7392	1.3806	2.7231	1.3485	2.5239	0.8842	1.9841

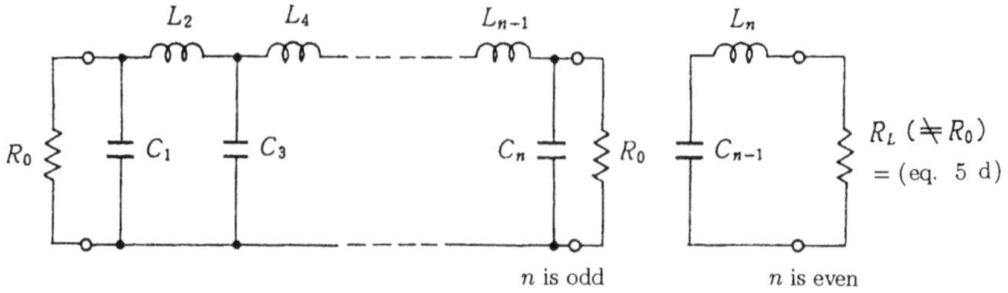

Figure 5. L and C used for the LPF with a source impedance of R_0 Ω.

BEF

The BEF can be realized by replacing C'_k of Fig. 6 by the parallel resonant circuit with C'_{pk} and L'_{pk}, and L'_k by the series resonant circuit with L'_{sk} and C'_{sk}; this construction is shown in Fig. 8. In this figure, the subscripts p and s and the prime were omitted. The values for C'_{pk}, L'_{pk}, and L'_{sk} are obtained from Eq. (8):

$$C'_{pk} = \frac{1}{w\omega_0 R_0 g_k}, \qquad C'_{sk} = \frac{wg_k}{\omega_0 R_0}$$
$$L'_{pk} = \frac{wR_0 g_k}{\omega_0}, \qquad L'_{sk} = \frac{R_0}{w\omega_0 g_k} \tag{8}$$

The frequency performance of BEF becomes as shown in the solid line of Fig. 8. When the resonant frequencies of the parallel and series resonant circuits are deviated in the proper values, we can realize the frequency performance with the equal rippled attenuation in the specified frequency band as shown in the dotted line of Fig. 8a.

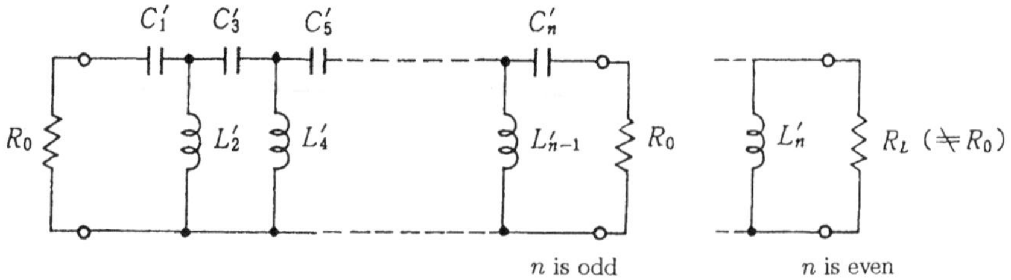

Figure 6. Construction of the lumped element HPF.

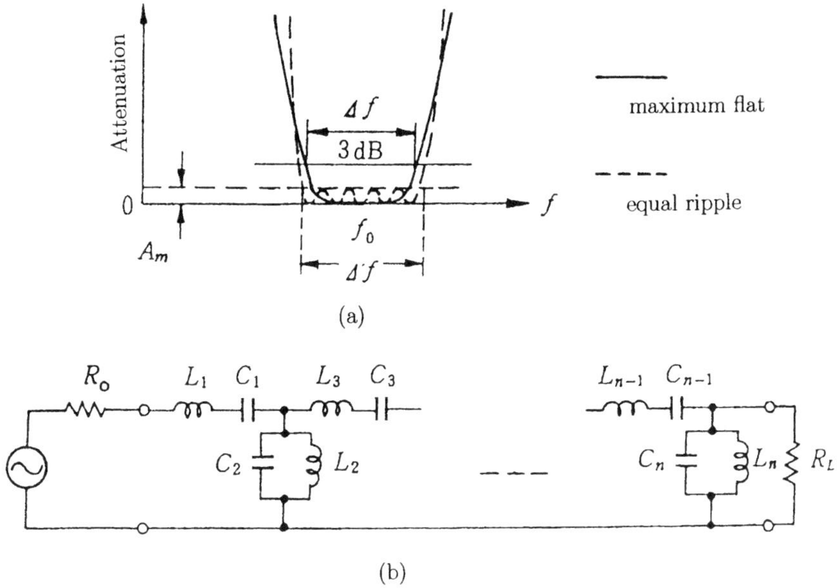

Figure 7. The frequency performance (a) and the construction of the lumped element BPF (b).

2.2 Construction with Distributed Constant Lines

LPF and HPF

The inductance and capacitance as shown in Figs. 5 and 6 can be easily realized by the distributed constant lines with a much shorter length than the wavelength, as mentioned in Chapter 3, Section 1.

When such elements are used in the filters, the characteristics of the filter is realized more accurately in the case of production, especially in higher-frequency regions, because they have good reproducibility and they can avoid the effect of floating capacity around the elements.

The parallel inductance and capacitance can be made by the shortened-end and the open-end distributed constant lines with a short length and their values take the values as shown in Eqs. (1e)–(1h) in Chapter 3. The series inductance and the parallel capacitance, however, can be also realized by making a higher and lower characteristic impedance of the line at the portion where we want to insert the series elements in the following explanation. When the characteristic impedance of the line is Z_c, we have the capacity C and inductance L per meter from Eq. (9a), as explained in Eq. (4e) of Chapter 1:

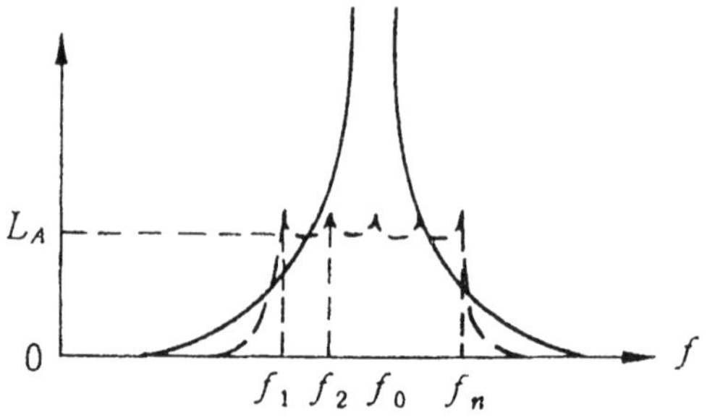

(a)

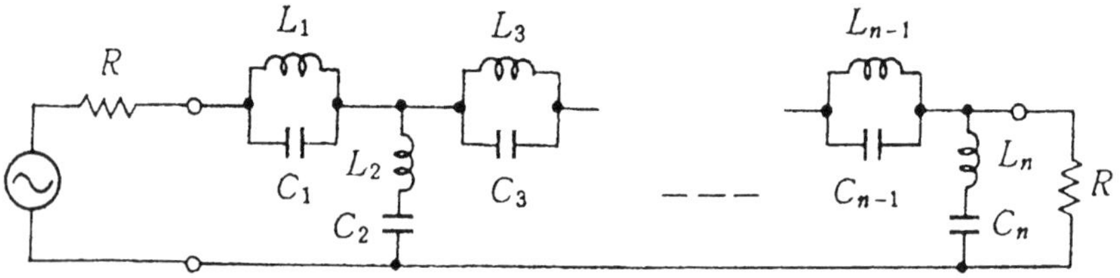

(b)

Figure 8. Frequency performance (a) and the construction of the lumped element BEF (b).

$$C = \frac{1}{Z_c v_p}, \qquad L = \frac{Z_c}{v_p} \tag{9a}$$

where v_p is the phase velocity. In the case of the characteristic impedance R_0, the inductance L_0 and the capacitance C_0 are calculated from

$$C_0 = \frac{1}{R_0 v_p}, \qquad L_0 = \frac{R_0}{v_p} \tag{9b}$$

Therefore, the decreasing and increasing of the characteristic impedance from R_0 to Z_c, replaces the increase of C and L, respectively, with the values from

$$(C - C_0)l = \left(\frac{1}{Z_c} - \frac{1}{R_0}\right)\frac{l}{v_p} \quad \text{for } Z_c < R_0$$

$$(L - L_0)l = \left(\frac{1}{Z_c} - \frac{1}{R_0}\right)\frac{l}{v_p} \quad \text{for } Z_c > R_0 \tag{10a}$$

This leads to the series inductance L_{eq} and the parallel capacity C_{eq} as

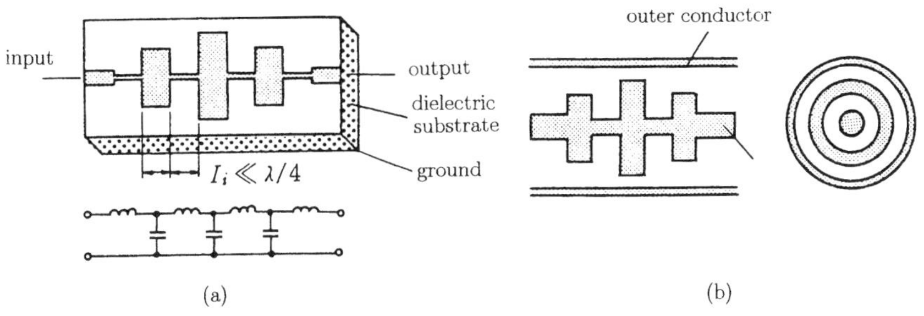

Figure 9. Examples of LPF. (a) Microstrip LPF; (b) coaxial LPF.

$$C_{eq} \simeq \left(\frac{1}{Z'_c v'_p} - \frac{1}{R_0 v_p}\right) l, \qquad L_{eq} \simeq \left(\frac{Z''_c}{v''_p} - \frac{R_0}{v_p}\right) l \tag{10b}$$

where we take account of the change of the phase velocity from v_p to v'_p and v''_p corresponding to Z'_c and Z''_c, respectively. This occurs in the microstrip line mentioned in Eqs. (24a)–(24c) of Chapter 2, the values of which are connected to v'_p and v''_p.

Based on this concept, we can make a LPF from the microstrip and the coaxial line, as shown in Figs. 9a and 9b, respectively. As shown in Fig. 9, the thicker conductor in the middle is caused by the required C with larger values, which is understood by the g values of the middle portion being larger than the end portions.

The series inductance can be adjusted by the width of the line and also the length. In the practical design, the equivalent inductance existing at the portion where the width of the line is rapidly changed and the evanescent higher mode is generated should be considered in series.

HPF

One of the examples of HPF made by triplate is shown in Fig. 10a, and the equivalent network is shown in Fib. 10b. Since the inductances L_1 and L_2 are made by the distributed constant lines of short length l and the shortened end, L_1 and L_2 are calculated from Eq. (1g) of Chapter 3; that is,

$$L_{1,2} \simeq \frac{Z_{1,2}}{v_p} l = \frac{Z_{0\,1,2}}{v_0} l \tag{10c}$$

where $Z_{0\,1,2}$ and v_0 show the characteristic impedance and the phase velocity in the case of $\varepsilon_r = 1$.

In the HPF, the large C, and small C, and the middle C are shown in Figs. 11a, 11c, and 11d, respectively.

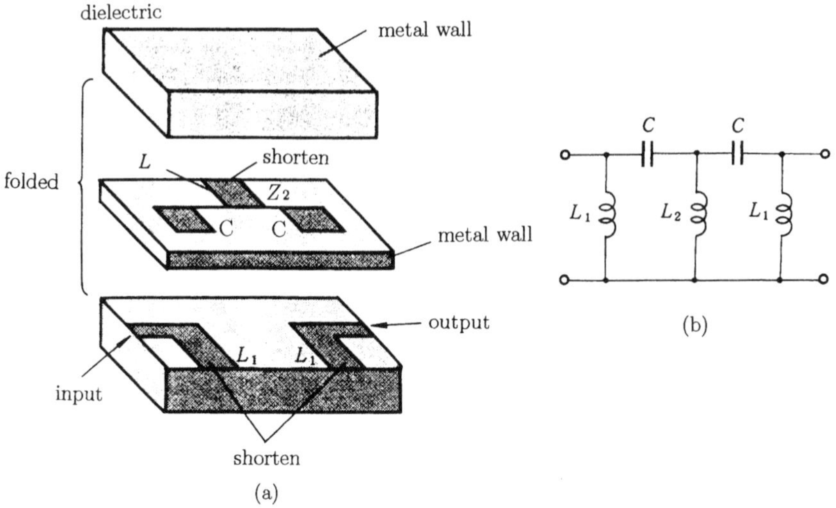

Figure 10. HPF made by a triplate. (a) Construction; (b) equivalent network of (a).

As mentioned at Chapter 1, Section 3, the waveguide is HPF, with cutoff frequency f_c. In the case of the $TE^{\square}_{10}$ waveguide,

$$f_c = \frac{v_0}{\lambda_c} = \frac{v_0}{2a\sqrt{\varepsilon_r \mu_r}} = \frac{v_p}{2a} \tag{11}$$

Therefore, the attenuation constant, α, is calculated from Eq. (12a) and the wave attenuation, A_u, for a length l is calculated from Eq. (12b):

$$\alpha = \omega\sqrt{\mu\varepsilon}\sqrt{\left(\frac{f_c}{f}\right)^2 - 1} = \frac{2\pi\sqrt{(f_c/f)^2 - 1}}{\lambda} \tag{12a}$$

$$A_u = \log_e e^{\alpha l} = \alpha l = \frac{2\pi l}{\lambda}\sqrt{\left(\frac{f_c}{f}\right)^2 - a} \quad \text{(Np)} \tag{12b}$$

where λ is the wavelength in the infinite medium with ε and μ.

When the attenuation is expressed by A (dB), it is calculated from

$$A = 54.6\,\frac{l}{\lambda}\sqrt{\left(\frac{f_c}{f}\right)^2 - 1} \quad \text{(dB)}$$

$$\lambda = \frac{v_0}{f\sqrt{\mu_r \varepsilon_r}} \tag{13}$$

where F_c is the cutoff frequency [Eq. (54)].

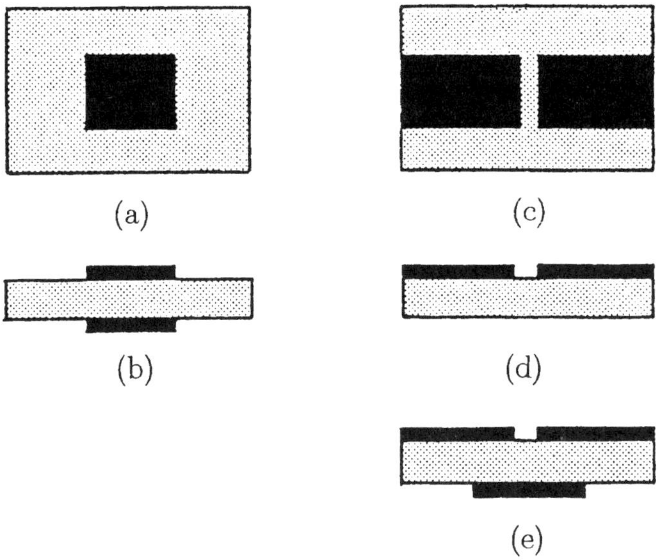

Figure 11. Construction to make series capacitor by using a high-dielectric material. (a) Top view; (b) side view of (a); (c) top view; (d) side view of (c); (e) side view of the construction to increase the C values of (c).

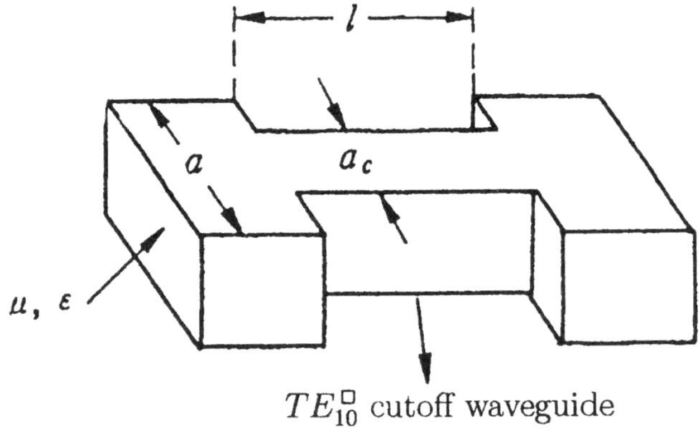

Figure 12. HPF made by a $TE_{10}^{\square}$ waveguide.

Example The $TE_{10}^{\square}$ waveguide with width $a = 1$ cm and length 2 cm is filled by material with the relative constant $\varepsilon_r = 10$. We will obtain the attenuation of 0.5 GHz. f_c and λ are obtained as follows:

$$f_c = \frac{v_0}{\lambda_c} = \frac{3 \times 10^8}{2a\sqrt{\mu_r \varepsilon_r}} = \frac{3 \times 10^8}{2 \times 10^{-2} \times \sqrt{10 \times 10}} = 1.5 \times 10^9 = 1.5 \quad \text{(GHz)}$$

$$\lambda = \frac{3 \times 10^8}{0.5 \times 10^9 \times \sqrt{10 \times 10}} = 6 \times 10^{-2}\ \text{m} = 6\ \text{cm}$$

Therefore, the attenuation A (dB) is obtained from Eq. (13) as

$$A = 54.6 \times \frac{2}{6} \sqrt{\left(\frac{1.5}{0.5}\right)^2 - 1} = 51.48\ \text{dB}$$

BPF and BEF

The construction of Fig. 7 can be transformed to that of Fig. 13a, where $D_1, \ldots, D_n$ are resonators and they are coupled to each other with coupling coefficient k_{ij}. D_1 and D_n are also coupled to the input and output ports, which result in the external Q, Q_{e1} and Q_{e2}. (See Appendix 2 of the volume.)

On the other hand, the BEF is constructed with the transmission line coupled to the several resonant resonators, as shown in Fig. 13b, which is obtained by applying the equivalent transform as shown in Fig. 13c to the construction in Fig. 8.

The values of Q_{e1}, Q_{e2}, and $k_{i,i+1}$ are obtained from Eqs. (14) and (15). (See Appendix 2 of this volume.)

$$Q_{e1} = \frac{g_1}{w}, \qquad Q_{en} = \frac{g_n}{w} \tag{14}$$

$$k_{i,i+1} = \frac{w}{\sqrt{g_i g_{i+1}}} \tag{15}$$

The bandwidth ratio w is provided by Eq. (16) as mentioned in the earlier subsection on BPF.

$$\begin{aligned} w &= \frac{\Delta f_{3\,\text{dB}}}{f_0} \quad \text{(maximum flat)} \\ &= \frac{\Delta f}{f_0} \quad \text{(equal ripple)} \end{aligned} \tag{16}$$

In the following subsections we will show several examples of Q_{e1}, Q_{e2}, and $k_{i,j}$.

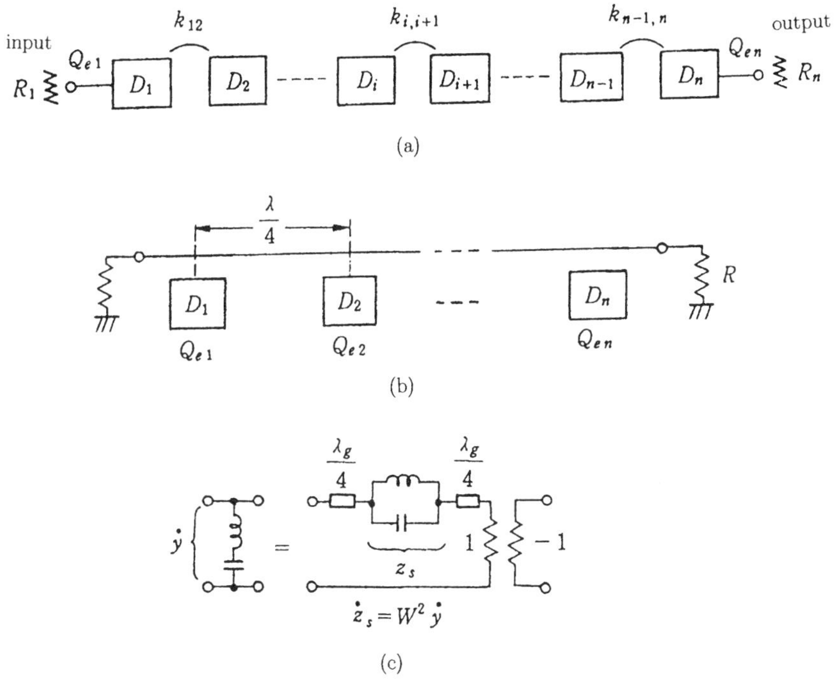

Figure 13. BPF and BEF with n resonators (a and b) and equivalent transform with $\lambda_g/4$ lines (c).

Coupling Between the Resonator and the Source or the Load

Coupling by the Reactance of So-Called Capacitive Coupling and Inductive Coupling The capacitance or the inductance with the proper values are connected between the input and D_1 and also between the output and D_n of Fig. 13a. The position of the input and the output is properly selected, for example, at the open end of a $\lambda_g/4$ resonator and sometimes at the proper portion of $\lambda_g/4$.

Coupling by Coupled $\lambda_g/4$ Distributed Constant Lines When a resonator is made by a $\lambda_g/4$ distributed constant line, it can be coupled with another $\lambda_g/4$ distributed constant line as shown in Fig. 14a. Denoting the inductances and capacities of the coupled distributed constant lines by L_{11}–L_{22} and C_{11}–C_{22}, the equivalent network of Fig. 14b becomes that in Fig. 14b, where the values of C_p, L_p, and n are obtained as shown in Eq. (17).

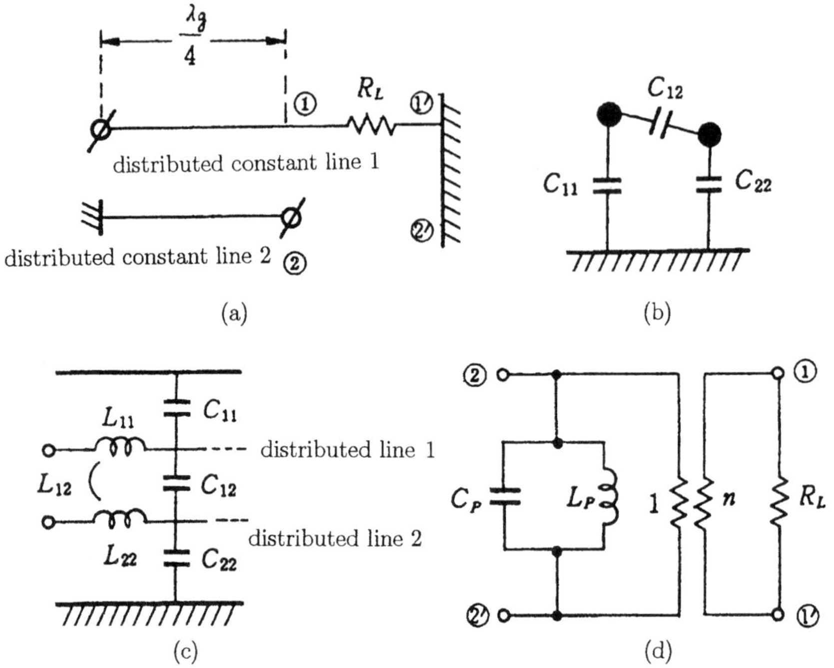

Figure 14. Coupling between $\lambda_g/4$ resonator and load R_L through $\lambda_g/4$ coupled distributed constant lines and the equivalent network. (a) External load R_L is coupled to $\lambda_g/4$ resonator of line 2 through $\lambda_g/4$ line 1; (b) sectional view of distributed constant lines; (c) inductance and capacitance of the coupled distributed lines; (d) equivalent network of (a).

$$C_p = \frac{\pi}{4\omega W_2'}, \qquad L_p = \frac{1}{\omega^2 C_p}$$
$$n = \frac{L_{12}}{L_{22}} = \frac{C_{12}}{C_{11} + C_{12}}$$
$$W_2' = \frac{1}{v_p C_{22}} \tag{17}$$

Therefore Q_e is obtained from

$$\frac{Q_e}{\sqrt{\varepsilon_{r,\text{eff}}}} = \frac{3\pi}{4}\left(\frac{C_{11}^{\circ} + C_{12}^{\circ}}{C_{12}^{\circ}}\right)^2 C_{22}^{\circ}\,(\text{pF})R_L \times 10^{-4}$$
$$C_{ij}^{\circ} = \frac{C_{ij}}{\varepsilon_r} \tag{18}$$

In Eq. (17), C_{ij} and L_{ij} $(i, j = 1, 2)$ denote the capacities and the inductances per meter, measured as C_{ij} F/m and H/m, respectively. When the parallel coupled lines have a symmetrical sectional view, as shown in Figs. 15a and 15b, we have

$$C_{11} = C_{22} \tag{19}$$

Denoting the characteristic impedance of the even mode by Z_e and the odd mode by Z_0, we get

$$Z_e = \frac{1}{C_{11}v_p} \tag{20a}$$

$$Z_0 = \frac{1}{(C_{11} + 2C_{12})v_p} \tag{20b}$$

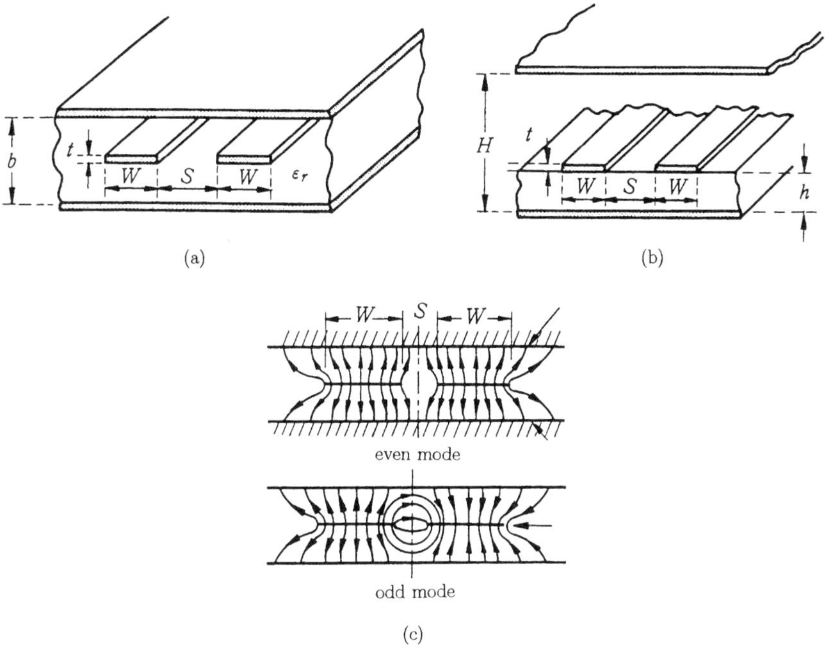

Figure 15. Coupled parallel lines with symmetrical sectional view and the electrical field distributions for even and odd modes. (a) Symmetrical coupled triplates; (b) symmetrical coupled microstrips; (c) electrical field distribution of even and odd modes of (a).

In Fig. 14d, because Q_e is calculated from

$$Q_e = \frac{\omega C_p R_L}{n^2}$$

substituting Eqs. (17), (19), (20a), and (20b) into the above equation, we get

$$Q_e = \frac{\pi}{4}\left(\frac{1 + Z_0/Z_e}{1 - Z_0/Z_e}\right)^2 \frac{R_L}{Z_e} \tag{21a}$$

Especially, in the case of $Z_e = R_L$, we get

$$Q_e = \frac{\pi}{4}\left(\frac{1 + Z_0/Z_e}{1 - Z_0/Z_e}\right)^2 \tag{21b}$$

On the other hand, if Q_e is given, we obtain the ratio Z_0/Z_e from

$$\frac{Z_0}{Z_e} = \frac{2\sqrt{Q_e Z_e/\pi R_L} - 1}{2\sqrt{Q_e Z_e/\pi R_L} + 1} \tag{21c}$$

However, Z_e and Z_0 in Fig. 15a for $t = 0$ are obtained by conform mapping [2]:

$$Z_e = \frac{30\pi}{\sqrt{\varepsilon_r}} \frac{K(k_e')}{K(k_e)}$$

$$Z_0 = \frac{30\pi}{\sqrt{\varepsilon_r}} \frac{K(k_e')}{K(k_e)}$$

$$k_e = \tanh\left(\frac{\pi}{2}\frac{W}{b}\right) \tanh\left(\frac{\pi}{2}\frac{W + S}{b}\right), \qquad k_e'^2 = 1 - k_e^2$$

$$k_0 = \tanh\left(\frac{\pi}{2}\frac{W}{b}\right) \coth\left(\frac{\pi}{2}\frac{W + S}{b}\right), \qquad k_0'^2 = 1 - k_0^2 \tag{22}$$

where $K(k)$ is the first-kind complete elliptical integral, which is defined in Note 1 in the Appendix to this chapter.

In Eq. (22), $K(k)/K(k')$ is expressed by [3].

$$\frac{K(k)}{K(k')} = \begin{cases} \dfrac{\pi}{\ln 2 + 2 \tanh^{-1}\sqrt{k'}}, & k^2 \le 0.5 \\[2ex] \dfrac{\ln 2 + 2 \tanh \sqrt{k}}{\pi} & 0.5 \le k^2 < 1 \end{cases} \tag{23}$$

When the thickness is taken account, Z_e and Z_0 are shown by [4].

$$Z_e = \frac{30\pi(b - t)}{\sqrt{\varepsilon_r}[W + (bC_f/2\pi]A_e]}$$

$$Z_0 = \frac{30\pi(b - t)}{\sqrt{\varepsilon_r}[W + bC_f/2\pi]A_0}$$

$$A_e = 1 + \frac{\ln(1 + \tanh\theta)}{\ln 2}$$

$$A_0 = 1 + \frac{\ln(1 + \coth\theta)}{\ln 2}$$

$$\theta = \frac{\pi S}{2b}$$

$$C_f = 2\ln\left(\frac{2b - t}{b - t}\right) - \frac{t}{b}\left(\frac{t(2b - t)}{(b - t)^2}\right) \tag{24}$$

As an example, the calculated values of Z_e and Z_0 for $t = 0$ are shown in Fig. 16 [6].

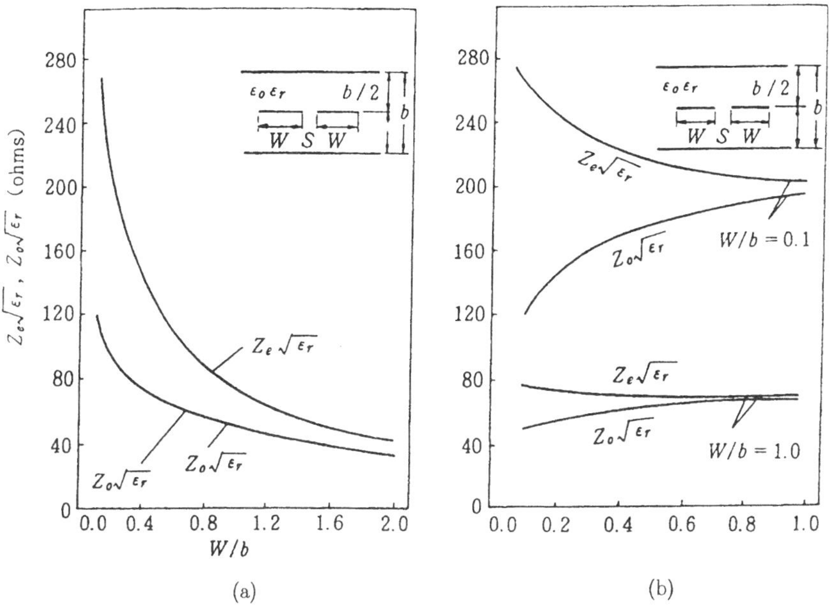

Figure 16. (a) $Z_e\sqrt{\varepsilon_r}$ and $Z_0\sqrt{\varepsilon_r}$ for a $S/b = 0.1$ corresponding to W/b; (b) $Z_e\sqrt{\varepsilon_r}$ and $Z_0\sqrt{\varepsilon_r}$ for $W/h = 0.1$ and 1.0 corresponding to S/b. (From Ref. 6.)

It is understood that Z_e is larger than Z_0. This is caused because the capacity of the odd mode is larger than that of the even mode with the additional capacity of $2C_{12}$, as in Eq. (20b).

Next, for $H = 2h$ and $t = 0$ in Fig 15b, Z_e and Z_0 are obtained as [7]

$$Z_e = \frac{30\pi}{\sqrt{(\varepsilon_r + 1)/2}} \frac{K(k_e')}{K(k_e)}$$

$$Z_0 = \frac{30\pi}{\sqrt{(\varepsilon_r + 1)/2}} \frac{K(k_0')}{K(k_0)}$$

$$k_e = \tanh\left(\frac{\pi}{4}\frac{W}{h}\right) \tanh\left(\frac{\pi}{4}\frac{W + S}{h}\right), \qquad k_e'^2 = 1 - k_e^2$$

$$k_0 = \tanh\left(\frac{\pi}{4}\frac{W}{h}\right) \tanh\left(\frac{\pi}{4}\frac{W + S}{h}\right), \qquad k_0'^2 = 1 - k_0^2 \tag{25}$$

As understood by comparing Eqs. (25) and (22), Eq. (25) is obtained by replacing the values of ε_r and b by the following values, considering the symmetry between the upper and lower limits for the electric fields:

$$\varepsilon_r \rightarrow \varepsilon_{r,\text{eff}} = \frac{\varepsilon_r + 1}{2}$$

$$b \rightarrow 2h$$

A practical example to obtain the size satisfying the required values of Q_e will be shown in the following:

Step 1. When Z_e and R_L are given, Z_0/Z_e is obtained from Eq. (21a). As a reference, we show the relationship between Z_0/Z_e and $Q_e(Z_e/R_L)$ in Fig. 17.

Step 2. W/b and S/b to obtain Z_0/Z_e required at step 1 are calculated from Eq. (22) or (25). For this purpose, it is best that the calculations of Z_e and Z_0/Z_e corresponding to several values of S/b and W/b be made in advance. As a reference, the values of $\sqrt{\varepsilon_r}Z_e$ and Z_0/Z_e were obtained corresponding to S/b with the parameters $W/b = 1.0$, 0.5, and 0.1 as shown in Fig. 18. As an example, when we need a Q_e of 40, Z_0/Z_e should be 0.77, as in the case of $Z_e = R_L$ from Fig. 17. From Fig. 18, we get $S/b = 0.48$ and $\sqrt{\varepsilon_r}Z_e = 198\ \Omega$ for $W/b = 0.1$, $S/b = 0.31$, and $\sqrt{\varepsilon_r}Z_e = 112\ \Omega$ for $W/b = 0.5$, and $S/b = 0.19$ and $\sqrt{\varepsilon_r}Z_e = 72\ \Omega$ for $W/b = 1.0$. In the case $R_L = 50\ \Omega$, $\varepsilon_r = (198/50)^2$ for $W/b = 0.1$, $\varepsilon_r = (112/50)^2 = 5$ for $W/b = 0.5$, and $\varepsilon_r = (72/50)^2 = 2.074$ for $W/b = 1.0$ are required. Therefore, the preparation of the many groups for the curve corresponding to values of W/b by Eq. (24) is convenient in designs. For example, when $b = 3.2$ mm and $\varepsilon_r = 2.074$, we get $W = 3.2$ mm and $S = 0.609$ mm.

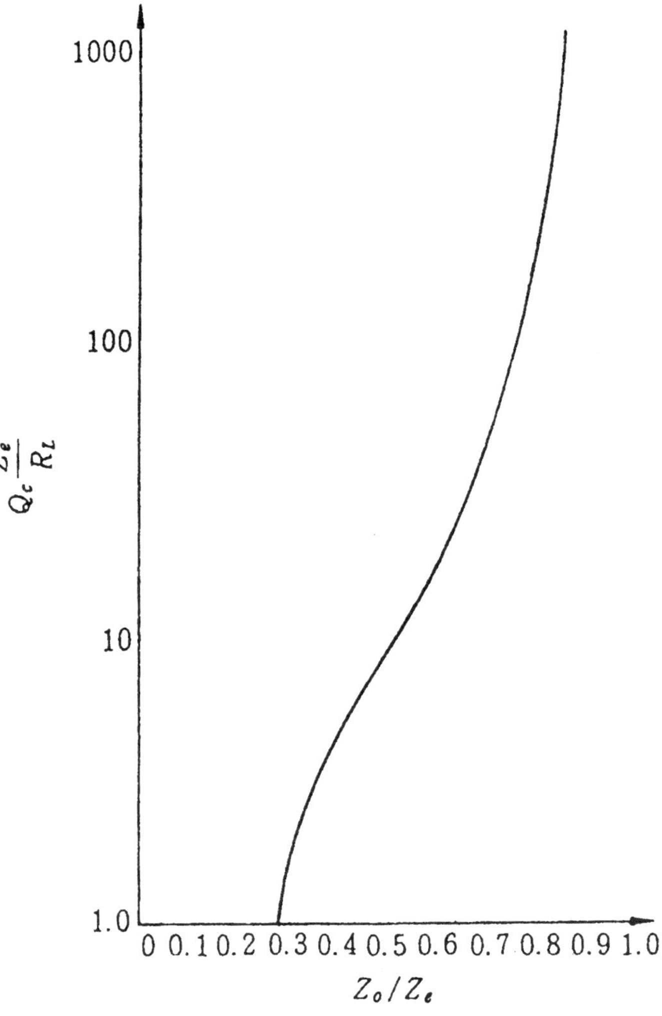

Figure 17. Relationship between $Q_e(Z_e/R_L)$ and Z_0/Z_e [Eq. (21a)].

Next, we consider the BEF as shown in Fig. 19a. When the sectional view takes the symmetrical construction as shown in Fig. 19b, the equivalent network becomes as shown in Fig. 19c. [In Eq. (56i) of Chapter 3, we substitute the values $R_c = 1$, $W_c = Z_e$, and $W_\pi = Z_0$.]

Since the construction is transformed to that of Fig. 19d, we get the external Q, Q_e, obtained from

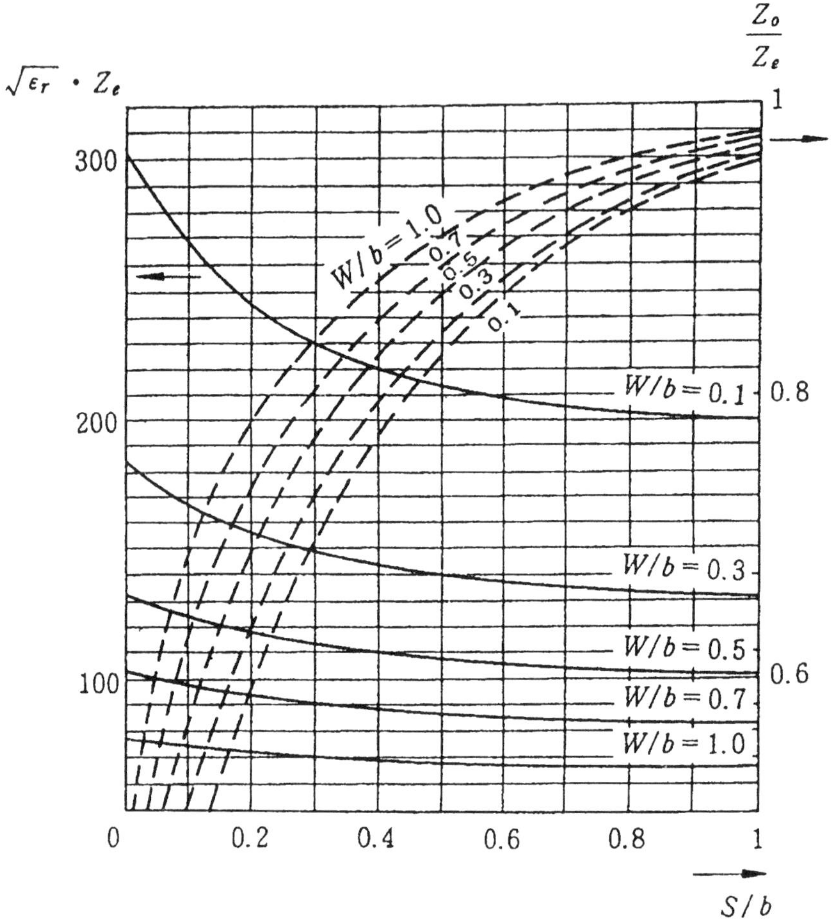

Figure 18. $\sqrt{\varepsilon_r}Z_e$ and Z_0/Z_e for W/h = 1.0, 0.5, and 0.1.

$$Q_e = 2R\omega C_p = \frac{\pi R}{2}\frac{1 + Z_0/Z_e}{1 - Z_0/Z_e} = 1 + \frac{C_{11}}{C_{12}} \tag{26}$$

when we used $C_p = \pi/4\omega W_2$ as in Eq. (7) of Chapter 3. In this case, the frequency performance becomes as shown in Fig. 19e, where Q_e is related to

$$Q_e = \frac{f_0}{\Delta f} \tag{27}$$

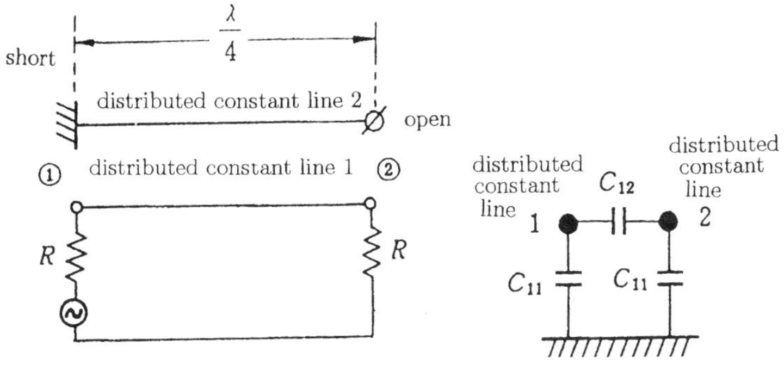

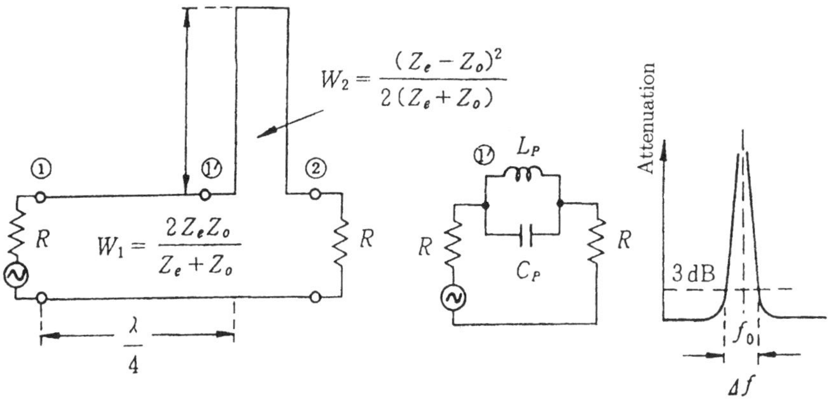

Figure 19. BEF with one resonator, the equivalent network, and the frequency performance. (a) BEF with one λ/4 coupled distributed constant line; (b) sectional view of coupled symmetrical distributed constant line and the capacities; (c) equivalent circuit of (a); (d) equivalent circuit between 1′ and 2 in (c) ($R = W$); (e) frequency performance of (a).

Coupling Between a Dielectric Resonator and a Microstrip Line Because $TE^{\circ}_{01\delta}$ DR (dielectric resonator) has magnetic fields as shown in Fig. 29a, it can be coupled to a microstrip line, with the construction shown in Fig. 20a.

When the source is applied to terminals 2 and 2′ of Fig. 20b, DR coupled to the microstrip line induces the magnetic flux, generates the reverse electric motive force and chokes the transmission of the wave to the load connected at terminals 3 and 3′. The equivalent network is shown in Fig. 21a, where

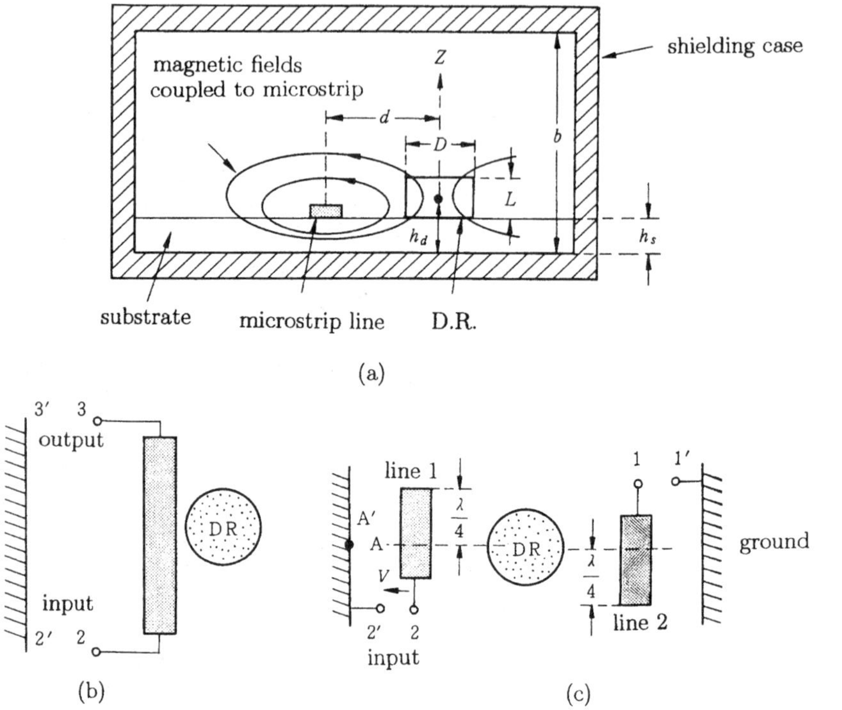

Figure 20. Coupling construction between a DR and a microstrip. (a) Sectional view; (b) top view of BRF; (c) top view of BPF.

L_rC_r is the parallel resonant circuit with the same resonant frequency as the DR, and the equivalent network becomes Fig. 21b. Therefore, the construction of Fig. 20b is used for BRF.

Next, the construction of Fig. 20c is used for BPF with one resonator, where the DR is placed at a distance $\lambda_g/4$ from the open end of the microstrip lines. Here, the magnetic fields of the microstrip lines become maximum, which makes it possible to couple strongly with DR. The equivalent network of the input port becomes as shown in Fig. 21c. When terminals A and A' are connected by the source with the internal impedance of R, the resonant circuit with C_r and L_r is loaded by the source impedance through the coupling coefficient of L_m, which results in the external Q, Q_e. To obtain the values of Q_e, we have to calculate the magnetic flux generated by the DR and to get the consumed power in the source and the load impedances, P_d. Then, Q_e can be obtained from

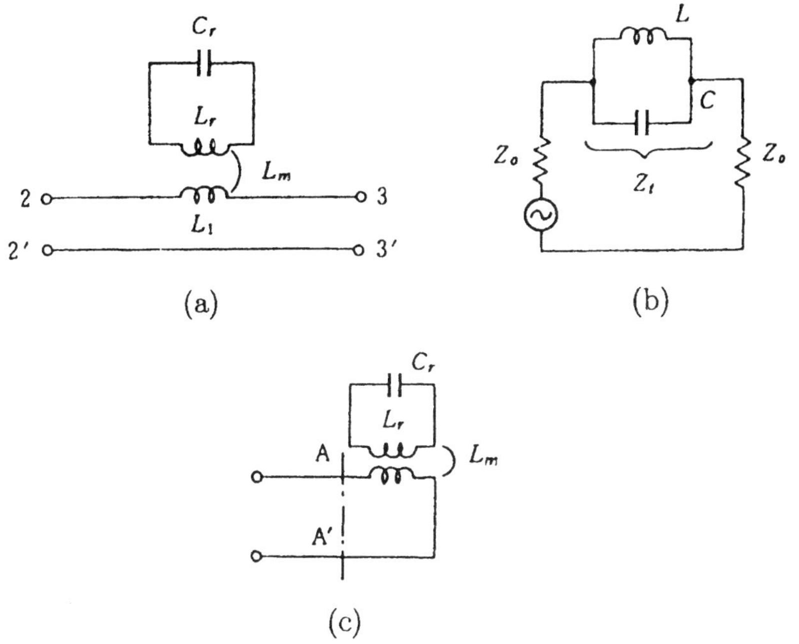

Figure 21. Equivalent network of the coupling between DR and the microstrip line. (a) Equivalent network of Fig. 20a; (b) equivalent network of Fig. 20b with a source and load; (c) equivalent network of the input coupling of Fig. 20c.

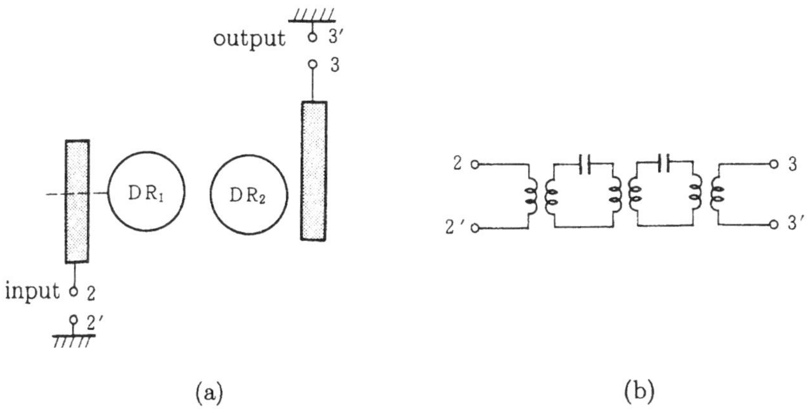

Figure 22. BPF and two DRs (a) and the equivalent network (b).

$$\frac{\omega \tilde{W}_t}{P_d} \tag{28}$$

where $\tilde{W}_t$ is the time average of the total reactive energies included in the DR.

We can, however, consider the DR and the terminal of the microstrip line as the two-port network. The method, as mentioned before, to calculate Q_e is by flowing the current at the port corresponding to the DR. On the other hand, by the reciprocal theory, we can obtain Q_e by flowing the current at the port corresponding to the terminal of the microstrip line, which is another way to calculate the values of Q_e. We will explain it step by step.

1. The electromagnetic field outside the DR can be obtained by replacing the DR with the magnetic dipole moment M (or the magnetic current element) placed at the center of the disk parallel to the axis of the $TE^{\circ}_{01\delta}$ DR.

   The values of M are obtained from

   $$M = IS \tag{29a}$$

   where S is the area of a one-turn loop antenna and I is the current flowing to the antenna.
2. Cutting the loop antenna, the cut point will be denoted by port 1, and the terminal of the microstrip used for the connection of the load impedance will be denoted by port 2. Then, we get a two-port network. When the current, I, is flowing in port 1, the voltage V appears between the impedance R connected at port 2, which produces the consumed power P_d as

   $$P_d = \frac{V^2}{R} \tag{29b}$$

   On the other hand, the value of M is related to $\tilde{W}_t$, which is the time average of the total reactive energies of the DR; then, we can obtain the values of Q_e from Eq. (28). This is really the method, which is already described before, for using Eq. (28).
3. The voltage V is the same voltage that appeared at port 2 when port 1 is derived by the current I on the basis of reciprocal theory.

Indicating the assumed area of the loop corresponding to the DR by S in Fig. 23, the voltage appearing at port 1 should be calculated from Eq. (29c) by Faraday's Law:

$$|V| = \omega\mu_0 \int_S |H(I)| \, ds \tag{29c}$$

When the values of V of Eq. (29c) is substituted into Eq. (29b), we get

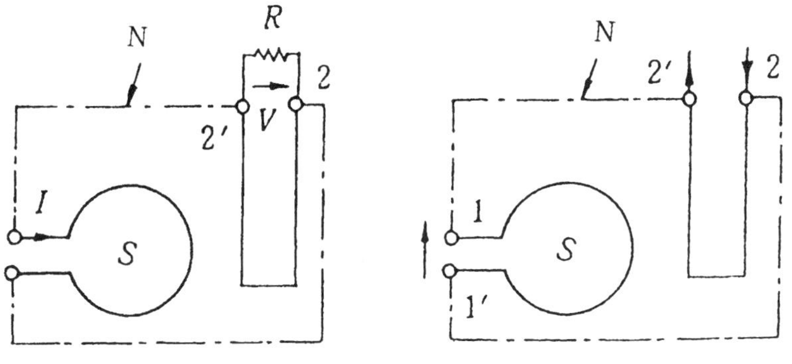

Figure 23. The voltage V and the current I in the equivalent two-port network consisting of port 1, corresponding DR, and port 2, corresponding to the microstrip line terminal. (a) The voltage V appeared at port 2 when the current is flowing to the loop corresponding to the DR; (b) the voltage V appeared at the cut point of the equivalent loop of the DR when the current I is flowing to port 2 of the microstrip.

$$Q_e = \frac{R\tilde{W}_t}{\omega\mu_0^2[\int_s H\, ds]^2} \tag{29d}$$

where $\tilde{W}_t$ is the time average of the total reactive energies corresponding to the magnetic current moment M of Eq. (29a). In the case of Fig. 20c, the Q_{e1} caused by the coupling between the DR and line 1 can be obtained from Eq. (29d).

However, the values Q_e in the construction of Fig. 29b take twice the values of Eq. (29d), because the impedance becomes $2R$ as the result of the series connection of the load and the source; that is, the circuit in Fig. 20b can be shown in Fig. 24a, where we have the symmetrical plane S'. Because the $TE^{\circ}_{01\delta}$ mode is the odd mode about the S' plane, the S' plane becomes the electrical wall. Then, we can consider the Fig. 24a in light of Fig. 24b. In the case of Fig. 24b, the integral region S in Eq. (29d) becomes half, which results in Eq. (29e). It shows the values in Eq. (29d):

$$Q_e = \frac{2RW_t}{\omega\mu_0^2[\int_{ss} H\, ds]^2} \tag{29e}$$

When the area of the loop is small enough, H takes the same values in the loop. In this case, the integral parts of Eqs. (29d) and (29e) become

$$\int_s H\, ds = HS, \tag{29f}$$

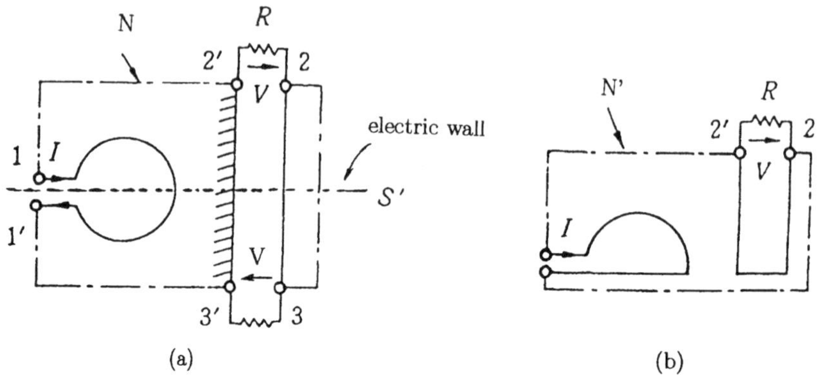

Figure 24. The construction of the DR coupled with the microstrip line using 2 ports (a) and the equivalent circuit considering the image (b).

where S is the area of the loop which results in Eqs. (29g) and (29h), corresponding to Figs. 20c and 20b, respectively.

In the case of Fig. 20c,

$$Q_e = \frac{RW_t}{\omega\mu_0^2 M^2}\left(\frac{I}{H}\right)^2 \tag{29g}$$

In the case of Fig. 20b,

$$Q_e = \frac{2RW_t}{\omega\mu_0^2 M^2}\left(\frac{I}{H}\right)^2 \tag{29h}$$

The expressions in Eqs. (29d) and (29e) are called the flux method and those of Eqs. (29g) and (29h) are called the H/I method [8].

W_t and M are, however, expressed by the maximum values of E_θ as follows [9]:

$$\tilde{W}_t = \frac{1}{4}\varepsilon_0\varepsilon_r\pi D^2 L E_0^2\left[\frac{1}{2}\left(1 + \frac{\sin k_z L}{k_z L}\right) + \frac{\cos^2(k_z L/2)}{\alpha_0 L\varepsilon_r}\right] \tag{30a}$$

$$F = \frac{\mu_0 M^2}{\tilde{W}_t} = \frac{0.927 D^4 L\varepsilon_r}{\lambda_0^2} \tag{30b}$$

By substituting Eq. (30b) into Eqs. (29g) and (29h), we get

$$Q_e = \frac{R}{\omega\mu_0}\frac{1}{F}\left(\frac{I}{H}\right)^2 \quad \text{(Fig. 20c)} \tag{31a}$$

$$Q_e = \frac{2R}{\omega\mu_0}\frac{1}{F}\left(\frac{I}{H}\right)^2 \quad \text{(Fig. 20b)} \tag{31b}$$

From the facts mentioned above, we can conclude the following:

(a) Q_e decreases for the larger h_d when the distance between the DR and the microstrip line and in the case of $h_d \ll b$ and d of Fig. 20a. This is caused because the coupling increases, because the values of the axial component of H of the DR increases for the larger h_d. The effect is shown in Fig. 25a. However, when h_d is greatly increased, the distance between the DR and the microstrip line increases, which results in increasing Q_e again.

When the DR and the microstrip are shielded as shown in Fig. 20a, the current of the microstrip, I, has the direction of the $\otimes$ sign of Fig. 26 and the electrical fields decrease exponentially by increasing the distance from the microstrip line, which results in decreasing H_z.

Because E distributes as $\sin(\pi h_d/b)$, H/I values in Eqs. (29g) and (29h) change in $\sin^2(\pi h_d/b)$. Therefore, Q_e is proportional to $1/\sin^2(\pi h_d/b)$, and Q_e decreases for increasing h_d when $h_d \ll b$.

(b) Q_e decreases for larger values of h_s. In Fig. 26a, when h_s increases, the distance between the microstrip line and the image line increases,

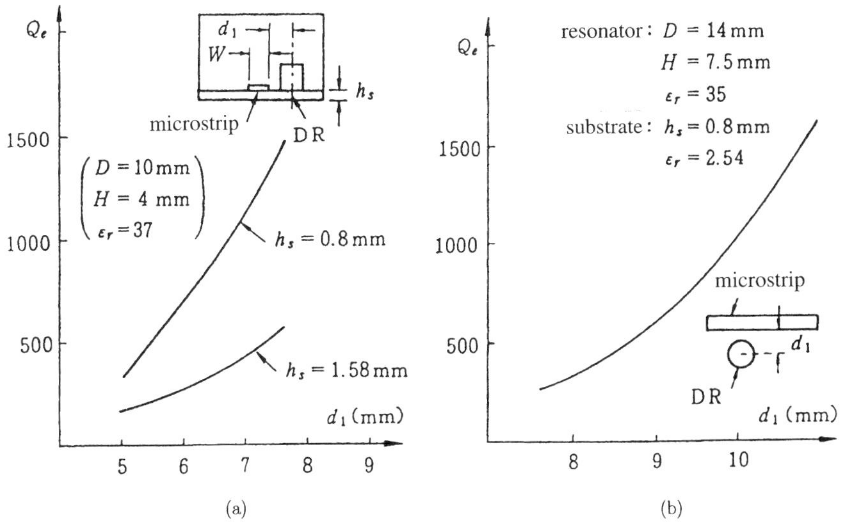

Figure 25. Relation between Q_e and d and h_s in the microstrip line coupled with the DR. (a) Relation between Q_e and d_1 obtained by the H/I method; (b) relation between Q_e and d_1 obtained by the flux method. (From Ref. 8.)

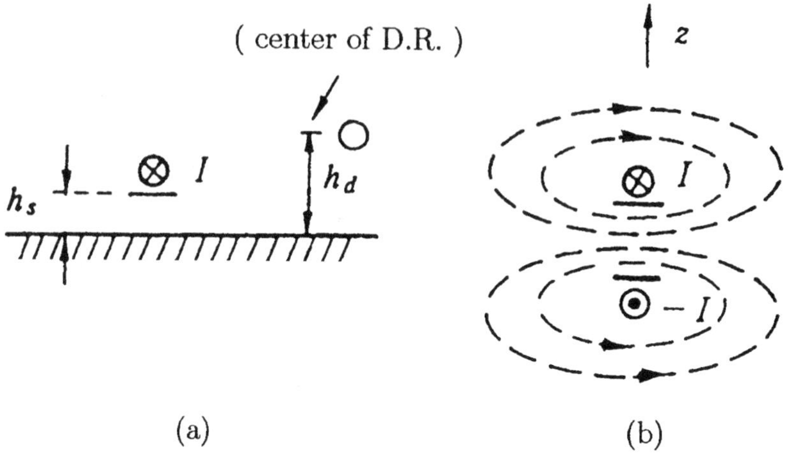

Figure 26. The magnetic field caused by current of the microstrip line. (a) The current I of the microstrip line; (b) the equivalent construction of (a) considering the image.

which results in decreasing the cancellation of I by the image current $-I$. Therefore, the magnetic fields coupled to the DR decrease and it decreases Q_e. Because $E \propto I \sin(\pi h_s/b)$, H_z is proportional to E, and increases for the larger h_s in $h_s \ll b$. Therefore, because

$$\left(\frac{H_z}{I}\right)^2 \propto \sin^2\left(\frac{\pi h_s}{h}\right)$$

Q_e decreases for increasing h_s.

(c) Q_e increases for increasing distance d between the DR and the microstrip line, because the magnetic fields caused by the current element on the microstrip decreases by d^{-2} by the Biot–Savar Law. In the case of Fig. 27, the evanescent TE mode between two parallel metal plates decreases as $e^{-\alpha d}$ (α is the attention constant of the evanescent wave), then $(H_z/I)^2 \propto e^{-2\alpha d}$. Q_e, therefore, increases proportional to $e^{2\alpha d}$.

Coupling Between Two Resonators Where two cavities with the same resonator angular frequency are electricomagnetically coupled, the two angular frequencies, ω_+ and ω_-, take place. In this case, we define the coupling coefficient k by Eq. (32), when $k \ll 1$:

$$k = \frac{\omega_+ - \omega_-}{\omega_0}, \quad \omega_+ > \omega_- \tag{32}$$

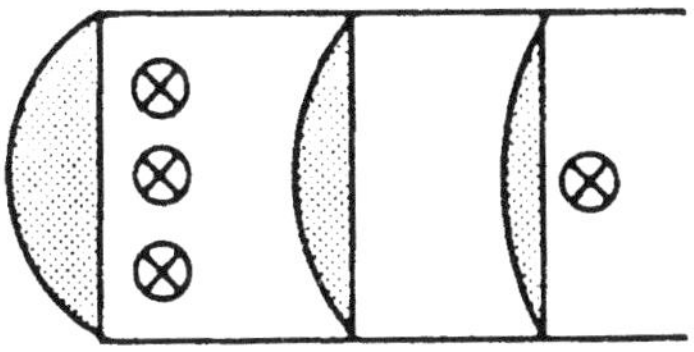

Figure 27. Evanescent TE wave decreasing toward right-hand side between two parallel metal plates.

Coupling by the Lumped Element Reactance Between Two Resonators with the Same Resonant Angular Frequency In the case of a symmetrical circuit with two resonators as shown in Fig. 28, it is easy to obtain the coupling coefficient. For example, the circuit in Fig. 28a has two resonant angular frequencies, ω_e and ω_o. The resonant currents for ω_e and ω_o, flow as shown in Figs. 29b and 29c, respectively.

This fact can be easily understood from the symmetrical two-port network obtained by cutting the symmetrical two ports of Fig. 28a. As in Eq. (16f)

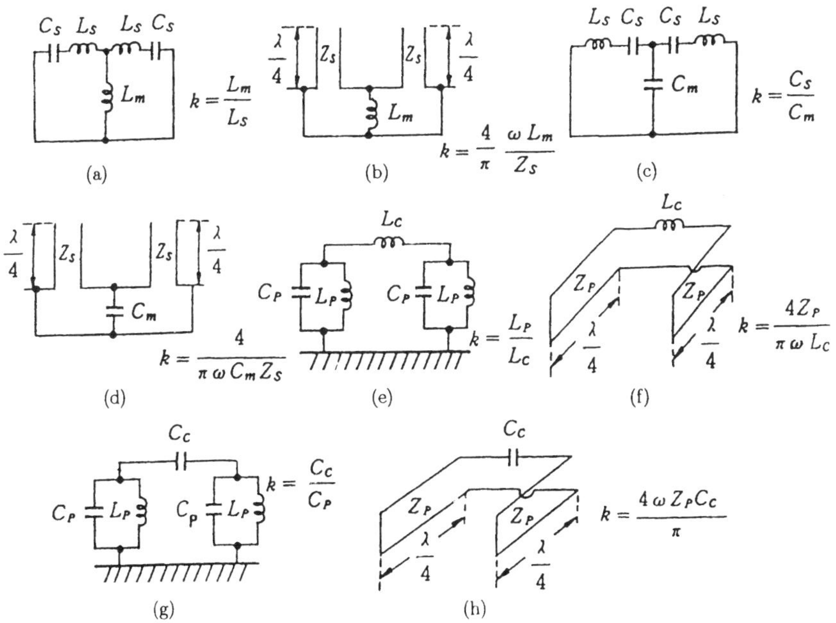

Figure 28. Several kinds of coupling by the lumped element reactance between two resonators.

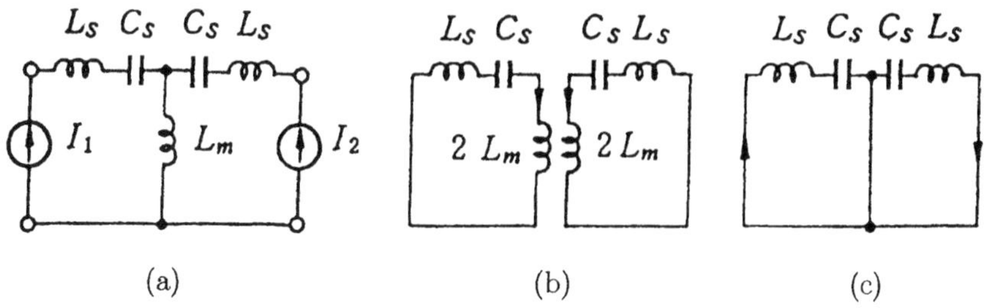

Figure 29. Two-port network obtained by cutting symmetrical two ports of Fig. 28a, and the two resonant currents. (a) The circuit of Fig. 28a is changed to a two-port network by cutting; (b) equivalent network of $I_1 = I_2$; (c) equivalent network of $I_1 = -I_2$.

of Chapter 2, the symmetrical two-port network has even mode and odd mode excitations corresponding their eigenimpedances, which are zero in this case because of the resonant condition. Therefore, the resonant angular frequency of the even and odd modes can be obtained from Eq. (33a), and the coupling coefficient k is obtained from Eq. (33b):

$$\omega_e = \frac{1}{\sqrt{(L_s + 2L_m)C_s}} = \omega_-, \qquad \omega_0 = \frac{1}{\sqrt{L_s C_s}} = \omega_+ \tag{33a}$$

$$k = 1 - \sqrt{\frac{1}{1 + 2L_m/L_s}} \simeq \frac{L_m}{L_s} \quad \text{(in the case } 2L_m \ll L_s\text{)} \tag{33b}$$

We can obtain values of k for each circuit of Fig. 28, and their values were shown in each part of the figure.

However, the LC series and the LC parallel resonator circuits as shown in Figs. 30a and 30c, respectively, can be replaced by the $\lambda_g/4$ line with the

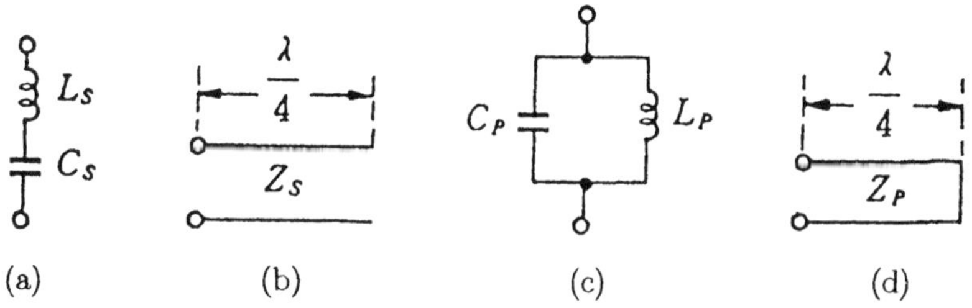

Figure 30. Narrow-band equivalent circuits of the series resonant circuit (a) and the parallel resonant circuit (c) by the $\lambda_g/4$ distributed line.

open end and the short end as shown in Fig. 30b and 30d, respectively. In this case, C_s, L_s, C_p, and L_p are calculated with respect to Z_s and Z_p, as shown in Eq. (34), which were described in Eqs. (7) and (8e) of Chapter 3.

$$C_p = \frac{\pi}{4\omega_r Z_p}, \quad L_p = \frac{4Z_p}{\pi\omega_r}$$

$$C_s = \frac{4}{\pi\omega_r Z_s}, \quad L_s = \frac{\pi Z_s}{4\omega_r} \tag{34}$$

Coupling by a $\lambda_g/4$ Distributed Constant Line We consider the circuit shown in Fig. 31. In the case of the even and the odd modes, the center point of the $\lambda_g/4$ line is open and short; therefore, the admittances jY_c and $-jY_c$ are connected to the parallel resonant circuit with the resonant angular frequency ω_r, respectively.

Showing the even- and odd-mode resonant angular frequencies by ω_e and ω_o, respectively, we get

$$\omega_e \simeq \frac{1}{\sqrt{L_p\,(C_p + Y_c/\omega_r)}} \tag{35a}$$

$$\omega_o \simeq \frac{1}{\sqrt{L_p\,(C_p - Y_c/\omega_r)}} \quad \left(Y_c = \frac{1}{Z_c}\right) \tag{35b}$$

Therefore, the coupling coefficient k can be obtained from Eq. (32) as follows:

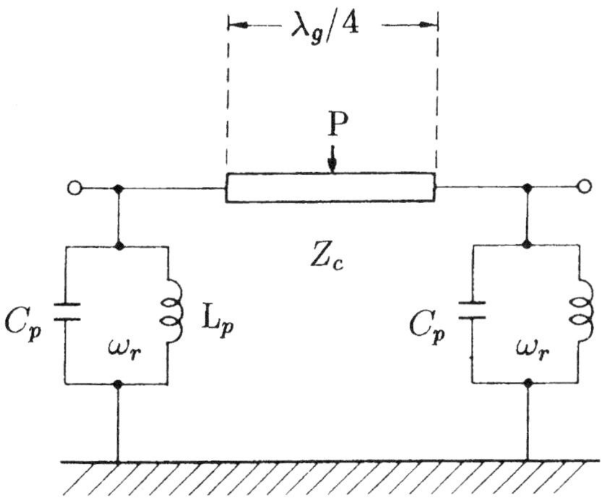

Figure 31. Coupling by a $\lambda_g/4$ distributed constant line with characteristic impedance Z_c.

$$k \simeq \frac{1}{\sqrt{1 - Y_c/\omega_r C_p}} - \frac{1}{\sqrt{1 + Y_c/\omega_r C_p}} \tag{36a}$$

In the case for $Y_c \ll \omega_r C_p$, we get

$$k \simeq \frac{Y_c}{\omega_r C_p} \tag{36b}$$

As mentioned above, the resonators with the same resonant angular frequencies are split by the coupling of the $\lambda_g/4$ line as ω_e and ω_o, where

$$\omega_e < \omega_r < \omega_o \tag{36c}$$

Interdigital Coupling Between Two TEM Resonators of $\lambda_g/4$ Lines We will consider the symmetrical interdigital $\lambda_g/4$ TEM resonators as shown in Figs. 32a and 32b. The equivalent network is shown in Fig. 32c, which is already described in Fig. 52 of Chapter 3. When ports 1 and 2 are excited by the antiphase, port 1 and 2′ of Fig. 32c are excited by the in-phase, and the center of the distributed lines, C and C', are opened.

On the other hand, when ports 1 and 2 are excited by the in-phase, ports 1 and 2′ are excited by the antiphase, and C and C' are shortened.

Therefore, the resonant angular frequencies corresponding to the even- and the odd-mode excitations, ω_e and ω_o, are obtained by the angular frequencies to satisfy the infinite impedance viewed from port 1 of Figs. 33a and 33b, respectively. By this method, we get the following results:

$$\begin{aligned} \omega_e &= \omega_r \frac{1}{\sqrt{1 + \Delta C/C}} \simeq \omega_r \left(1 - \frac{\Delta C}{2C}\right), \qquad \Delta C = \frac{Y_o - Y_e}{2\omega_r} \\ \omega_o &= \omega_r \frac{1}{\sqrt{1 - \Delta C/C}} \simeq \omega_r \left(1 + \frac{\Delta C}{2C}\right), \qquad C = \frac{\pi Y_e}{4\omega_r} \end{aligned} \tag{37}$$

Therefore, the coupling coefficient k are calculated from

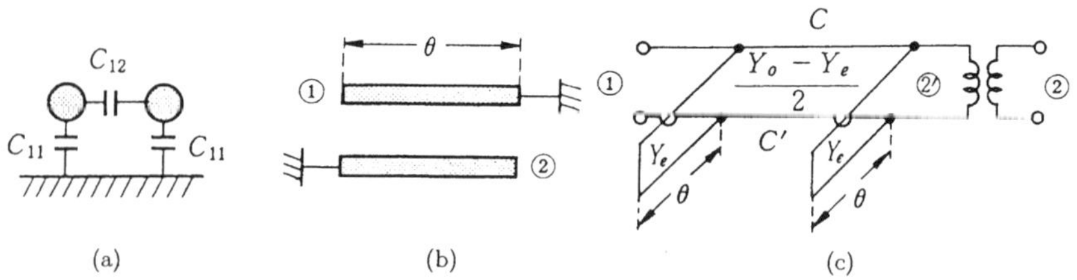

Figure 32. The construction of the symmetrical interdigital $\lambda_g/4$ TEM resonators and the equivalent network. (a) Sectional view; (b) top view; (c) equivalent network.

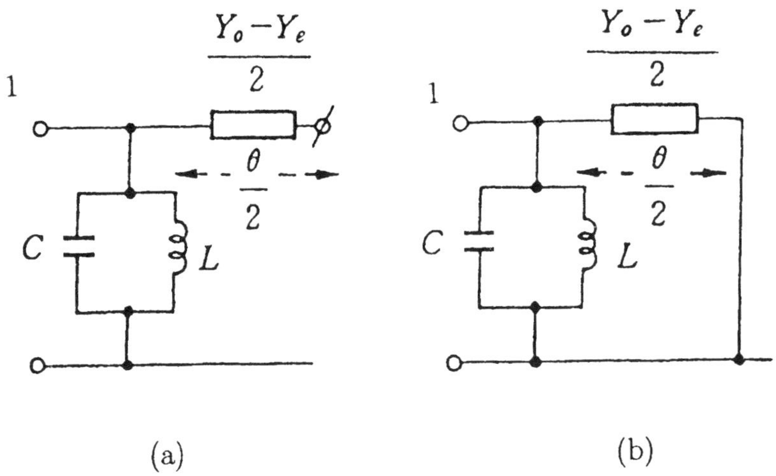

Figure 33. Equivalent networks when ports 1 and 2′ are excited by even (a) and odd (b) modes.

$$k = \frac{2}{\pi}\frac{Y_o - Y_e}{Y_e} = \frac{4}{\pi}\frac{C_{12}}{C_{11}} \tag{38}$$

Symmetrical Coupled Quasi-TEM $\lambda_g/4$ Resonators in an Inhomogeneous Medium Two parallel lines with open ends on the same side and short ends on the other side, placed in the homogeneous medium as shown in Fig. 34, do not couple each other. It is because the resonant frequencies of the even and the odd modes have the same values and results in $k = 0$.

On the other hand, when we consider the parallel lines in the inhomogeneous medium as shown in Fig. 35, the phase velocity of the quasi-TEM

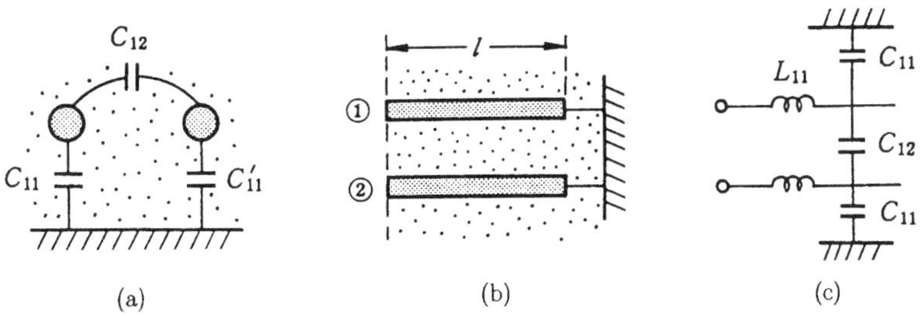

Figure 34. Symmetrical coupled lines in homogeneous medium. (a) Sectional view; (b) side view; (c) equivalent network.

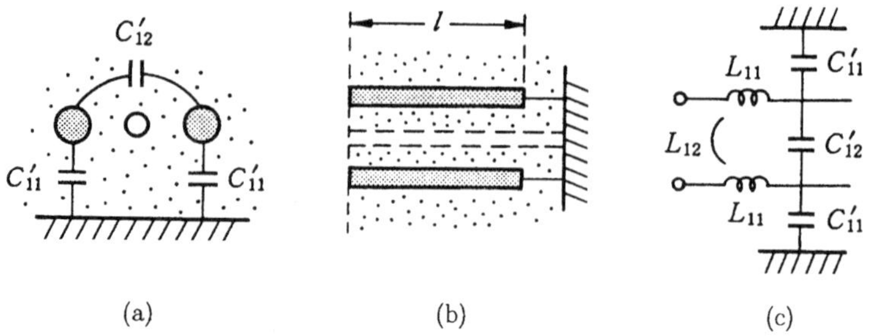

Figure 35. Symmetrical coupled lines in inhomogeneous medium. (a) Sectional view; (b) side view; (c) equivalent network.

mode for the odd mode is faster than that for the even mode. The reason is that the electrical fields for the odd mode cross the air portion of Fig. 35, which results in increasing the phase velocity. On the other hand, the electrical fields for the even mode do not cross the air portion, which results in keeping the same phase velocity as that in the case of homogeneous medium. Therefore, the resonant angular frequencies for the even mode and the odd mode, ω_e and ω_o, respectively, are $\omega_e < \omega_o$. The physical meaning can be explained in a quantitative way as follows.

When we denote the capacities per unit meter for the even and odd modes in the construction of Fig. 34, by C_e and C_o, and the inductances for their modes by L_e and L_o, respectively, we have

$$\frac{1}{v_p^2} = L_e C_e = L_o C_o, \qquad C_e = C_{11}, \quad C_o = C_{11} + 2C_{12} \tag{39}$$

In the case of Fig. 35, denoting the capacities for the even and the odd modes by C_e' and C_o', respectively, we get

$$L_e C_e' \neq L_o C_o' \tag{40}$$

based on the unchanging inductances between Figs. 34 and 35 in the case of the quasi-TEM mode.

Calculating C_e' and C_o' by

$$\begin{aligned} C_e' &= C_e - \Delta C_e \\ C_o' &= C_o - \Delta C_o \end{aligned} \tag{41a}$$

the phase velocities for the even and odd modes, v_e and v_o, respectively, can be obtained from

$$\frac{1}{v_e} = \sqrt{L_e(C_e - \Delta C_e)} \simeq \frac{1}{v_p}\left(1 - \frac{\Delta C_e}{2C_e}\right)$$
$$\frac{1}{v_o} = \sqrt{L_o(C_o - \Delta C_o)} \simeq \frac{1}{v_p}\left(1 - \frac{\Delta C_o}{2C_o}\right) \tag{41b}$$

where v_p is the phase velocity in the case of Fig. 65.

Denoting the length of the distributed constant line by l, the resonant frequencies of the even and the odd modes, f_e and f_o, respectively, can be expressed by

$$\frac{v_e}{4l} = f_e, \qquad \frac{v_o}{4l} = f_o \tag{41c}$$

Substituting Eq. (41c) into (41b), f_e and f_o are obtained, which results in obtaining values k by

$$k = \frac{\Delta C_o}{2C_o} - \frac{\Delta C_e}{2C_e} \tag{41d}$$

Because the values of C_e are not substantially affected by the air region, as mentioned earlier, Eq. (41d) becomes

$$k \simeq \frac{\Delta C_o}{2C_o} \tag{41e}$$

Substituting Eq. (39 into Eq. (41d), we get

$$k \simeq \frac{1}{2}\left(\frac{\Delta C_{11} + 2\Delta C_{12}}{C_{11} + 2C_{12}} - \frac{\Delta C_{11}}{C_{11}}\right) \simeq \frac{\Delta C_{12}}{C_{11} + 2C_{12}} \tag{42}$$

where the right-hand equation shows the values of k based on the approximation of $\Delta C_e \ll \Delta C_o$ and only in consideration of the effect ΔC_{12}.

The values of C_{ij} $(i, j = 1, 2)$ can obtained by the method of the resistive sheet, which will be shown in Appendix 4 and is provided in Ref. 10.

Partial Conductive Connection of Two Distributed Lines As shown in Figs. 36a and 36b, when the two distributed constant lines are partially connected, the symmetrical plane S is the magnetic wall for the even mode and the electrical wall for the odd mode. Therefore, the resonant angular frequencies for the even and the odd modes, ω_e and ω_o, respectively, are related by

$$\omega_e < \omega_o$$

The values of ω_e and ω_o in Fig. 36a can be obtained from Figs. 36b and 36c, respectively. In the same way, ω_e and ω_o for Fig. 36d can be obtained from Figs. 36e and 36f, respectively. In the case of Fig. 36e, we can consider

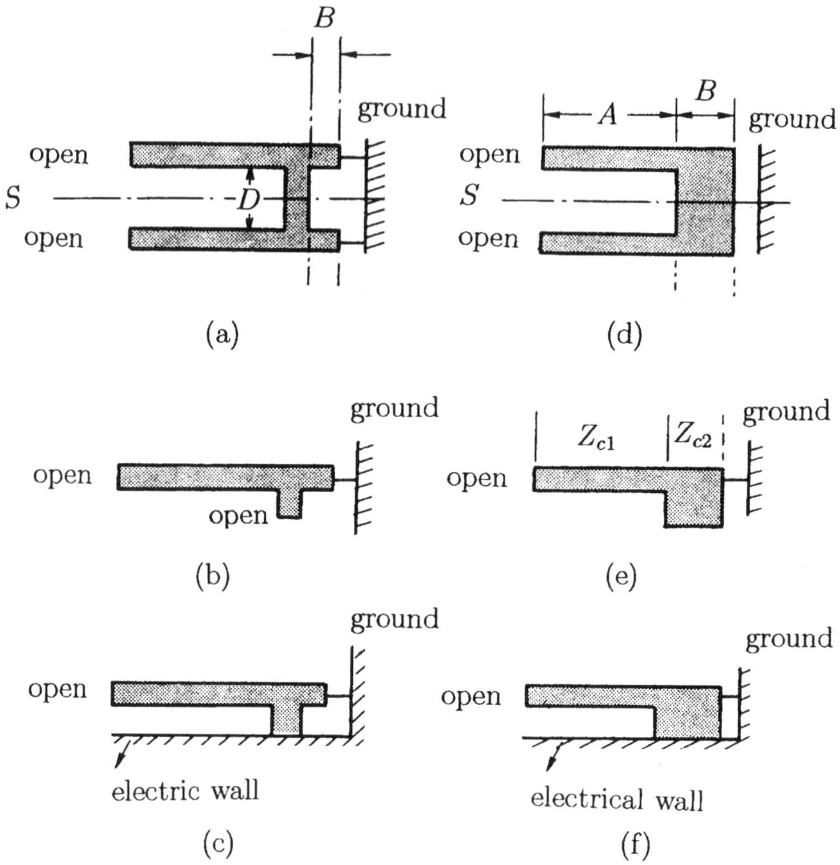

Figure 36. Construction of two distributed lines partially connected and their equivalent construction of the even and odd excitations. (a) Top view of construction; (b) even mode; (c) odd mode; (d) top view of construction; (e) even mode; (f) odd mode.

the distributed constant linc as thc cascadc connection of lines with characteristic impedances of Z_{c1} and Z_{c2} corresponding to regions A and B, together with the equivalent inductance generated by the discontinuity between regions A and B.

Coupling Between Neighboring Dielectric Resonators The two-stage BPFs are made by the cutoff waveguide, including two magnetically coupled DRs, as shown in Figs. 37a and 37b, or by two microstrip lines coupled

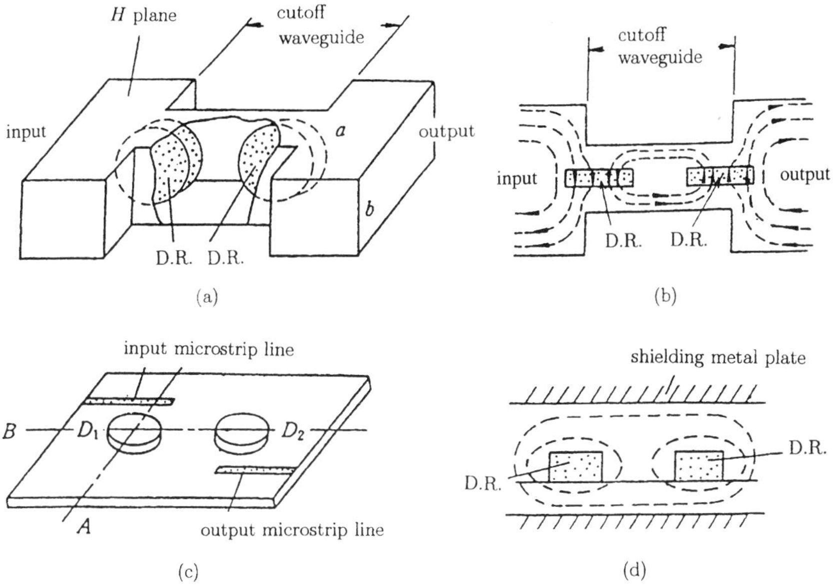

Figure 37. Waveguide and microstrip line BPF with two DRs of the $TE^{\circ}_{01\delta}$ mode. (a) Two-stage BPF constructed by the cutoff waveguide with two DRs; (b) top view of (a), parallel to the *H* plane (dots = *H* fields); (c) microstrip line filter with two DRs; (d) magnetic coupling in the plane of the broken line B in (c) (dashes = *H* fields).

through two DRs placed on the same substrate, as shown in Figs. 37c and 37d.

The coupling coefficient between neighboring resonators are obtained under the equivalent network, where the neighboring DRs are replaced by small loops with currents I [9]. As another method, the even- and odd-mode resonant frequencies are obtained by the perturbation method considering the effect of the neighboring DR. In this book, the principle of the former method is introduced, together with a practical example.

The magnetic dipole moment $M = IS$ is defined corresponding to the given electromagnetic field, as shown on the left-hand side of Fig. 38a, which induces the magnetic flux to the loop of the right-hand side. We denote the self inductance and the mutual inductance of the loops by L_r or L_m; the time average of the magnetic energy of the loop $\tilde{W}_m$ is calculated from

$$\tilde{W}_m = \frac{L_r}{2} I^2 \tag{43a}$$

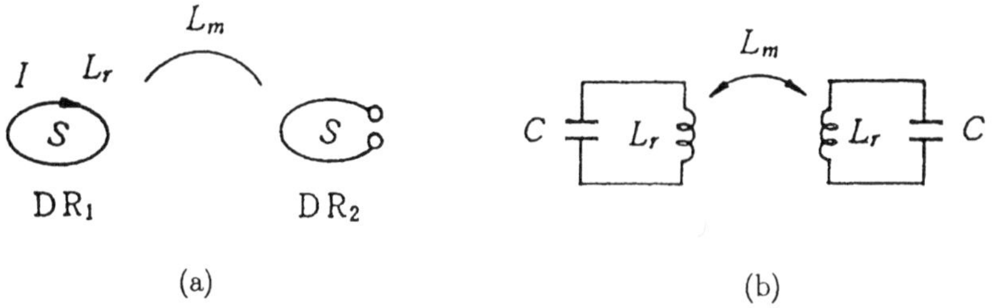

Figure 38. Small loops corresponding to two DRs (a) and their equivalent network (b).

which means that the total reactive energies of the resonator, $\tilde{W}_t$, of $L_r I^2$. L_m can be obtained by

$$L_m = \frac{\mu_0 H_2 S}{I} = \frac{\mu_0 H_2 M}{I^2} \tag{43b}$$

Because the coupling coefficient $k = L_m/L_r$, it is obtained from Eqs. (43a) and (43b) as follows:

$$k = \frac{\mu_0 M}{2\tilde{W}_m}, \qquad H_2 = \frac{\mu_0 M}{\tilde{W}_t} H_2 \tag{43c}$$

The values of H_2 in Eq. (43c), however, is obtained from the evanescent mode in the cutoff waveguide excited by the magnetic dipole moment M of DR_1, as shown in Fig. 38a. If we denote the pth normal mode of the waveguide including DRs by h_p^+ (+ shows the wave travels forward to DR_2 from DR_1), H_2 can be obtained from Eq. (43d) [11]:

$$H_z = \left[\sum a_p h_p^+ \right]_x \tag{43d}$$

where $[\ ]_x$ is the x (the axis of the loop) component. The values of a_p are calculated from Eq. (43e) [11]:

$$a_p = j \frac{\omega \mu_0}{2} h_p^- M \tag{43e}$$

where M is as large as M and is in the direction of the axis of the disk of $TE^{\circ}_{01\delta}$.

h_p^- of Eq. (43e) is the pth normalized mode traveling toward the opposite direction to h_p^+. Because $p = 1$ is the dominant mode, the attenuation of the evanescent mode is lower than that for $p > 1$. Therefore, when we only consider the case of $p = 1$, we get the approximate values of k from Eqs. (43c)–(43e) as follows (see Note 2):

$$k = \frac{\mu_0 M^2}{\tilde{W}_t} \frac{\alpha_1 e^{-\alpha_1 s}}{ab} = f\frac{\alpha e^{-\alpha_1 s}}{ab} \simeq \frac{0.927 D^4 L \varepsilon_r}{\lambda_0^2} \frac{\alpha_1 e^{-\alpha_1 s}}{ab} \tag{44}$$

where F is obtained from Eq. (30b).

Practical example. We will obtain the values of k, when two DRs with parameters

$$D = 9.825 \text{ mm}, \qquad L = 4 \text{ mm}, \qquad \varepsilon_r = 97.6, \qquad f_r = 3.388 \text{ GHz}$$

are installed at a distance of S (mm) in the cutoff waveguide with $a = 18.75$ mm and $b = 18.75$ mm.

From

$$\lambda_c = 2a = 37.5 \text{ mm}, f_c = 8 \text{ GHz}, \qquad \lambda_0 = \frac{3 \times 10^8 \times 10^3}{3.388 \times 10^9} = 88.548 \text{ mm}$$

we get

$$\alpha_1 = \frac{2\pi}{\lambda_0} \sqrt{\left(\frac{f_c}{f}\right)^2 - 1} = 0.1518$$

from Eq. (12a), where λ is replaced by λ_0. Therefore, we get

$$e^{-\alpha_1^s} = 0.0225 \quad \text{for } S = 25 \text{ mm}$$
$$e^{-\alpha_1^s} = 0.0037 \quad \text{for } S = 37.5 \text{ mm}$$

and

$$k = \frac{0.927 \times 9.825^4 \times 4 \times 97.6}{88.548^2} \frac{0.1518 \times e^{-\alpha_1^s}}{18.75 \times 18.75} = 0.1857 e^{-\alpha_1^s}$$

which results in

$$k = 0.00418 \quad \text{for } S = 25 \text{ mm}$$
$$k = 0.000686 \quad \text{for } S = 37.5 \text{ mm}$$

The results agree with those of Ref. 9.

Coupling Between Degenerate Modes by Perturbing the Shape First, we consider the $TE_{101}^{\square}$ cavity as shown in Fig. 39a, which has the modes of Fig. 39b and 39c, with the same resonant frequency [12].

Because the modes are orthogonal and have the same resonant angular frequency, ω_r, we get the modes shown in Figs. 39d and 39e by adding and subtracting the modes of Figs. 39b and 39d, respectively. When we deform the cavity as shown in Fig. 39f, the angular resonant frequencies of Figs.

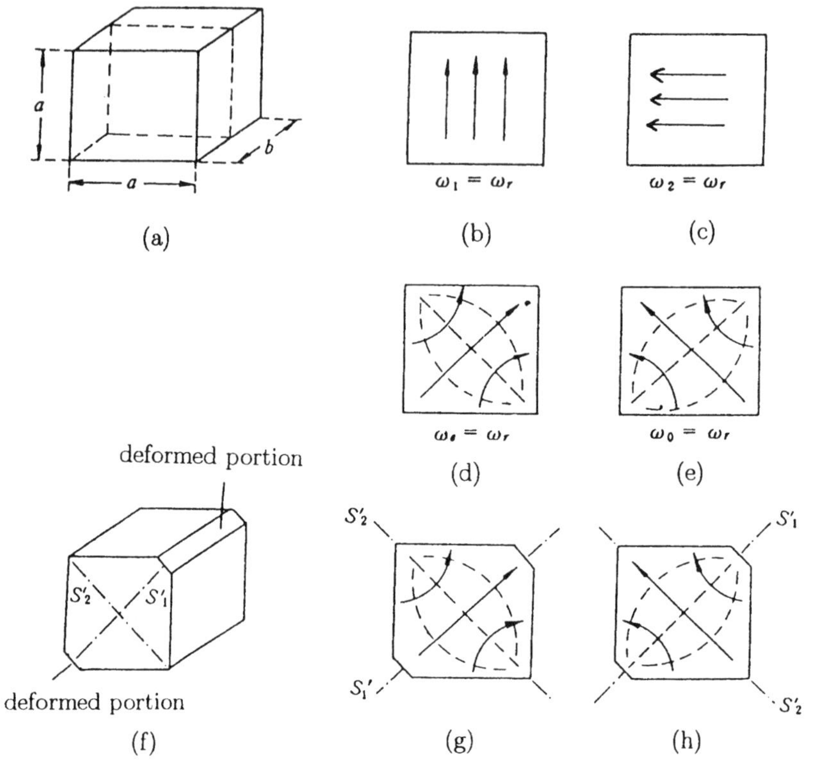

Figure 39. $TE^{\square}_{101}$ dual-mode resonator and its coupling. (a) TE_{101} resonator; (b) one of the modes of (a); (c) mode orthogonal to (b); (d) (b) + (c); (e) (b) − (c); (f) deformed cavity; (g) even mode for S'_1; (h) odd mode for S'_1.

39g and 39h becomes a little lower and higher, respectively, by the perturbation theory described in Appendix 2. This is understood from the physical meaning that the equivalent capacity between both ends of the electric fields increase even if the equivalent inductance does not substantially change and the resonant angular frequency, ω_e, decreases in the case of Fig. 39g. On the other hand, the equivalent inductance decreases even if the equivalent capacity does not substantially change and the resonant angular frequency of Fig. 39h. Therefore, ω_o increases in the case of Fig. 39h.

The notations ω_e and ω_o denote the resonant angular frequencies corresponding the even and odd modes, respectively, for the symmetrical plane S'_1. As mentioned earlier, the coupling coefficient k defined in Eq. (32) are obtained from

$$k = \frac{\omega_o - \omega_e}{\omega_r} \tag{45}$$

If the input $TE_{10}^{\square}$ waveguide is connected in such a way so as to couple the mode of Fig. 39b but not to the mode of Fig. 39c, the mode of Fig. 39b is decomposed to the two modes in Figs. 39d and 39e.

As we have the coupling between the modes in Figs. 39g and 39h through the deformed portions, we get the mode of Fig. 39c, which can excite the output $TE_{10}^{\square}$ waveguide coupled to the mode of Fig. 39c but not to that of Fig. 39b. Therefore, we get the two-stage BPF.

Such a coupling with dual modes can be realizing as follows, which are easily understood by the explanation given earlier.

Figure 40 shows the eight-stage BPF using four resonators and four HE_{11} dual-mode DRs, where the input port excites the first DR and the mode coupled to the other mode. The mode coupled to the next resonator includes the second DR. Such a process is repeated, until the eight-stage BPF is attained [13].

As another example, Fig. 41 shows the coupling between TM_{110} orthogonal dual modes by the slot mode on the surface of the DR at the cross corners [14].

Practical Examples of Construction

Microstrip Filters The top view of the microstrip-line two-stage BPF is shown in Fig. 42. As understood from Fig. 42, the input and the output microstrip lines with length $\lambda_g/4$ are coupled to the $\lambda_g/2$ resonators; then, the region surrounded by PQBA and EFRS are the coupling regions between the filter, to the source and the load. The regions can be designed by the method described in subsection 2.2, BPF and BEF. Because points C and D are open, points B and E are effectively shortened to the ground. Therefore, the region surrounded by BCED is the coupling region of the two resonators, for which the coupling coefficient is obtained by the method described in subsection 2.2, BPF and BEF.

Figure 43 shows the combline type BPF, where the second and third lines from the left are the $\lambda_g/4$ resonators and first and fourth lines are the input and the output lines. One of the coupling mechanisms of the resonators are based on the difference between even- and odd-mode resonant frequencies caused by the inhomogeneous medium, as mentioned at 5.1.2.(2)(c)(ii-iv); the other mechanism is based on the coupling taking place at the open ends caused by the evanescent E waves, as mentioned in 5.1.2.(2)(ii-ii). In the case of the combline made of the triplate, the evanescent E waves make their coupling. The input and the output lines coupled with the resonators by the mechanism of the interdigital line mentioned in 5.1.2.(1)(i-i).

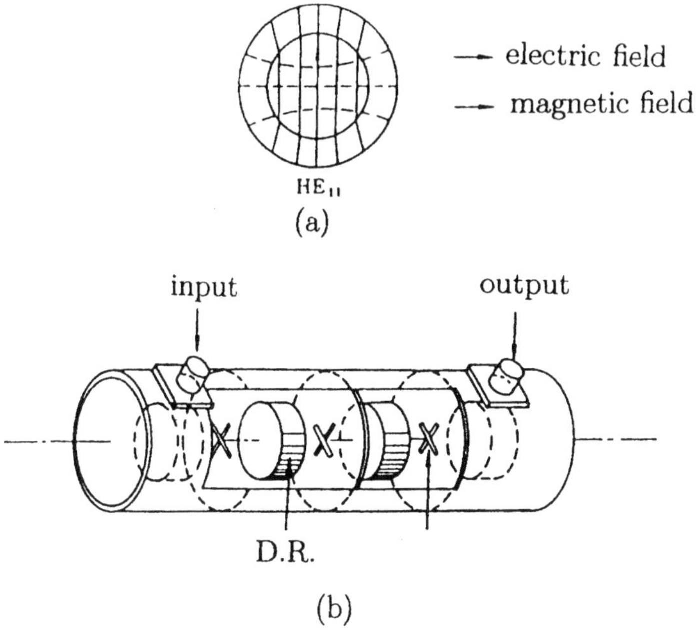

Figure 40. Eight-stage BPF using four DRs with dual modes.

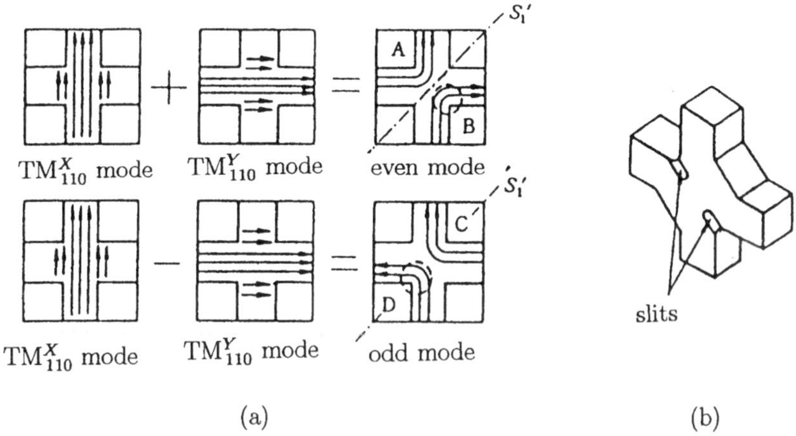

Figure 41. Orthogonal TM_{110} dual modes including the DR and their electric fields. (a) Electrical distribution of orthogonal dual modes and their combination; (b) slits used for the coupling of dual modes.

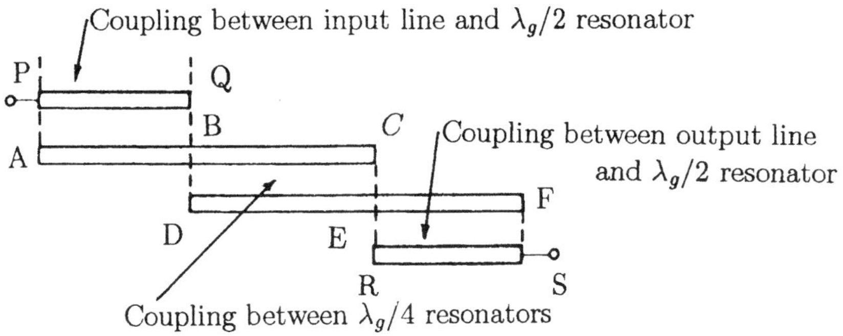

Figure 42. Top view of the microstrip-line two-stage BPF.

Figure 44 shows the construction of the capacitive coupling between resonators, and the inductive coupling between resonators and input (output). The line is wider near the open ends than near the short ends, which serve the length of the microstrip lines to be shorter and the second higher resonant frequency to be higher than three times the fundamental resonant frequency.

When the input and the output ports are coupled through capacities, the construction of Fig. 45 is available.

In Figure 46, we show the two-stage BPF with the construction of the triplate and the input connected to the resonators through inductive coupling. We can, however, the coupling between resonators and the input or the output port can be changed to the capacitive coupling by using the construction of Fig. 45 and also to the interdigital distributed constant line of length $\lambda_g/4$, as shown in Fig. 43.

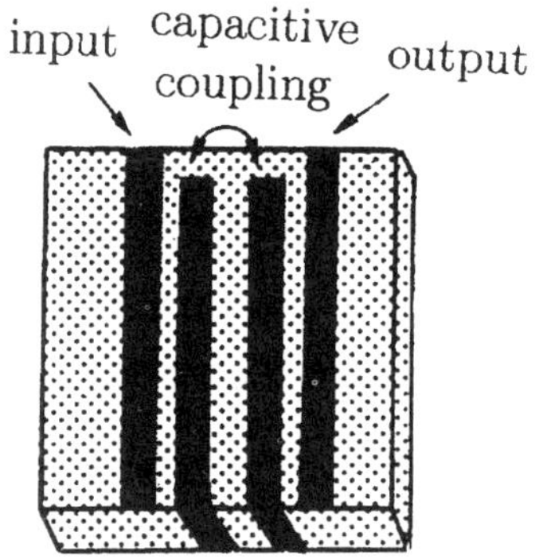

Figure 43. Combline-type BPF.

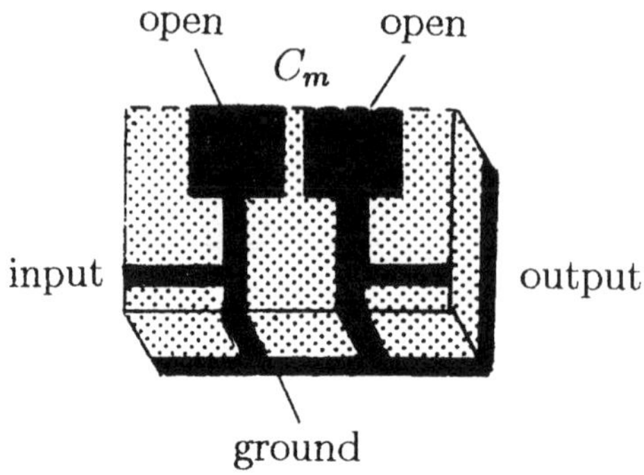

Figure 44. Capacitive coupling between resonators, and inductive coupling between input (output) and resonator.

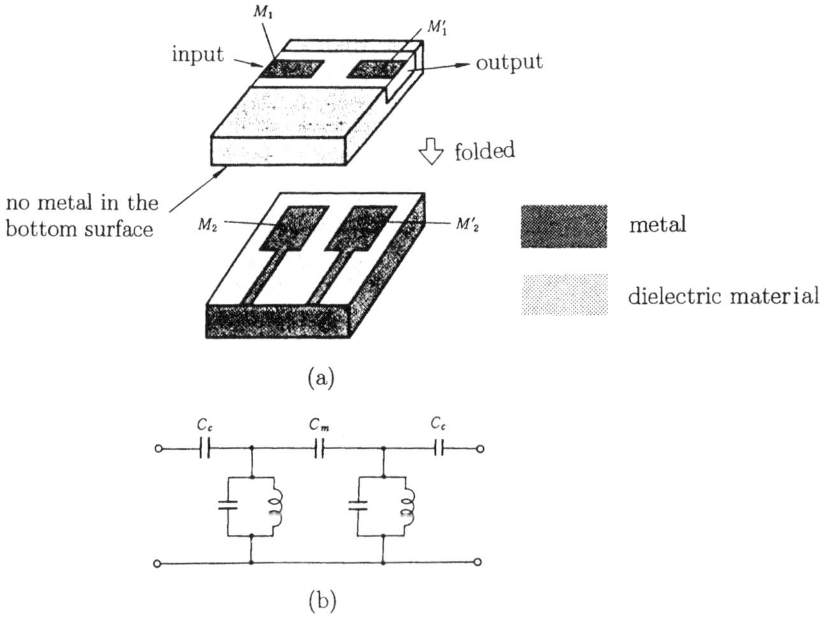

Figure 45. Construction (a) of capacitive coupled to the input and the output and the equivalent network (b), where the resonators are almost mode of the triplate except a part of the portion of the capacities.

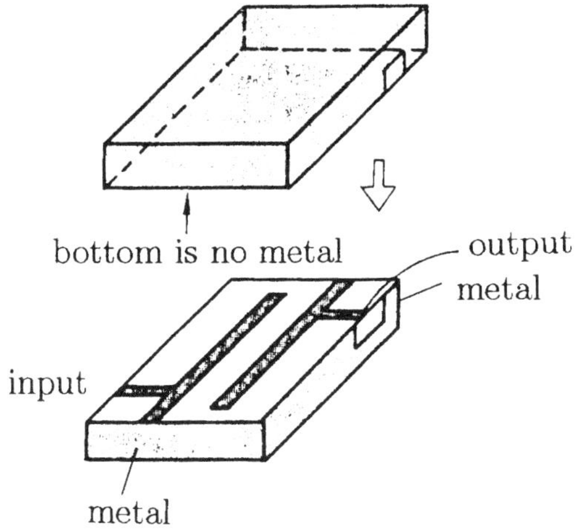

Figure 46. Interdigital two-stage BPF with the triplate construction.

By using the method of coupling between resonators as shown in Fig. 36, we can construct the BPF as shown in Fig. 47, where the coupling to the input and the output from resonators are made by the $\lambda_g/4$ distributed constant lines. The input and the output ports can be also coupled to the resonators by the capacitive coupling such as shown in Fig. 45 or by the inductive coupling as shown in Fig. 46.

There are many kinds of the construction using the combination of the coupling methods mentioned above, where we have to choose the method based on the size, the bandwidth, the attenuation characteristics out of the pass band, and so forth.

Next, the BEFs are constructed by the cascade connection of the construction as shown in Fig. 48 with a distance of $\lambda_g/4$ or $3\lambda_g/4$. The functions of BEF in Figs. 48a and 48b are based on the equivalent networks as shown in Chapter 3.2.3 (6) and (7) where the grounded point of Fig. 56a of Chapter 3 is replaced by the extended line of $\lambda_g/4$ in Fig. 48a. An example of the cascade connection is shown in Fig. 49. When λ_{g1}, λ_{g2}, λ_{g3}, . . . take a different value and d_1, d_2, d_3, . . . take proper values to give the proper external Q, we have the wideband BEF as shown in Fig. 8a by the dashed line. In the case of the BEF with a much wider elimination band, the $\lambda_g/4$ lines with open ends are directly connected to the main microstrip line between the input and the output. The distance between the neighboring $\lambda_g/4$ resonator is also $\lambda_g/4$ or $3\lambda_g/4$. The reason for using the distance $3\lambda_g/4$ is to avoid the interaction between the neighboring resonators.

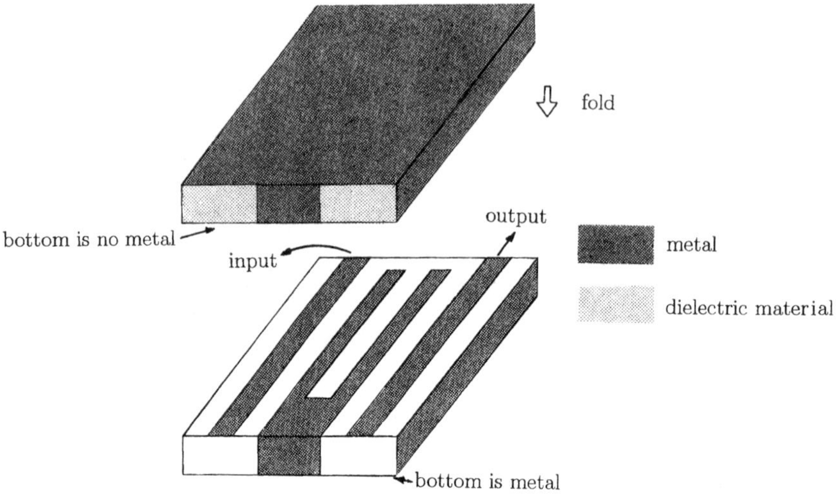

Figure 47. BPF with the coupled distributed constant lines using a partially conductive connection.

TEM or Quasi TEM Lines Made in the Ceramics with Low Loss and High Dielectric Constant The block of the ceramic material with high dielectric constant and low loss as shown in Fig. 50 is metalized on the bottom and the side surfaces.

When we have the holes parallel to the side wall and perpendicular to the bottom wall and the surface of the holes are metalized, TEM modes exist inside the block, which make the resonators for the case when the hole length is $\lambda_g/4$. In this construction the two TEM resonators do not couple

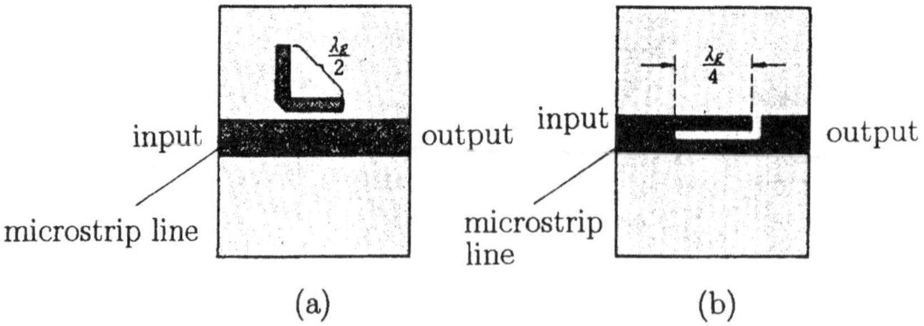

Figure 48. One-stage BEF (trap filter).

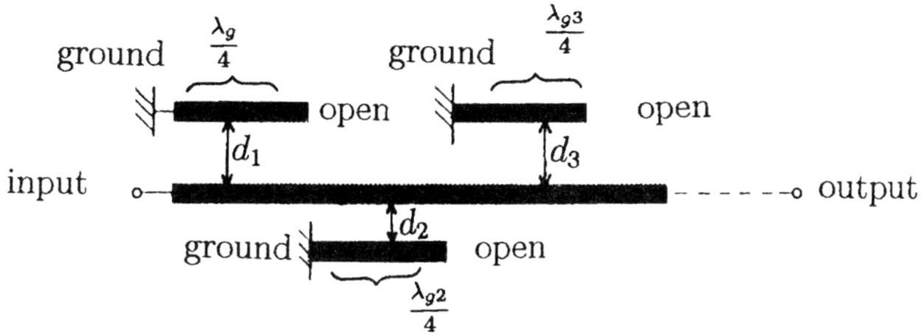

Figure 49. Cascade connection of the trap filter.

each other. Therefore, we need the reactance, such as the condenser or the inductor, to couple the resonators, as shown in Fig. 50.

We, however, make the coupling between the resonators by making the air hole or the air slot without the reactance, as shown in Figs. 51a and 51b. The principle is described as follows. Because the construction has the symmetrical plane, there are even and odd modes. In the case of the even mode, the electric fields are distributed as in Figs. 51c and 51d, which show that there are no electric fields in the portion with the air hole or the air slot. Therefore, the "almost" electric fields exist inside the high dielectric material.

On the other hand, in the case of the odd mode, the electric fields exist inside the air hole or the slot, which results in decreasing the electrical energies of the resonator even if the magnetic energies of the resonator do not change. This results in increasing the odd-mode resonant frequency; that is, we have the relation $\omega_o > \omega_e$. This is equivalent to the fact as shown in

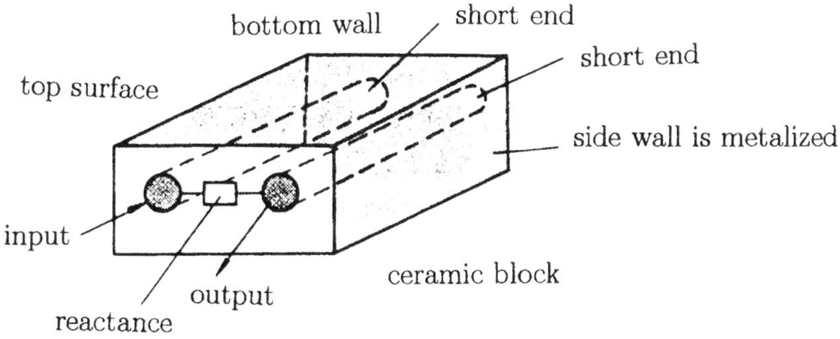

Figure 50. Coupling with a condenser between two TEM $\lambda_g/4$ resonators.

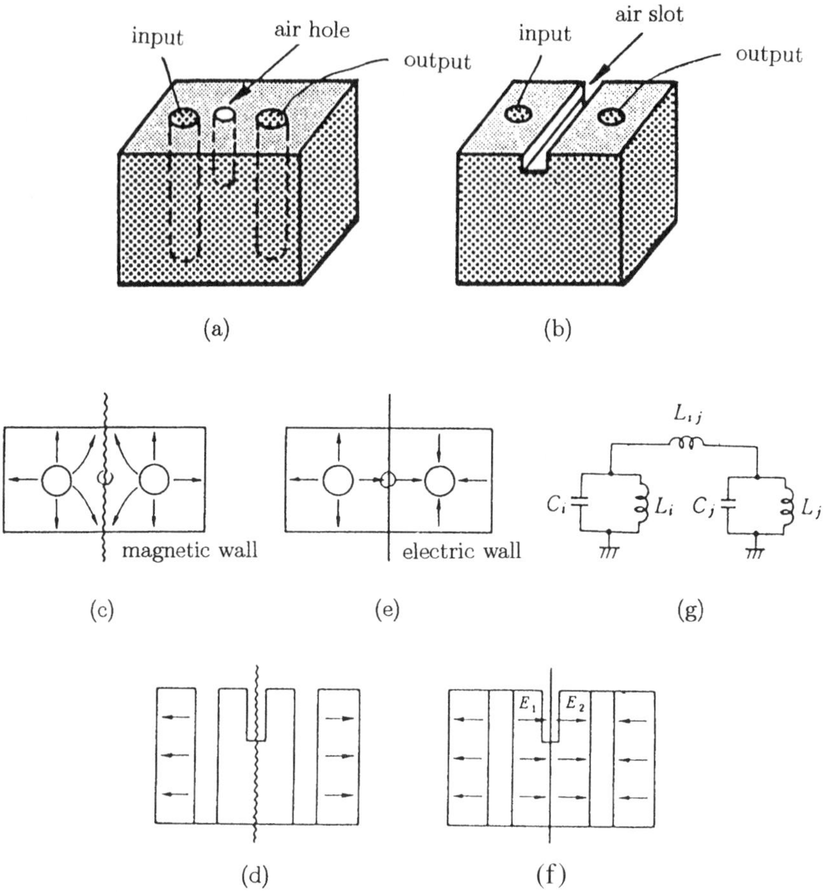

Figure 51. Coupling with an air hole or a slot between two $\lambda_g/4$ resonators. (a) Construction with an air hole on the top surface; (b) construction with an air slot in the top surface; (c) top view of electric fields of even mode; (d) side view of electric field of even mode; (e) electric field of odd mode; (f) side view of electric field of odd mode; (g) equivalent network of (a) and (b).

Fig. 28e, which shows that the reactance in Fig. 50 is inductive. It is realized by considering the fact that the coupling between the neighboring TEM resonators are made by the capacitive and inductive coupling with the same coupling coefficient as in Fig. 52b.

This matter is easily understandable from that the electric coupling coefficient $k_c[=C_p/(C_p + C_m)]$ in Fig. 52a takes the same values of the magnetic

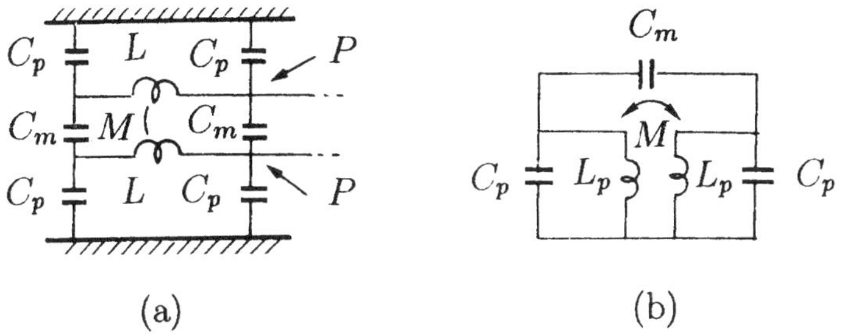

Figure 52. Equivalent network of two parallel $\lambda_g/4$ TEM resonators. (a) Equivalent network of two parallel lines in the homogeneous medium; (b) two $\lambda_g/4$ resonators by connecting p, p' to ground.

coupling coefficient $k_l(=M/L)$ in the parallel distributed lines in the homogeneous medium to make the TEM wave as explained in Chapter 3 by Eq. (39a).

However, when we make the air hole or the air slot, the capacity of C_m in Fig. 52b decreases the values, and the magnetic coupling becomes dominant over the capacitive coupling. This is the reason for the construction of Fig. 51g. Figures 51a and 52b takes the equivalent network of Fig. 51g.

On the other hand, when the inside of the hole and the slot is metalized, as shown in Fig. 53, the electric energies increase because the capacity between the both lines around their open ends. Therefore, the capacitive coupling becomes dominant over the inductive coupling, which results in the equivalent network of Fig. 53e.

Figures 54a and 54b show the construction of the filter which has the coupling mechanism between resonators through the inhomogeneous medium as in Figs. 35a and 35b.

Because the velocity of the odd mode is faster than that of the even mode based on the $\Delta C_e \ll \Delta C_o$ in Eq. (41b), the resonant angular frequencies of the even and the odd modes, ω_e and ω_o, have the relation of $\omega_e < \omega_o$.

Therefore, the equivalent network is as in Fig. 51c. This is also understood from Fig. 54 of Chapter 3 considering $B_2 < 0$, because $R_\pi = -1$, $R_c = 1$, and $\Delta < 0$ (as $k_e < k_0$) for the symmetrical construction in Eq. (56q) of Chapter 3.

Another construction, Fig. 55, is considered as an example using the principle explained in (ii-v) as the partial conductive connection of the neighboring lines.

The construction of Fig. 56 is a practical example for realizing the principle explained in (ii-v). In this case, regions B and D of Fig. 36a correspond to regions B and D of Fig. 56c.

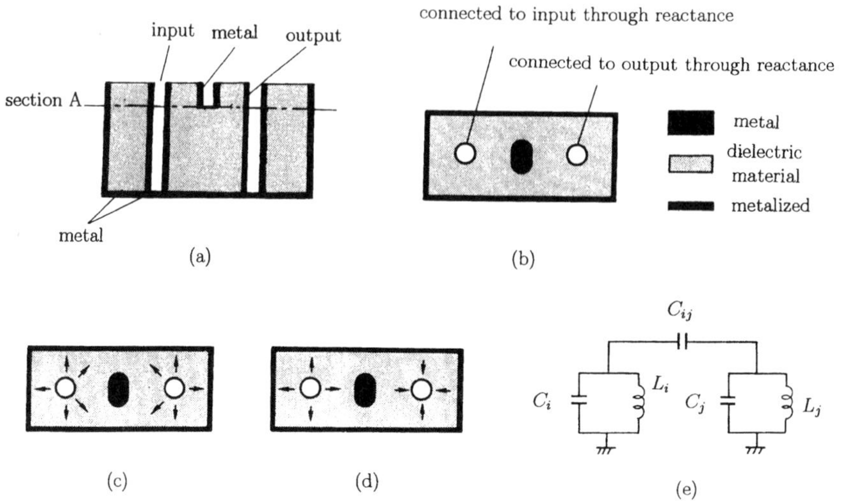

Figure 53. Coupling with a metalized hole between two $\lambda_g/4$ resonators. (a) Side view; (b) top view; (c) electric fields of even mode at section A of (a); (d) electric fields of odd mode of section A of (a); (e) equivalent network of (a).

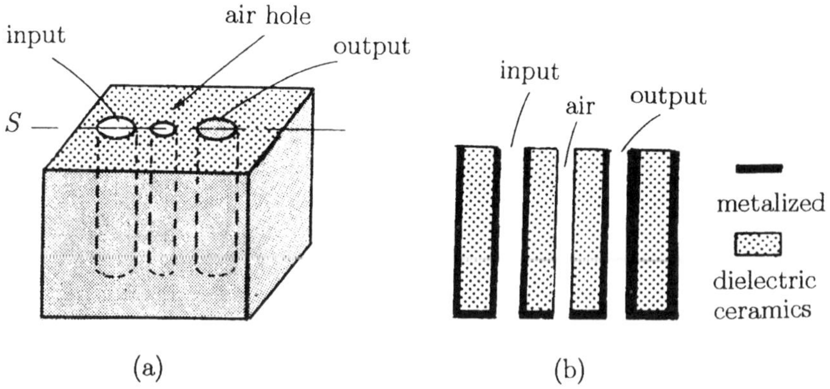

Figure 54. BPF with coupling between the resonators through the inhomogeneous medium. (a) Construction; (b) sectional view of S in (a).

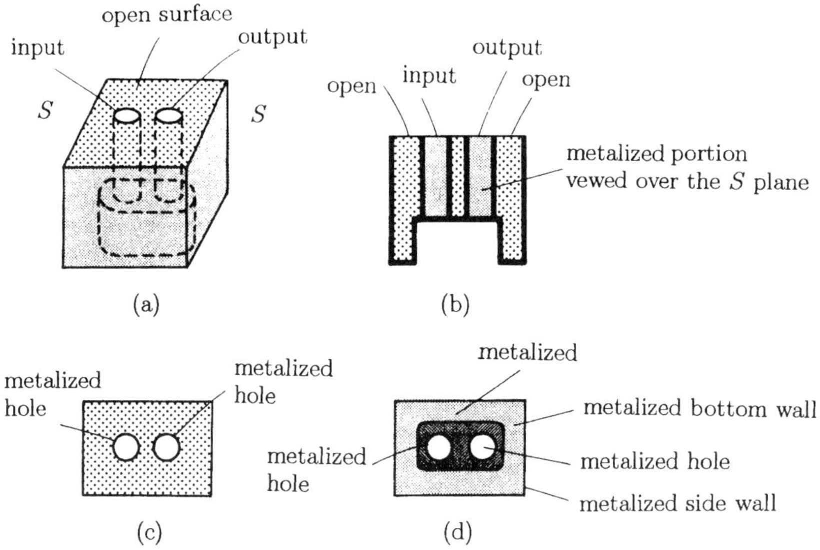

Figure 55. Example of the BPF with two parallel lines coupled by the partial conductive connective. (a) Construction; (b) sectional view of the *S* plane in (a); (c) top view; (d) bottom view.

For the coupling construction of the input and the output shown in Figs. 50, 51, and 53–55 to the source and the load, there are several constructions. One of the methods is the capacitive coupling described in (i-i).

The capacitive coupling is realized by the direct connection of condensers, as shown in Fig. 57a, by inserting the insulated metal post inside the hole as shown in Fig. 57b, or by the microstrip line made on the nearest wall from the open ends of the hole to make $\lambda_g/4$ resonator as shown in Fig. 57c.

Another method to couple the input and the output ports to the resonators is to use the $\lambda_g/4$ interdigital TEM lines as shown in Fig. 14 and using Eqs. (18) and (21a)–(21c).

The construction to realize such a coupling in the ceramic TEM filter is shown in Fig. 58, in which, the input and the output ports are connected to the top surface and they are opened at the bottom surface of the ceramic block. On the other hand, the resonator is open at the top surface and connected to the ground at the bottom surface. Figure 58 shows the one-stage BPF.

The two-stage BPF is constructed as shown in Fig. 59, where regions A are the coupling regions of the input and the output ports to the resonators,

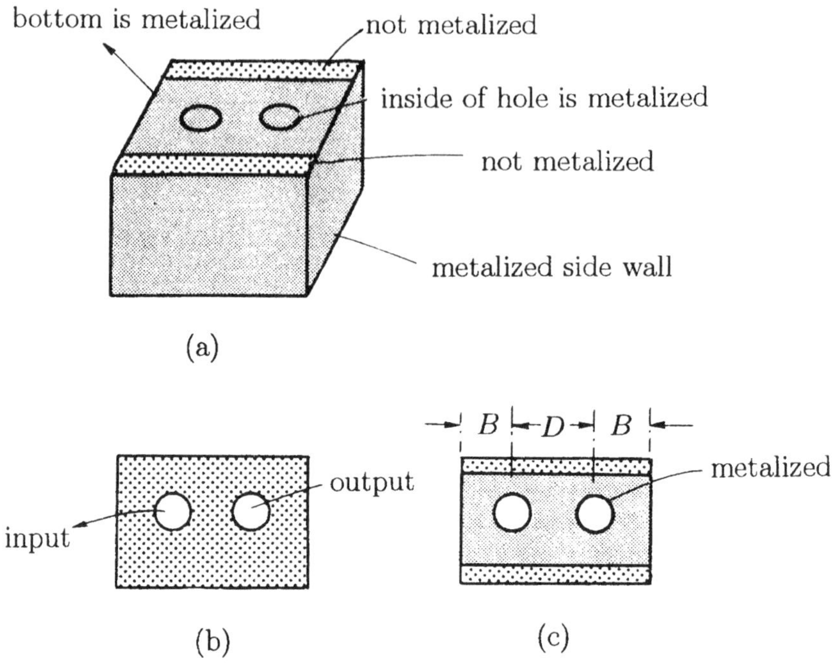

Figure 56. Example of the BPF with parallel lines coupled by the partial conductive connection. (a) Construction; (b) top view; (c) bottom view.

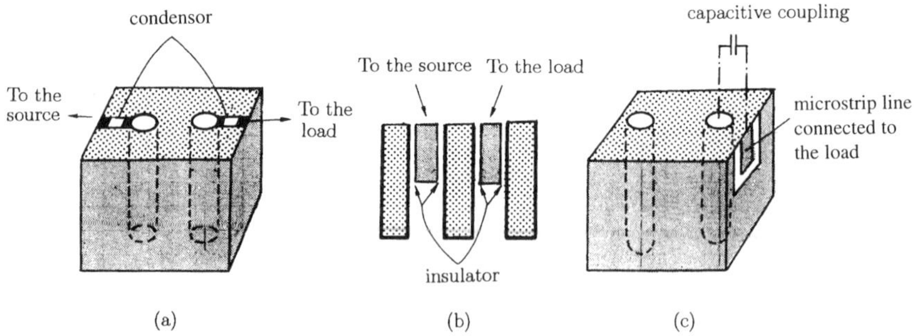

Figure 57. Several kinds of capacitive coupling. (a) Coupling condensers; (b) effective capacitive coupling by insulating metal post; (c) effective capacitive coupling by the microstrip made on the side wall.

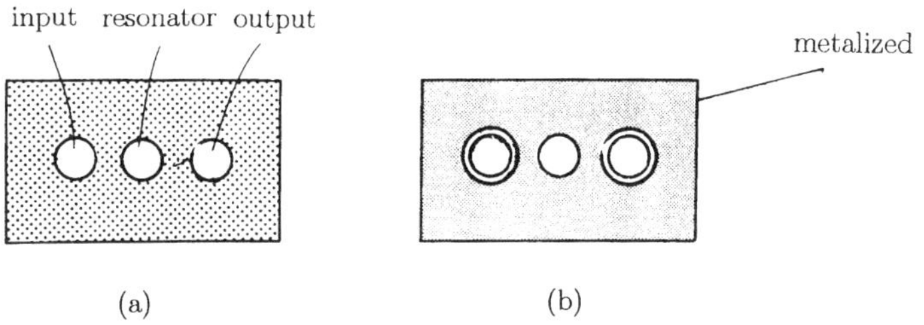

Figure 58. Input and output ports coupled to the resonator by interdigital $\lambda_g/4$ TEM lines. (a) Top view; (b) bottom view.

and region B is the coupling region between two resonators. However, the construction of the coupling mechanism is not shown in Fig. 59.

The proper couplings mentioned earlier (Figs. 50, 51, 53–56) can be used for the coupling of neighboring resonators.

Filter with Dielectric Resonators Figures 37a and 37c are examples of BPF used for the waveguide and the microstriplines, which are used for the small-size filters. On the other hand, for the purpose of the high-power filters, the loss components of the dielectric material generate the power loss, which increases the DR and destroys the DR. The main reasons are

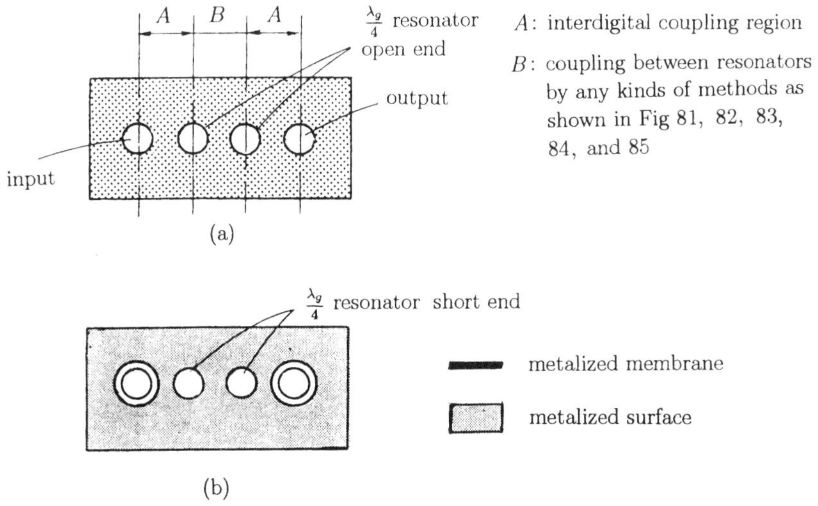

Figure 59. Construction of interdigital coupling between resonators and input or output of BPF. (a) Top view; (b) bottom view.

the poor heat conductivity of the dielectric material and the difficulty of moving the generated heat outside if the DR is installed in air.

To rescue such a bad condition, the contact of DRs with metal wall of shielding cases are preferable, and also using the large DRs of the dielectric material with suitably small dielectric constant, which should be compromized with the size of the filter and the available power.

An example is shown in Fig. 60 [15], where the TM_{110} DRs contact to the shielding boxes. The neighboring shielding boxes are magnetically coupled through the window. The input and the output ports are coupled to the shielding boxes by loops.

Combination of TEM Resonators and the Waveguide Resonator [16] As shown in Fig. 61a, the $TE^{\square}_{101}$ waveguide resonator filled by the high-dielectric material is extended to the direction of the wave propagation with the cutoff waveguide filled with the same material, and the post is installed in the cutoff waveguide parallel to the E plane of the waveguides. The input and the output ports are connected to the posts.

When the signal is applied to the input port, the magnetic flux takes place surrounding the post, and the flux excites the waveguide resonator through magnetic coupling. Because the magnetic flux is also coupled to the output post, the signal appears in the output. The even mode contributing to the signal transmission is shown in Fig. 61b. The filter is installed on the substrate used for the microstrip line, as in Fig. 61e. The measured performance is shown in Fig. 61d.

Dielectric Waveguide Filter [16] By using the high-dielectric material, we not only can decrease the size but also easily create the magnetic wall between the air and the ceramic. Therefore, a $\lambda_g/2$ waveguide resonator with

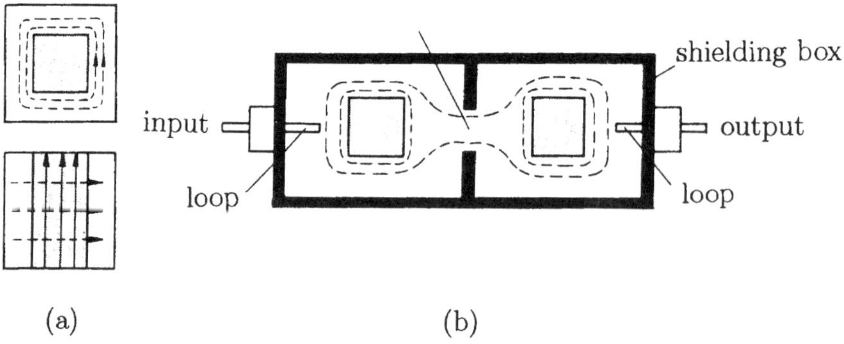

Figure 60. Two-stage BPF with TM_{110} DRs. (a) Electromagnetic field of TM_{110} mode; (b) two-stage BPF using TM_{110} resonators with magnetic coupling.

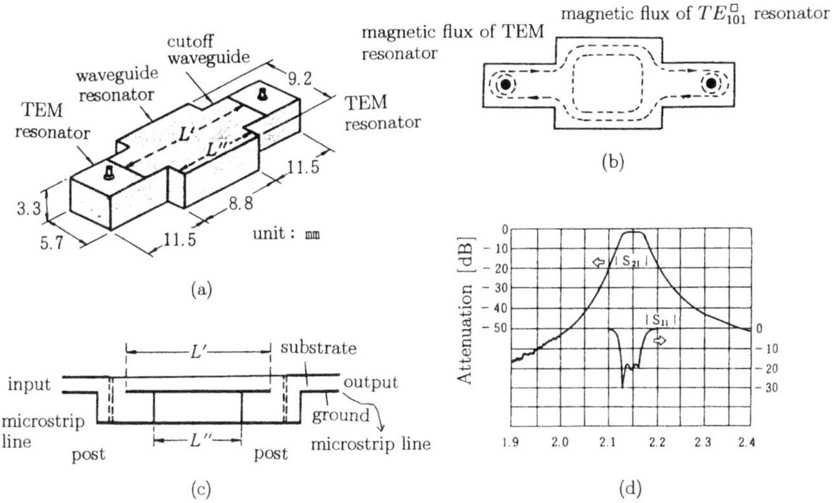

Figure 61. BPF using a $TE_{101}^{\square}$ resonator and a TEM resonator with dielectric material. (a) Construction; (b) even-mode magnetic flux; (c) installation on the substrate used for microstrip lines; (d) frequency (GHz).

short ends can be replaced by a $\lambda_g/4$ waveguide resonator with open and short ends.

The coupling of such a $\lambda_g/4$ waveguide resonator can be done by inductive coupling through the usually used inductive window, or by the capacitive coupling by making the slow without metal. The coupling of the input and output to the $\lambda_g/4$ resonator can be made by making the microstrip on

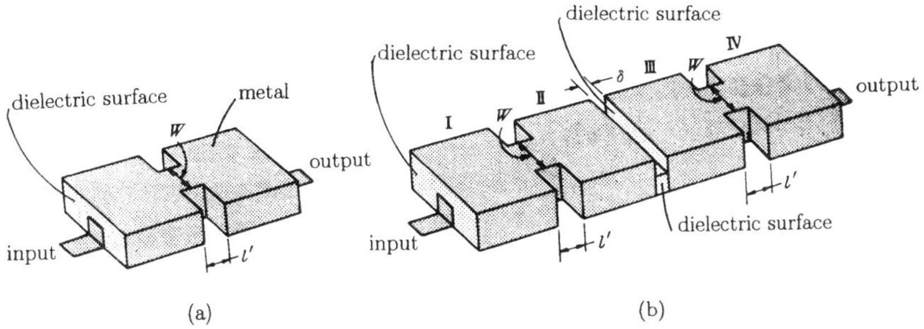

Figure 62. BPF by a cascade connection of a $\lambda_g/4$ waveguide resonator. (a) Two-stage BPF; (b) four-stage BPF.

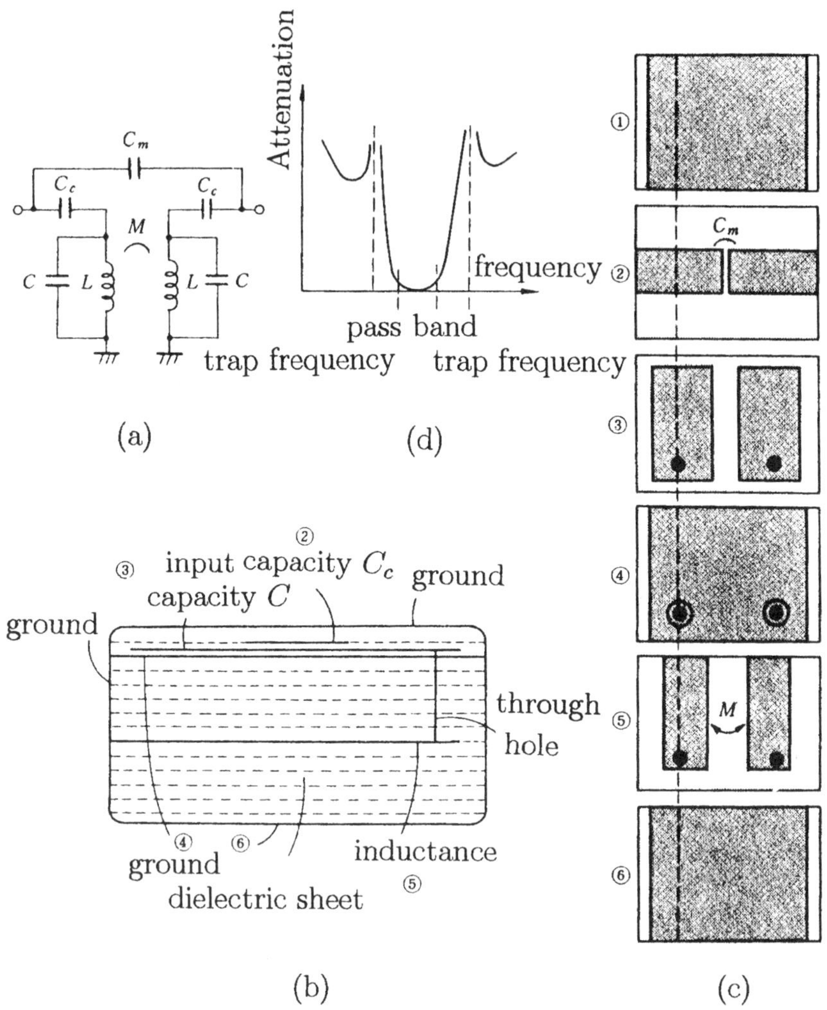

Figure 63. The construction of the laminated *LC* BPF, the equivalent network, and the performance. (a) Equivalent network; (b) sectional view at dotted line of (c); (c) metal pattern at each dielectric sheet; (d) frequency performance.

the open surfaces, which are connected to the microstrip lines on the substrate.

Practical examples are shown in Figs. 62a and 62b. Figure 62a shows the two-stage BPF, where the coupling between resonators is made by the magnetic coupling through a inductive window. Figure 62b shows the four-stage

BPF, where the coupling between resonators are made by two magnetic couplings and one capacitive coupling at the center. In Fig. 62 the coupling between resonators decreases by using a smaller W and a larger δ.

Laminated Ceramic Filter Recently, dielectric ceramics made by low sintering temperatures, such as 850–900°C lower than the melting temperature of metals were developed. By using such a ceramic, we can make a thin ceramic with metal patterns on the surface of the ceramics. After folding several such sheets of thin ceramics, as shown in Fig. 63c, and sintering, we can get the laminated ceramic filter as shown in Fig. 63b. The patterns on each ceramic sheet are the examples for realizing the BPF with the equivalent network of Fig. 63a, and the performance is shown in Fig. 63b. The ordinal construction of the waveguide filters can be certainly realized only by filling the high-dielectric material in the region of the air of the ordinal construction. By such a technique, the size can be small, proportional to $1/\sqrt{\varepsilon_r}$.

APPENDIX

Note 1

$$F(\phi, k) = \int_0^{\phi} \frac{d\phi}{\sqrt{1 - k^2 \sin^2 \phi}} = \int_0^{z} \frac{dz}{\sqrt{(1 - z^2)(1 - k^2 z^2)}} \tag{1}$$

is called the elliptic integral of the first kind.

In the case of $\phi = \pi/2$ or $z = 1$ in Eq. (1),

$$F\left(\frac{\pi}{2}, k\right) = K(k) = \int_0^{\pi/2} \frac{d\phi}{\sqrt{1 - k^2 \sin^2 \phi}} = \int_0^{1} \frac{dz}{\sqrt{(1 - z^2)(1 - k^2 z^2)}} \tag{2}$$

is called the complete elliptic integral of the first kind. In Eqs. (1) and (2), k is called the modulus and $k' = \sqrt{1 - k^2}$ is called the complementary modulus.

Note 2

Normal modes are chosen as

$$\iint_{s_0} \boldsymbol{e}_p \times \boldsymbol{h}_p \, d\boldsymbol{s}_0 = 1$$

where $\boldsymbol{e}_p$ is the normal of the electric field, S_0 is the section's area, and $\boldsymbol{s}_0$ is the normal vector to the section of the waveguide.

Because the wave impedance at the evanescent wave is $j\omega\mu/\alpha$, substituting this value into the above equation, we get

$$\frac{\omega\mu}{\alpha}\int\int_{s_0} |\boldsymbol{h}_{pt}(z = 0)|^2 \, ds = 1$$

where t is the sectional component of $\boldsymbol{h}_p$.

If the axial direction of DR is x,

$$\boldsymbol{h}_{pt}^{\pm} = \boldsymbol{i}_x A_x \sin\left(\frac{\pi x}{a}\right) e^{\mp\alpha_t z}$$

in the case of $TE_{10}^{\square}$, where z is the direction of wave propagation in the $TE_{10}^{\square}$, waveguide. Therefore, we get

$$\frac{\omega\mu}{\alpha} A_x^2 \frac{ab}{2} = 1$$

$$A_x^2 = \frac{\alpha}{\omega\mu}\frac{2}{ab}$$

From Eqs. (43d) and (43e), therefore, H_z is obtained from

$$|H_2| = \frac{\omega\mu M}{2}|\boldsymbol{h}_{pt}|^2 e^{-\alpha_1 z}$$

If DR is placed at $x = a/2$, we get $|\boldsymbol{h}_{pt}| = A_x^2$ and $|H_2| = (\omega\mu M/2)A_x^2 e^{-\alpha_1 z}$. Substituting this relation into Eq. (43c), we get

$$k = \frac{\mu_0 M^2}{\tilde{W}_t}\frac{\alpha_1}{ab}e^{-\alpha_1 z}$$

REFERENCES

1. Matthai, G. L., Young, L., and Jones, E. M. T., *Microwave Filters, Impedance-Matching Networks, and Coupling Structures*, McGraw-Hill, New York, 1964.
2. Chon, S. B., Shielded transmission lines, *IRE Trans. Microwave Theory Tech.*, *MTT-3*(10), 29–38 (1955).
3. Hillberg, W., From approximations to exact relations for characteristics impedances, *IEEE Trans. Microwave Theory Tech.*, *MTT-17*, 259–265 (1965).
4. Bahl, I. and Bhartia, P., *Microwave Solid State Circuit Design*, Wiley–Interscience, New York, p. 28.
5. Cohn, S. B., Shielded coupled-strip transmission line, *IRE Trans. Microwave Theory Tech.*, *MTT-3*, 29–38 (October 1955).
6. Bhat, B. and Kohl, S. K., *Stripline-Like Transmission Lines for Microwave Integrated Circuits*, John Wiley and Sons, New York, pp. 216 and 219.

7. Gladwell, G. M. L. and Cohen, S., A Chebysheu approximation method for microstrip problems, *IEEE Trans. MIT-23*, 865–870 (1975).
8. Guillon, P. and Mekerta, S., A bandstop dielective resonator filter, *IEEE MTT-SInt, Microwave Symp. Dig.*, 170–173 (June 1981).

   In this reference, the loop constructed by the L_rC_r series resonant circuit is coupled to the microstrip line with the mutual inductance L_m. By the equivalent network, Q_e is obtained. However, in this book, it was obtained directly by the reciprocal theory.
9. Cohn, S. B., Microwave bandpass filters containing high-Q dielectric resonators, *IEEE Trans. Microwave Theory Tech.*, *MTT-16*(4), 218–227 (1968).
10. Konishi, Y., Determination of capacitances and inductances of the uniform coupled lines in the anisotropic inhomogeneous medium by resistive sheet, *IEEE Trans. Broadcasting* (June 1996).
11. Collin, R. E., *Field Theory of Guided Waves*, McGraw-Hill, New York, pp. 200–204.
12. Ragan, G. L., *Microwave Transmission Circuits*, McGraw-Hill, New York, pp. 675–676.
13. Fiedziusko, S. J., Dual-mode dielectric resonator loaded cavity filter, *IEEE Trans. Microwave Theory Tech.*, *MTT-30*(9), 1311–1316 (1982).
14. Nishikawa, T., Ishikawa, Y., Hattori, J., Wakino, K., and Kobayashi, Y., 4-GHz band bandpass filter using an orthogonal array coupling TM_{110} dual mode dielectric resonator, in *Proceedings of the 19th European Microwave Conference*, Sept. 1989, pp. 886–891.
15. Nishikawa, T., et al., 800 MHz band high-power bandpass filter using TM_{110} mode dielectric resonators, *Denshi*, *28*, 115–118 (1989).
16. Konishi, Y., Novel dielectric waveguide components: microwave applications of new ceramic material. *Proc. IEEE*, *79*(6) (June 1991).

6

Principle of Practical Circuits

1 REACTANCE ELEMENTS

1.1 Construction of Inductors

Lumped Element Inductors

For lumped element inductances, the straight ribbon type, the circular spiral type, and the rectangular spiral type as shown in Figs. 1a, 1b, and 1c, respectively, are mainly used for the microwave integrated circuits.

The inductances of the inductors shown in Fig. 1 are shown in Table 1, where K is the correcting coefficient based on the current concentrating effect to the edge and it is obtained as shown in Fig. 2 [4]. K' is also the same correcting effect; $K' \simeq 1.5$ [3].

The calculated values and measured values of the circular spiral coil are shown in Table 2.

Distributed Constant Lines with the Short End and the Length $l << \lambda_g/4$

As described in Eqs. (1c) and (1g) of Chapter 3, we get

$$L_{\mathrm{eq}} \simeq Ll, \qquad L = \frac{Z_c}{v_p} \tag{1}$$

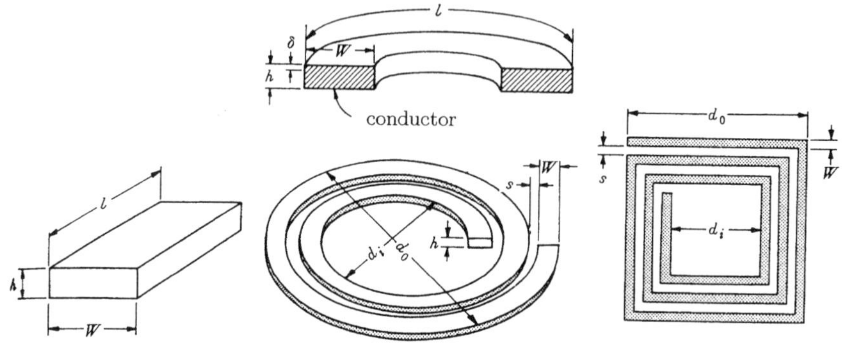

Figure 1. Several lumped element inductors: (a) straight ribbon type; (b) circular spiral type; (c) rectangular spiral type.

where v_p is the phase velocity and L is the inductance of the distributed constant line per unit length.

This technique is used for connecting the inductance to guided wave in parallel. The examples are shown in Figs. 3a–3c.

Examples of the series inductance connecting to the coaxial line are shown in Figs. 3e–3g, which show the series inductor to connected inner or outer conductor of the coaxial line. In Figs. 3e and 3f, the inductors are made by the coaxial lines. In Fig. 3g, it is made by the radial transmission line.

Distributed Constant Lines with the High Characteristic Impedance and the Length $l \ll \lambda_g/4$

Considering of the construction of Fig. 4a, an increase of the inductance at the region of the distributed line with the high characteristic impedance of Z' ($>Z_0$) takes place, as the inductance per unit length of the distributed constant line in the region between T_1 and T_2 is higher than others. Denoting the inductance per unit in the case of Z' and Z_0 of the characteristic impedance by L' and L_0, respectively, the increase of the inductance L_{eq} is calculated from

$$L_{eq} = (L' - L)l = \left(\frac{Z'}{v_p'} - \frac{Z_0}{v_p}\right) \tag{2}$$

Table 1. Constant of Inductors

Kinds of inductor	Inductance	*Q* Values	Ref.
Straight ribbon type	$L_R = 5.08 \times 10^{-3} l \left[\ln \left(\frac{l}{W + h} \right) + 1.193 + 0.2235 \frac{W+h}{l} \right]$ (1)	$Q_R = 2.15 \times 10^3 \frac{L_R\,(\text{nH})}{K} \frac{W}{l} \times \sqrt{\frac{\rho\,(\text{Cu})}{\rho}} \sqrt{\frac{f\,(\text{GHz})}{2}}$ (2)	2, 3
Circular spiral type	$L_s = 1 \frac{a^2 n^2}{8a + 11c}$ (nH) (3) $a = \frac{1}{4}(d_0 + d_i)$ $c = \frac{1}{2}(d_0 - d_i)$ n = number of turns d_0, d_i = unit is mils	$\frac{Q_s d_0^{1/2}}{L_s^{1/2}} = \frac{1.25 \times 10^2 W}{K'(1 - 7d_i/15d_0)^{1/2}} \times \sqrt{\frac{\rho\,(\text{Cu})}{2}} \sqrt{\frac{f(\text{GHz})}{\rho}}$ (4) $\frac{Q_{s\max} d_0^{1/2}}{L_s^{1/2}} = \frac{1.3 \times 10^2 W}{K'} \times \sqrt{\frac{f\,(\text{GHz})}{2}} \sqrt{\frac{\rho\,(\text{Cu})}{\rho}}$ (5) (maximum Q values for $d_0 = 5d_i$)	5, 6
Rectangular spiral type	$L\,(\mu\text{H}) = 8.5 \times 10^{-10} d_0\,(\text{cm}) \times n^{5/3} = 8.5 \times 10^{-10} \frac{A}{W^{5/3}}$ (6) $d_0 = \sqrt{A}$, $W = s$ A = area of coil (cm^2) n = number of turns W = width of coil (cm) s = space between coils (cm)	$Q_0 = 12.4 \frac{f}{f_0} \frac{t^2}{\rho} \frac{1}{n^{4/3}} \times 10^4$ (7) t = thickness of thin coil ρ = resistivity of conductor	7

Source: Ref. 3.

where v_p' and v_p are the phase velocities of the corresponding distributed lines. Therefore, L_{eq} becomes large in the case of $Z' > Z$ under the condition $v_p' \simeq v_p$.

Because in the practical case, the evanescent E wave and H wave are generated at the portion of the discontinuity of the line, the parallel capacities C_e caused by the E wave and the series inductance L_e caused by the H wave are added to the equivalent network of Fig. 4b, which results in that shown in Fig. 4c.

The practical constructions corresponding to Fig. 4 are shown in Fig. 5.

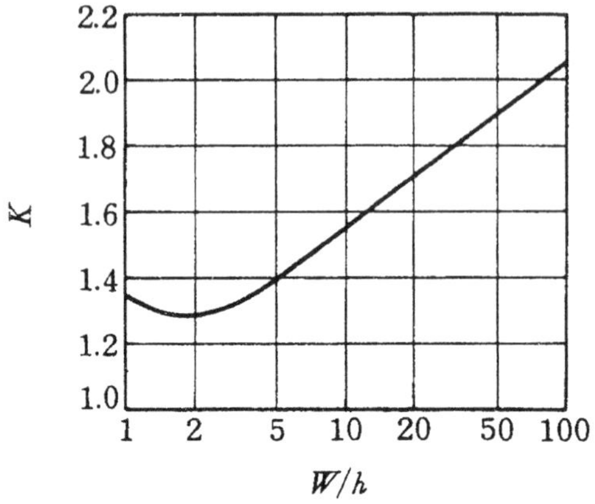

Figure 2. The values of K of Table 1. (From Ref. 1.)

Evanescent H Wave

As mentioned in Chapter, 1, Section 3.3, the magnetic energy is dominated by the electric energy in the evanescent H wave. By using this fact, we can obtain the parallel inductance in a waveguide with the construction of the so-called inductive window, as shown in Fig. 6a.

When we consider the even-mode excitation of the inductive window, the magnetic fields take the distribution of Fig. 6b, which can be decomposed to the dominant mode of $TE^{\square}_{10}$, the evanescent higher mode of $TE^{\square}_{30}$, $TE^{\square}_{50}$,

Table 2. Q and Inductance L of Thin Circular Spiral Coils

n	d_0 (mil)	W (mil)	s (mil)	d_i (mil)	L (nH)	Q $f = 0.5$ GHz	Q $f = 1$ GHz	Q $f = 2$ GHz
2.5	48	4	2	11	4.5[a]	81[a]	115[a]	162[a]
					4.4[b]	43[b]	61[b]	100 ± 20[b]
2.5	49	5	1	11	4.3[a]	97[a]	137[a]	194[a]
					4.4[b]	52[b]	137[b]	120 ± 25[b]
3.0	32	2.5	1.0	10	4.8[a]	67[a]		
					4.6[b]	36[b]		

Note: n = Number of turns; h = 12 μm.
[a]Measured values.
[b]Calculated values from Eqs. (3) and (4).
Source: Ref. 3.

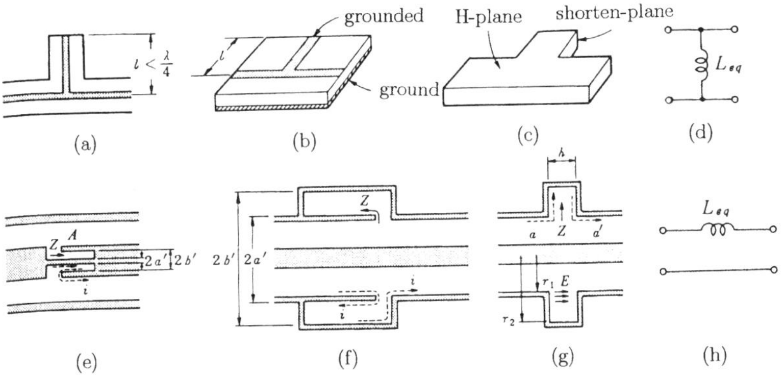

Figure 3. Inductors made by the distributed constant line with the short end. (a) Coaxial line; (b) microstrip line; (c) $TE_{10}^{\square}$ waveguide; (d) equivalent circuit of (a), (b), and (c); (e) series inductor in the inner conductor of the coaxial line; (f) series inductor connected in the outer conductor of the coaxial line; (g) series inductor made by the radial transmission line connected in the outer conductor of the coaxial line; (h) equivalent circuit of (e) (f), and (g).

and so forth. Such a higher mode takes place the inductive energies. In the case of the odd-mode excitation, the magnetic field has the same distribution for the case $TE_{10}^{\square}$ mode, which means that the inductive window has no effect. For the reason mentioned above, the equivalent network of the inductive window becomes that shown in Fig. 6c.

The values of the susceptance B in Fig. 6c is obtained as

$$\frac{B}{Y} = -\frac{\lambda_g}{a} \cot^2 \left(\frac{\pi d}{2a} \right) \tag{3}$$

where Y is the wave admittance of the $TE_{10}^{\square}$ mode [8,9].

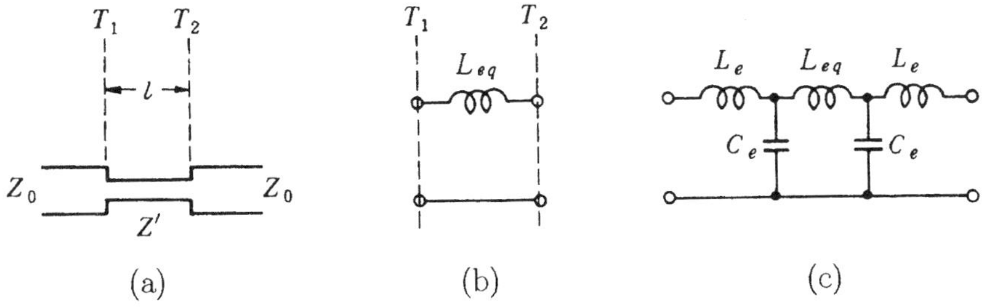

Figure 4. Construction and equivalent network: (a) construction; (b) equivalent network; (c) equivalent network considering the discontinuity of conductor.

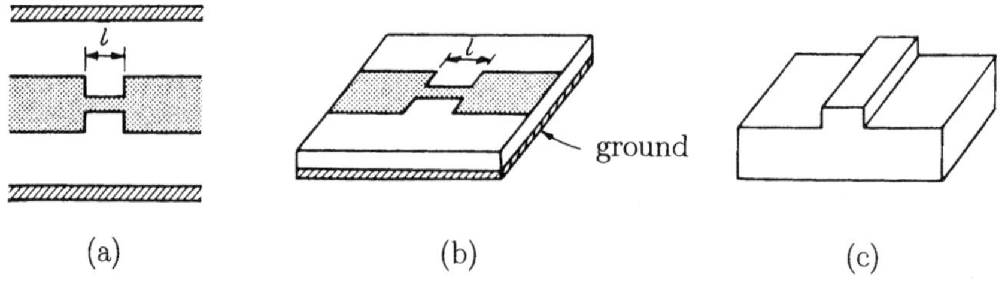

Figure 5. Constructions to obtain series inductance: (a) coaxial line; (b) microstrip line; (c) $TE_{10}^{\square}$ waveguide.

In the case of the construction of Fig. 7a, the higher mode of the evanescent H waves are also generated for the even-mode exitation. When it is excited by the odd-mode excitation, the magnetic energies at the region of the post are removed, which results in the capacitive susceptance B_b. The equivalent network, therefore, becomes as shown in Fig. 7c. The construction of Fig. 7a, is called the inductive post. The values of B_a and B_b of Fig. 7c are obtained as shown in Fig. 8 [9].

The equivalent network is changed by the positions of the reference plane. The network of Fig. 7c is that of the reference planes of T_1 and T_2, which is situated in the center of the post. When the reference planes are chosen in T_1' and T_2' as shown in Fig. 7b, the values of B_b of Fig. 7c become negative when B_b is inductive.

1.2 Construction of Capacitor

Multilayer Condenser

As shown in Fig. 9, the capacitor is obtained from the sandwich construction of the bottom electrode, a dielectric material, and the opposite electrode on the substrate. The capacitance is

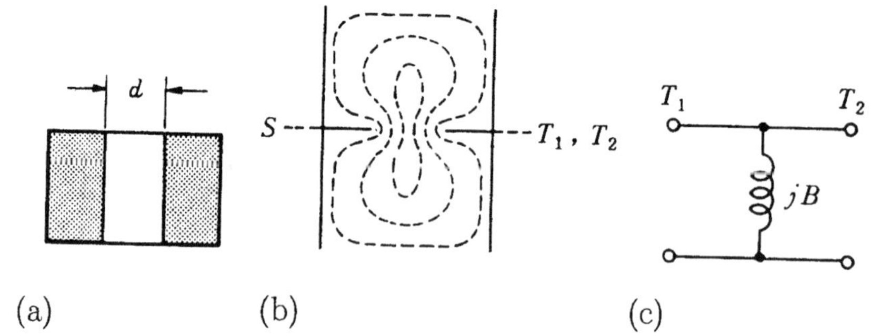

Figure 6. Inductive window: (a) construction; (b) magnetic fields for even mode; (c) equivalent network of (a).

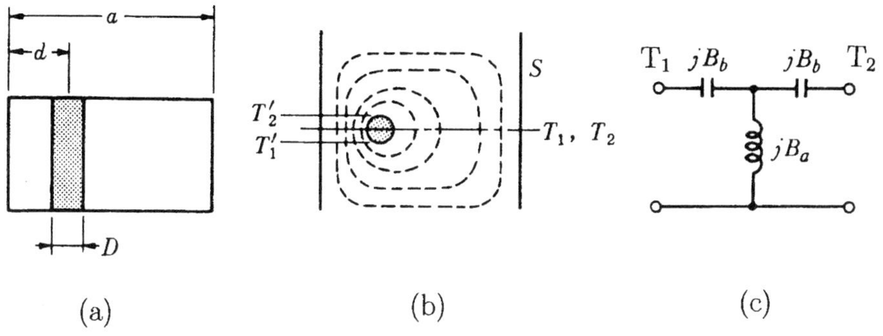

Figure 7. Inductive post: (a) construction; (b) magnetic fields for even-mode excitation; (c) equivalent network of (a).

$$C[\mu F] = (0.0885 \times 10^{-6})\, \varepsilon_r \frac{Wl}{d} \tag{4}$$

where the size of the electrode is W (cm) $\times$ l (cm) and ε_r is the relative dielectric constant of the dielectric material.

The surface of the substrate should be smooth and less than 3000 Å thick to avoid deterioration of Q values caused by conductive loss. The total Q values of the condenser, Q_{cap}, are determined by

$$\frac{1}{Q_{cap}} = \frac{1}{Q_c} + \frac{1}{Q_\varepsilon} \tag{5}$$

where Q_c is caused by the conductive loss and Q_ε is caused by the dielectric loss. The conductive loss is taken place by the current flowing on the elec-

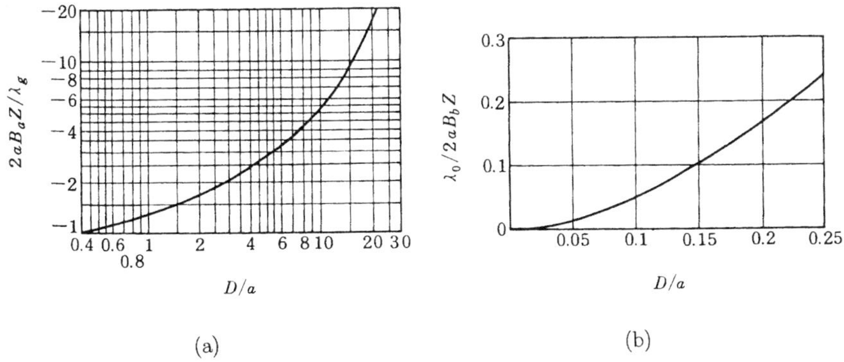

Figure 8. Equivalent constant of the inductive post.

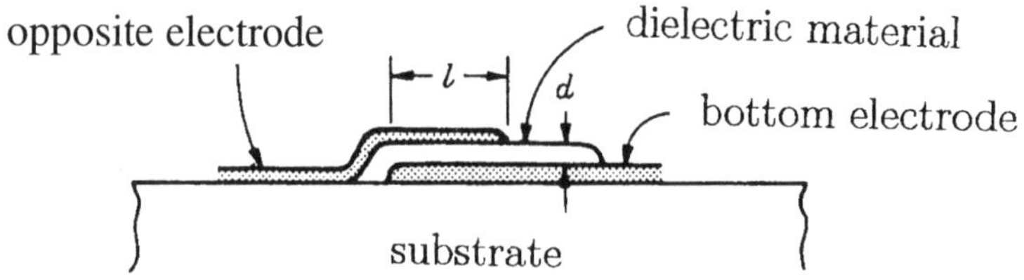

Figure 9. Construction of an element of the thin multilayer condenser.

trode, which is continued to the displacement current in the dielectric material. The currents become maximum at the feed points to the electrodes. When the electrode is copper, Q_c takes the values

$$Q_c = \frac{3W}{2\omega C R_s l} = \frac{2.9 \times 10^4 W}{f^{3/2}(\text{GHz}) \times (\text{mpF}) \times l} \tag{6}$$

In the case for Q_c less than Q_ε, Q_c is the important factor of Q_{cap}. In the case of Q_ε higher than 5000, a Q_{cap} of 1 pF is about 700 at 2 GHz; it is over 100 for 1 pF at 10 GHz [10].

Next, for the planar-type condenser, the interdigital construction as shown in Fig. 10 is used. The Q_{cap} values of about 400 for 0.1–10 pF are reported in Ref. 11.

Distributed Constant Line with Open End and Length l << $\lambda_g/4$, or Coupling Capacitor Between Two Conductors

In the case of the distributed constant line with open end and length less than a quarter wavelength, the capacity is calculated from

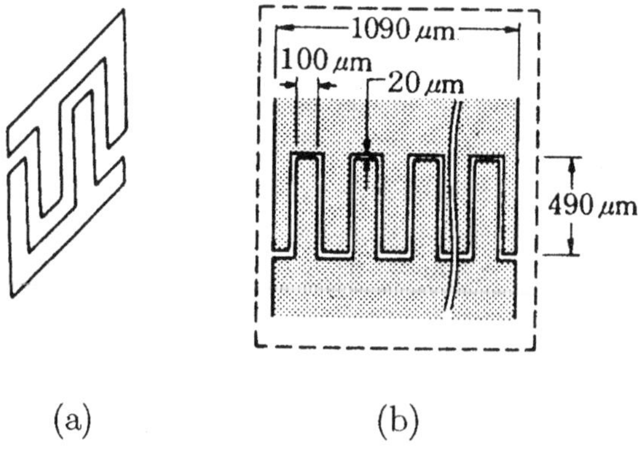

Figure 10. Examples of interdigital constructions: (a) basic construction; (b) practical construction.

$$C_{eq} = Cl, \qquad C = \frac{1}{Z_c v_p} \tag{7}$$

where C is the values of the distributed constant line per unit length. Several constructions are shown in Fig. 11.

Distributed Constant Line with Low Characteristic Impedance and Length $l << \lambda_g/4$

Because the region of low characteristic impedance increases the capacity, the equivalent capacity C_{eq} is obtained from Eq. (8) in the same way as for Eq. (2):

$$C_{eq} = \left(\frac{1}{Z' v_p} - \frac{1}{Z_0 v_p}\right) \tag{8}$$

where Z' and Z_0 are characteristic impedances of the region with C_{eq} and the characteristic impedance of the line, respectively. As an example, in the case of the construction of Fig. 12, because we have relation

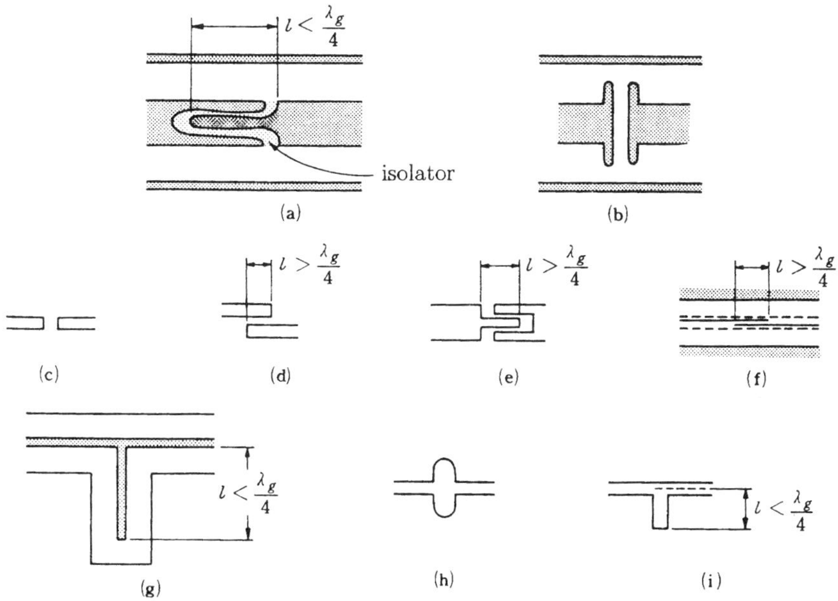

Figure 11. Several constructions for capacitors: (a) and (b) are used for the series capacity in the coaxial line; (c)–(f) are used for the series capacity in the microstrip line; (g) is used for the parallel capacity in the coaxial line; and (h) and (i) are used for the parallel capacity in the microstrip line.

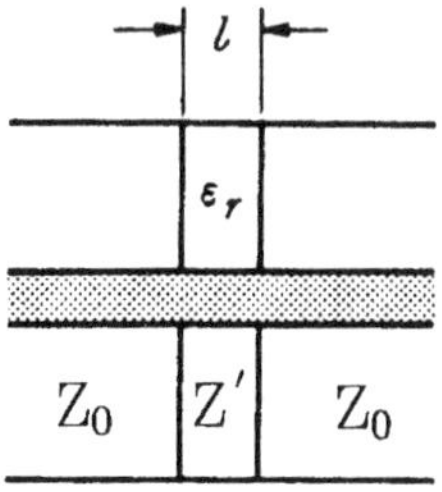

Figure 12. Method to partially fill the dielectric material.

$$Z' = \frac{Z_0}{\sqrt{\varepsilon_r}}, \qquad v_p' = \frac{v}{\sqrt{\varepsilon_r}}$$

where v_0 is a light velocity, we get

$$C_{eq} = \frac{l}{v_0 Z_0}(\varepsilon_r - 1) \tag{9}$$

Evanescent E Wave

As mentioned in Chapter 1, Section 3.3, the electric energy is dominated by the magnetic energy in the evanescent E mode. By using this fact, we can obtain the parallel capacitance in a coaxial line and waveguide. Practical examples are shown in Figs. 13–16. In these figures, we showed the constructions and distributions of the electric fields for the even-mode excitation. The electric fields are the evanescent E wave because the coaxial line and TE_{10} waveguide are designed to pass the dominant mode only. Therefore, in the same way as stated in Chapter 5, Section 1.1, the equivalent capacity, C_{eq}, is connected to the transmission line in parallel, as shown in Fig. 17a for the case of the constructions of Figs. 13 and 14.

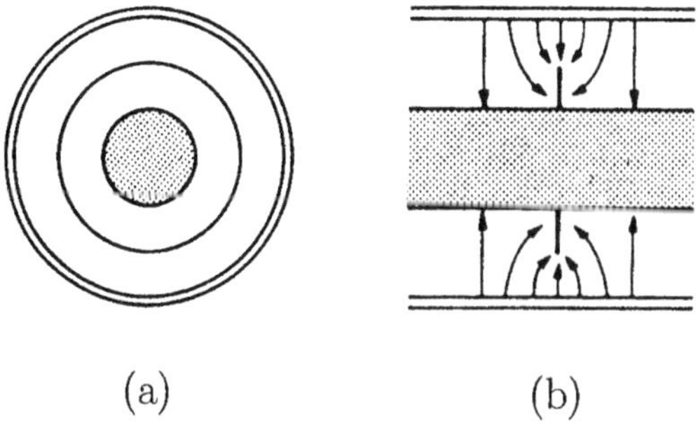

Figure 13. Coaxial line with circular thin disk: (a) construction; (b) electric field.

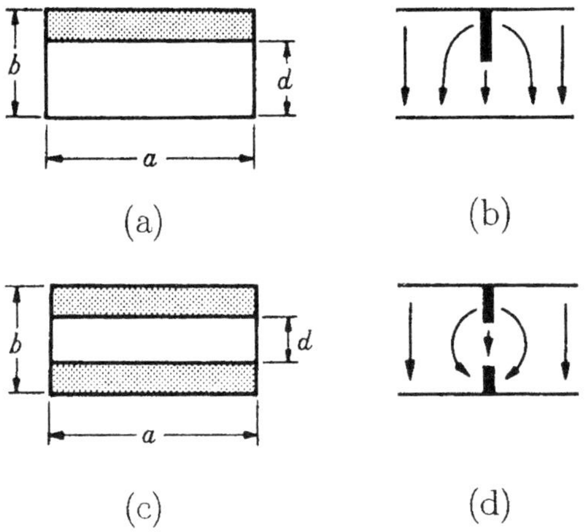

Figure 14. Capacitive window in the $TE^{\square}_{10}$ waveguide: (a) construction (asymmetric); (b) electric; (c) construction (symmetric); (d) electric field.

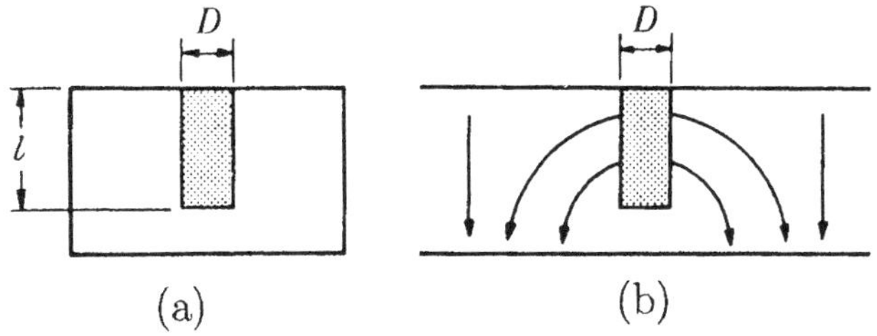

Figure 15. Capacitive post in the $TE^{\square}_{10}$ waveguide (parallel to the E plane): (a) construction; (b) electric field.

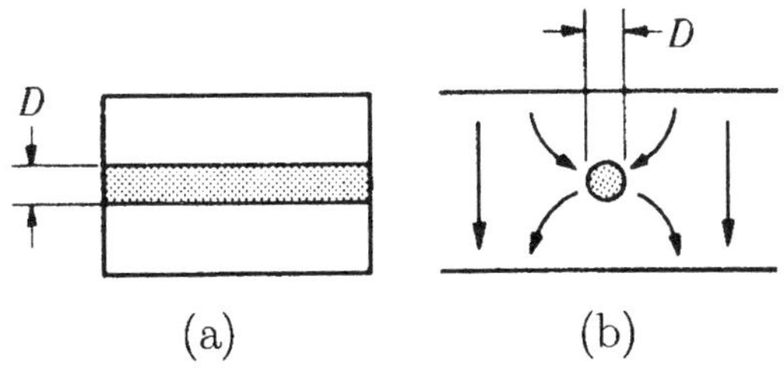

Figure 16. Capacitive bar in the $TE^{\square}_{10}$ waveguide (parallel to the *H* plane): (a) construction; (b) electric field.

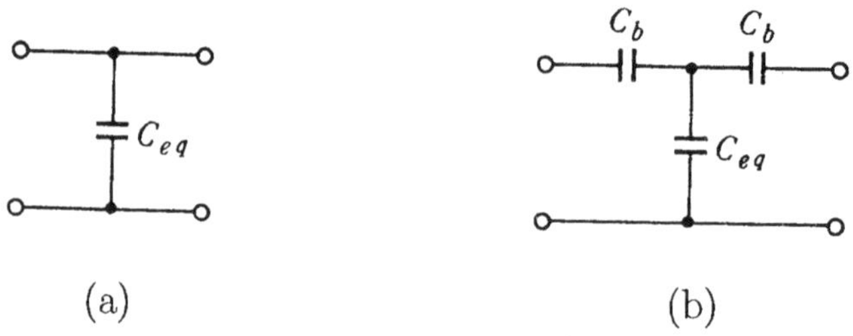

Figure 17. Equivalent network of (a) Figs. 13 and 14 and (b) Figs. 15 and 16.

In the case of the constructions in Figs. 15 and 16, however, the post and the bar have diameter D. Therefore, when we choose the reference plane of the equivalent network in the center of the post and bar, the equivalent network becomes as shown in Fig. 17b by the reason explained in Chapter 5, Section 1.1.

The values of C_{eq} in the case of Fig. 13, is calculated and shown in Fig. 18. The values of the suceptance of C_{eq}, $B(=\omega C_{eq})$ in the case of the Fig. 14, are obtained from Eq. (10) [9]:

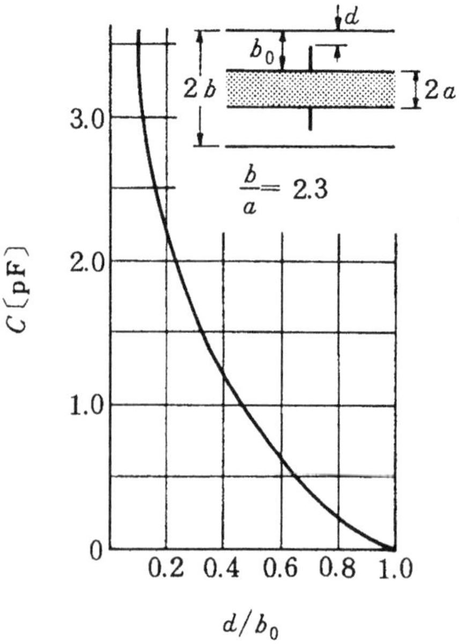

Figure 18. The values of C_{eq} of Fig. 13.

In the case of Fig. 14a

$$\frac{B}{Y} = \frac{8b}{\lambda_g} \ln\left(\frac{\pi d}{2b}\right) \quad (10)$$

in the case of Fig. 14c

$$\frac{B}{Y} = \frac{4b}{\lambda_g} \ln\left(\frac{\pi d}{2b}\right)$$

2 POWER DIVIDER AND COMBINER

2.1 Principle of Wilkinson Power Divider

Because the power divider is a reciprocal circuit, it is also used for the power combiner, for which an example is shown in Fig. 19. The power divider should have the following characteristics:

1. The supplied power to the input port should be divided into n output ports with the required power ratio.
2. The impedances of the input and the output ports should be matched to the source and the load impedances, respectively.
3. When the power is supplied to the one of n output ports, the power should not appear at any other output ports.

The power divider with such characteristics was proposed by Wilkinson [12]; the equivalent network and the coaxial construction is shown in Fig. 20.

At first, the principle is explained for a two-way power divider as shown in Fig. 21a. In Fig. 21, T_1 and T_2 are the quarter-wave distributed constant lines with characteristic impedance of W, and R is the resistance between ports 1 and 2. When the power is supplied to port 0 as shown in Fig. 21b,

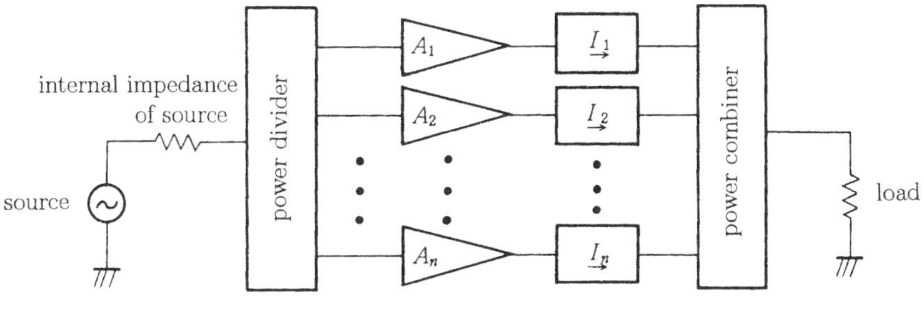

Figure 19. A system of the high-power amplifier: A_i = ith amplifier, I_i = ith isolator.

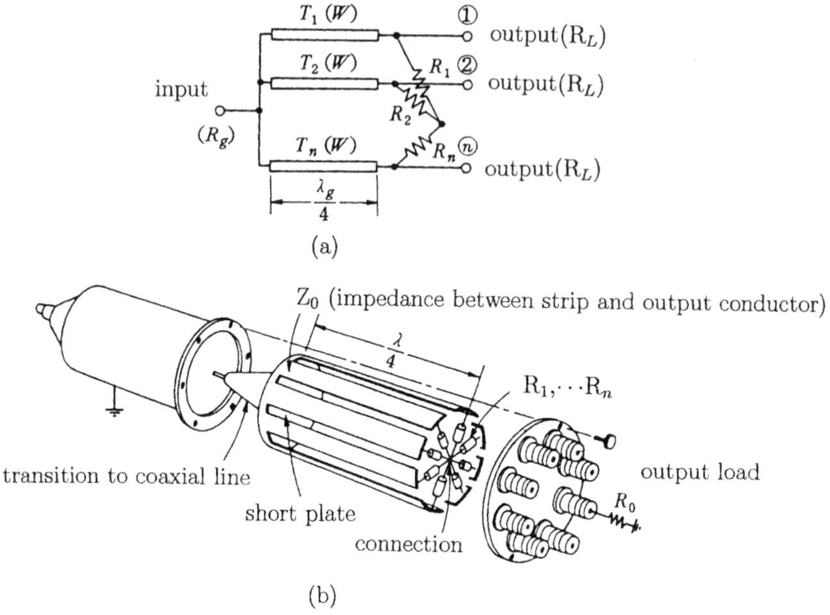

Figure 20. Wilkinson power divider: (a) equivalent network; (b) a realized construction of (a).

it is divided between ports 1 and 2 and current is not flowing in R because the potentials of ports 1 and 2 take the same values.

Next, when the power is supplied to port 1, as shown in Fig. 21c, the current is not flowing in R_L connected at port 2. The principle is explained as follows:

(i) From the matching condition at port 0, we must have the relation

$$W = \sqrt{2R_l R_g} \tag{11}$$

because the impedance viewed from the left-hand side of a quarter-wave line should be $2R_g$ to satisfy the matching condition at port 1, and W must take the values of the quarter-wave transformer as given by Eq. (2b) of Chapter 3.

(ii) From the isolated condition between ports 1 and 2 as shown in Fig. 21c, we must have the relation

$$R = 2R_L \tag{12}$$

Equation (12) can be easily introduced as follows. Because the power supplied to port 1 is not supplied to port 2, the voltage at port 2 must be

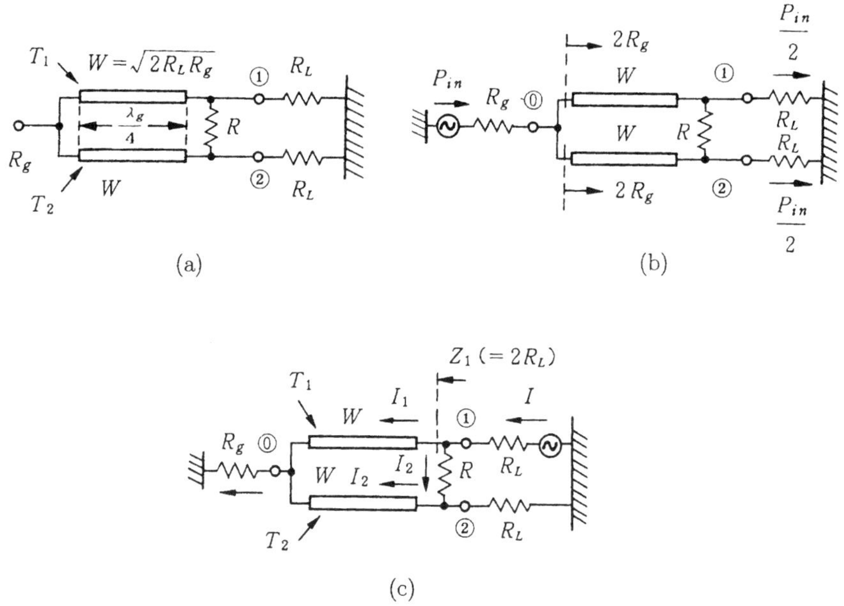

Figure 21. Two-way Wilkinson power divider (a) explanation of the operation: (b) power supplied to port 0, divided between ports 1 and 2; (c) power supplied to port 1 but not to port 2.

zero, which results in the infinite impedance viewed from the left-hand side of T_2 toward port 2. Then for the impedance viewed from port 1 toward T_1, Z_1 is obtained from

$$Z_1 = \frac{W^2}{R_g} = \frac{2R_L R_g}{R_g} = 2R_L \tag{13}$$

Next, the voltage at port 0 as in Fig. 21c, V must be

$$V = \frac{WI_1}{j} \tag{14a}$$

from the characteristics of a quarter-wave distributed constant line. (See Note 1 at end of chapter.) We get the current of T_2 on the right-hand side, I_2, from

$$I_2' = \frac{V}{j\omega} = \frac{1}{jW}\left(\frac{WI_1}{j}\right) = -I_1 \tag{14b}$$

Therefore, the current flowing into T_1 from port 1 must have the same values and the same direction as the current flowing into T_2 from port 2. Then, the impedance of R in Fig. 21c should have the same values as the impedance viewed from port 1 toward T_2, which results in

$$R = 2R_L \tag{14c}$$

The two-way power divider can be made on the dielectric substrate with the thin metal membrane, as shown in Fig. 22, where the other surface of the substrate is covered by a metal membrane.

Next, the construction of an n-way power divider is shown in Fig. 23a and the principle of the operation is shown in Fig. 23b. The constant can be obtained in the same way as for the two-way power divider and is described as follows. When the power is supplied to port 0 in Fig. 23a, it is divided among ports 1, 2, . . . , n, where each port is connected to the load impedance R_L. The impedance viewed from the left-hand side of each $\lambda_g/4$ line of $T_1, T_2, \ldots, T_n$ toward the right-hand should be nR_g under the matching condition of the port 0, where R_g is the source impedance. The characteristic impedance of the $\lambda_g/4$ distributed line, W, therefore, is calculated from

$$W = \sqrt{nR_LR_g} \tag{15a}$$

Next, when the signal is supplied to port 1 as shown in Fig. 23b, the signal should not appear at the other output ports. Because the voltage of the ith port ($i \neq 1$) is zero, the impedance viewed from the right-hand side of T_1 toward the left, Z_1, is calculated from

$$Z_1 = \frac{W^2}{R_g} = nR_L \tag{15b}$$

Because the voltage of the ith port ($i \neq 1$) is zero, the current does not flow through R_L connected to the $i(i \neq 1)$th port. Therefore, if we denote the current flow into T_1 from port 1 by i_1, the current with same values as i_1 should flow into $T_2, \ldots, T_n$ as shown in Fig. 23b. This is caused for the

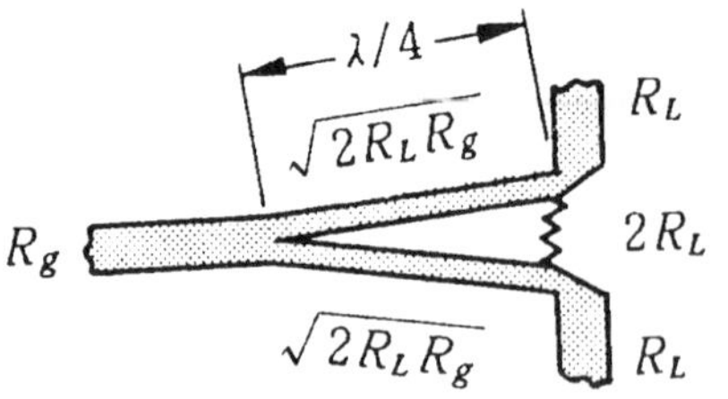

Figure 22. Pattern of two-way power divider made on a dielectric substrate.

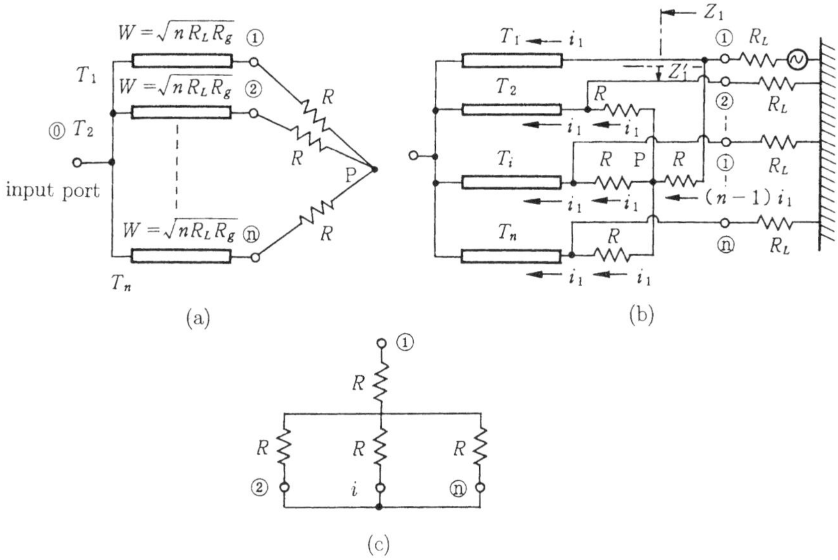

Figure 23. An n-way Wilkinson power divider: (a) construction; (b) explanation of isolation between different output ports; (c) impedance viewed from port 1 toward R.

same reason described for the two-way power divider. The current flowing through R between port 1 and point P in Fig. 23b, is $(n - 1)i_1$.

On the other hand, considering the voltage of port i ($i \neq 1$) is zero, the impedance viewed from port 1 toward the side where R is connected, Z_1', takes the values corresponding to that of Fig. 23c. Therefore, we get

$$Z_1' = R + \frac{R}{n - 1} \tag{15c}$$

Z_1', however, should be

$$Z_1' = \frac{Z_1}{n - 1} \tag{15d}$$

because the current through R connected to port 1 is $n - 1$ times of that flowing into T_1 as mentioned earlier.

From Eqs. (15b)–(15d), we get

$$R = R_L \tag{16}$$

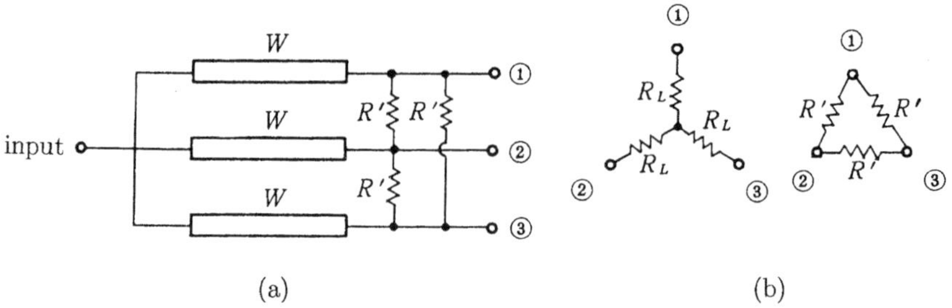

Figure 24. Radial type of $n = 3$ power divider (a) and Y-Δ transform (b), where $R' = 3R_L$.

In the case of $n = 2$, as shown in Fig. 21, point P of Fig. 23b can be considered to be at the center of R in Fig. 21a. Therefore, we can get the result of Eq. (14c).

When n increases to 2, we encounter the difficulty of obtaining point P by planar circuits.

In the case of $n = 3$, the resistance network of the Y connection of Fig. 24a can be transformed to the Δ connection by the $Y - \Delta$ transformation as shown in Fig. 24b. The construction where the resistances are connected between neighboring output terminals, as shown in Fig. 24a, is called the radial type. In the radial type, the construction where one resistance is removed is called the fork type. The fork type for $n = 3$ and 4 take the constructions of Fig. 25a and 25b, respectively.

The reason for using radial and fork types is to construct planar circuits for the microwave integrated circuits more easily. The radial and fork types can take the infinite isolation between the different output ports in the case of $n \leq 3$ and $n < 2$, respectively, even if the Wilkinson type takes the infinite

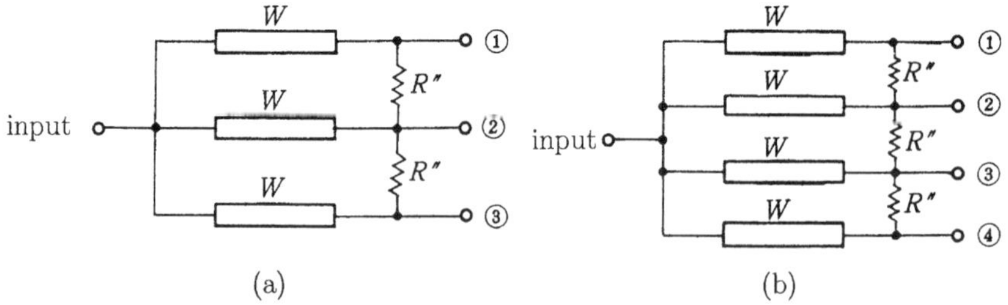

Figure 25. Fork type for $n = 3$ (a) and $n = 4$ (b).

isolation for any n. The values of the isolation for the optimum design in the case of the R type (radial type) are shown in Table 3 [13].

For the purpose of constructing the power divider by planar circuits, the series resistances connected between the port and the center point P are sometimes replaced by the equivalent network, as shown in Fig. 26a, which results in the construction of Fig. 26b [14].

By combining the two-way power dividers of Fig. 23a we get a four-way divider and a three-way divider as shown in Figs. 27 and 28, respectively.

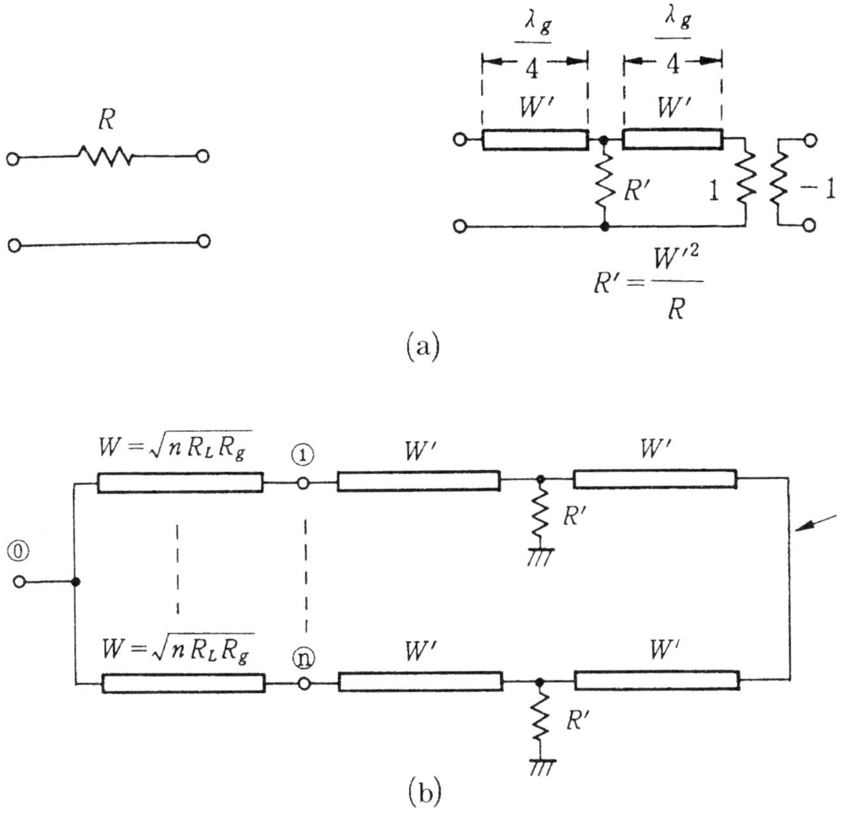

Figure 26. Wilkinson power divider with parallel resistances in place of series resistances: (a) equivalent network; (b) R of Fig. 23a is replaced by two $\lambda_g/4$ lines and parallel resistance R' as shown in (a).

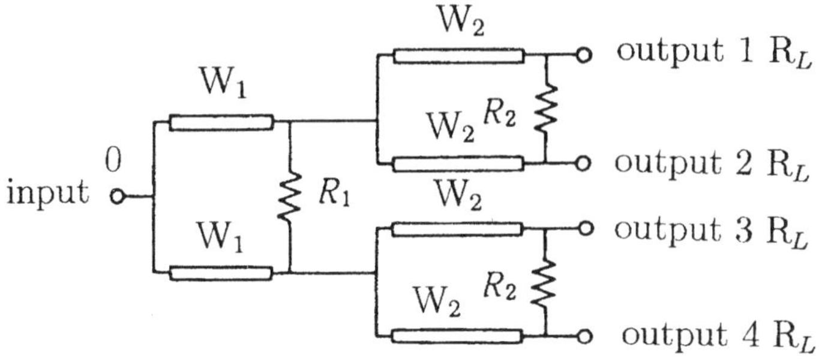

Figure 27. A four-way divider.

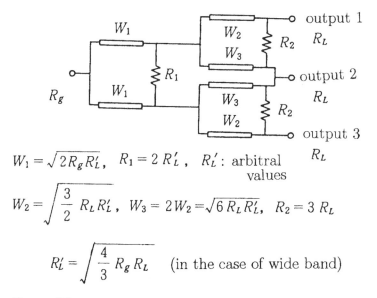

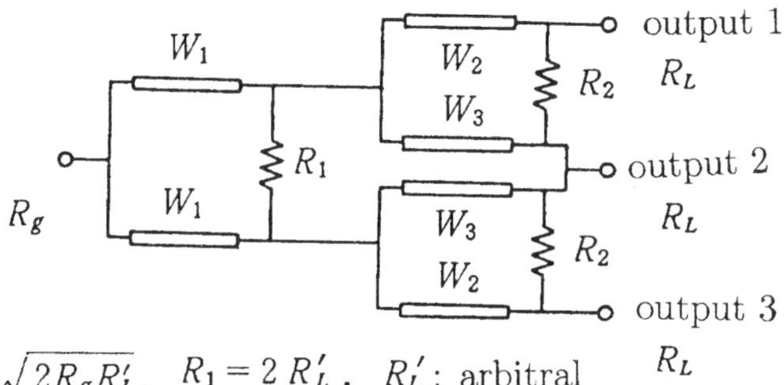

$W_1 = \sqrt{2R_g R_L'}$, $R_1 = 2R_L'$, R_L': arbitral values

$$W_2 = \sqrt{\frac{3}{2} R_L R_L'}, \quad W_3 = 2W_2 = \sqrt{6 R_L R_L'}, \quad R_2 = 3R_L$$

$$R_L' = \sqrt{\frac{4}{3} R_g R_L} \quad \text{(in the case of wide band)}$$

Figure 28. A three-way divider.

The explanation given above is only concerned about the operation at the center frequency. The $\lambda_g/4$ line shown in Fig. 22, however, does not have an electrical angle of $\pi/2$ radian when the frequency deviates from the center frequency. In such a case, the isolation between the different output ports becomes finite because the electrical angle of the pass from port 1 to port 2 via port 0 becomes no more π radian and the wave cannot cancel out with the wave through R from port 1. At the same time, the input and the output ports deviate from match with the source and the load impedance.

The frequency performance can be easily obtained as shown in Fig. 29. As understood from Fig. 29, isolation of more than 14 dB and the VSWR (voltage standing wave ratio) of less than 1.42 are kept in the frequency range $f_h/f_l = 2$. To get the wideband performance, the constructions of Fig. 30 are used and the design is established [15]. As simple examples, the constructions and the frequency performances for $n = 2$ and $n = 3$ are shown in Figs. 31a and 31b, respectively.

The explanation given above is concerned with the case of equal output power. The power divider with different output powers, however, can be realized from the construction in Fig. 32. When the output impedances viewed from ports 2′ and 3′ toward the loads, R_2 and R_3 are obtained from Eq. (17a), the output powers at ports 2 and 3 take the relative power ratio of $1:K^2$ under the same potential at ports 2′ and 3′.

In this case, Z_2, Z_2', Z_3, Z_3', and R are obtained from Eq. (17b). (See Appendix 5.)

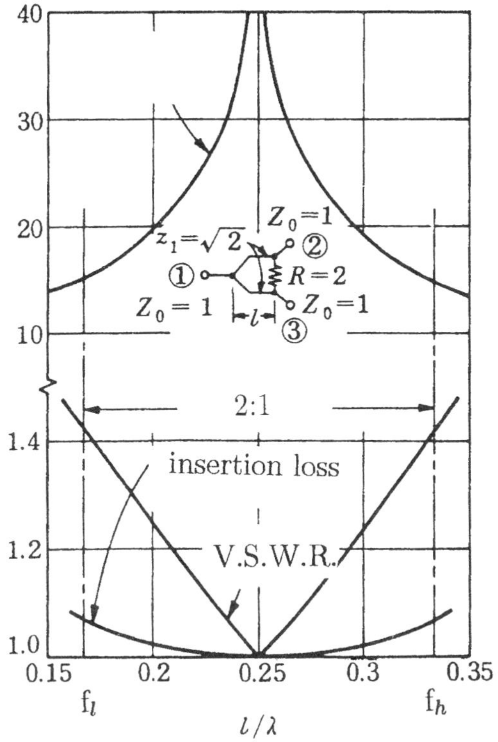

Figure 29. Frequency performance of the two-way power divider.

Table 3. Values of the Isolation and Insertion Loss of an R-type Power Divider for Several Values of n

n	Values of Isolation of R-type for Optimum Design (dB)	Insertion Loss
2	∞	0.25
3	∞	0.333 . . .
4	21.6	0.5
5	19.5	0.4472
6	17.6	0.4279
7	17.2	0.35955
8	16.1	0.3596
10	14.9	0.3596
12	14.1	0.3590

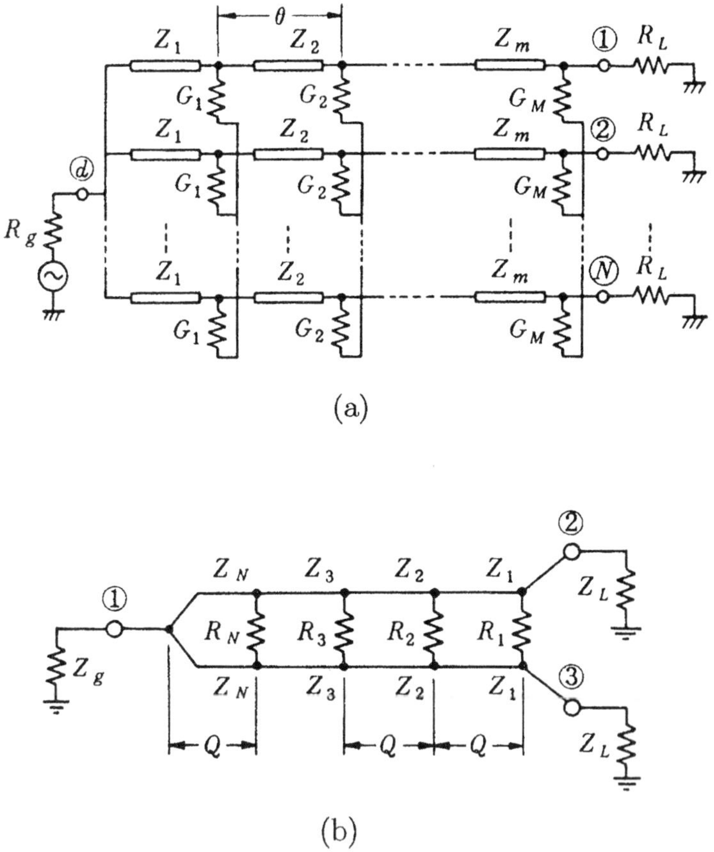

Figure 30. Construction of wideband power dividers: (a) wideband power divider with n output ports; (b) wideband power divider with two output ports.

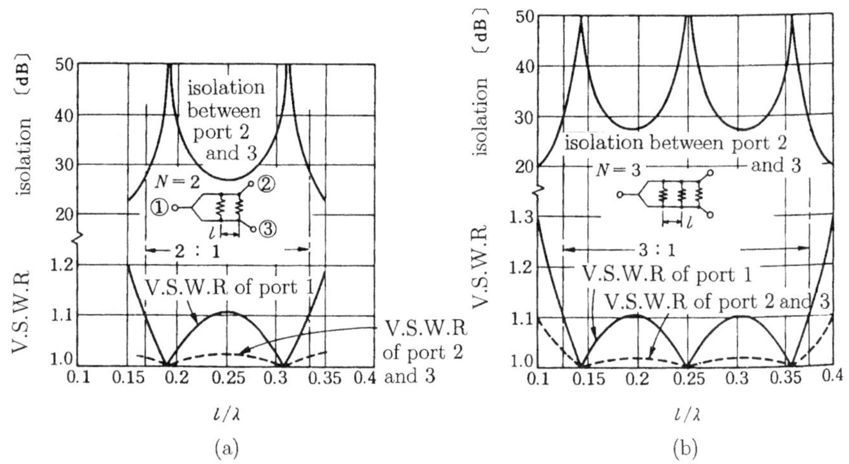

Figure 31. Wideband power dividers construction and frequency performance for n = 2 (a) and n = 3 (b).

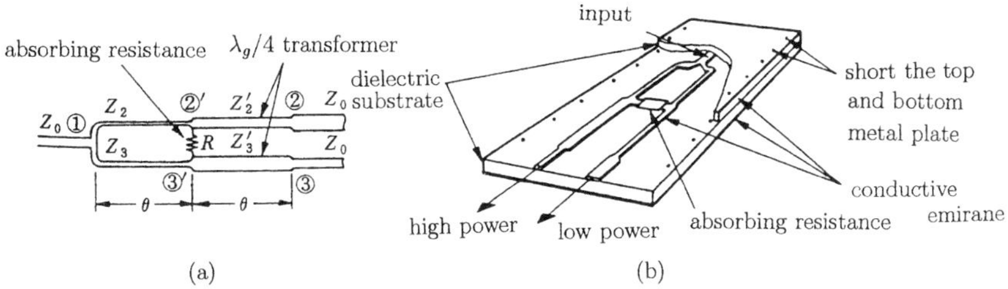

Figure 32. Power divider supplying different output powers: (a) construction; (b) realization from strip line.

$$R_2 = KZ_0, \qquad R_3 = \frac{Z_0}{K} \tag{17a}$$

$$Z_2 = Z_0\sqrt{K(1 + K^2)}, \qquad Z_2' = Z_0\sqrt{K}$$

$$Z_3 = Z_0\sqrt{\frac{1 + K^2}{K^3}}, \qquad Z_3' = \frac{Z_0}{\sqrt{K}}$$

$$R = Z_0\frac{1 + K^2}{K} \tag{17b}$$

In this case, we can choose the voltage at ports 2′ (and 3′) with an arbitrary value, which makes it possible to design for a wide band performance. The results are shown in Eq. (17c):

$$Z_2 = Z_0(1 + K^2)^{3/4}, \qquad Z_2' = Z_0(1 + K^2)^{1/4}$$
$$Z_3 = Z_0\frac{(1 + K^2)^{3/4}}{K^2}, \qquad Z_3' = Z_0\frac{(1 + K^2)^{1/4}}{K}$$
$$R = Z_0\frac{(1 + K^2)^{3/4}}{K} \tag{17c}$$

2.2 Lumped Element Power Divider with Ferrite

Two-Way Power Divider

The two-way lumped element power divider has the construction shown in Fig. 33a, where T_1 and T_2 are the transformers for the impedance matching and the power splitting, respectively.

When the signal is supplied to input port 1, it appears at port 1′ and splits to ports 2 and 3, as shown in Fig. 33b. Because the current flowing into port 1 is divided to ports 2 and 3 equally, the potentials of ports 1′, 2, and 3 take the same values and the impedance viewed from port 1′ toward the load becomes $W/2$, where W is the load impedance at ports 2 and 3, as shown in Fig. 33c. Therefore, a transformer with the ratio $m:m' = \sqrt{2}:1$ is required for the impedance matching at port 1.

Next, when the signal is supplied to port 2, we assume that the current i_2 flowing into port 2 is divided to i_2 in R and i_2'' in T_2. If the current does not appear at port 3, the current should not flow in W connected at port 3, and the current of i_2' should flow into T_2.

When T_2 is the ideal transformer, we have the condition

$$ni_2'' - ni_2' = 0 \tag{18a}$$

Considering the condition

$$i_2'' + i_2' = i_2 \tag{18b}$$

we get the values of i_2' and i_2'' as

$$i_2'' = i_2' = \frac{i_2}{2} \tag{18c}$$

Therefore, the voltage difference between both ends of R is $(i_2/2)R$, and the voltage between port 1 and the ground becomes $(i_2/4)R$ because port 1′ is the center of T_2.

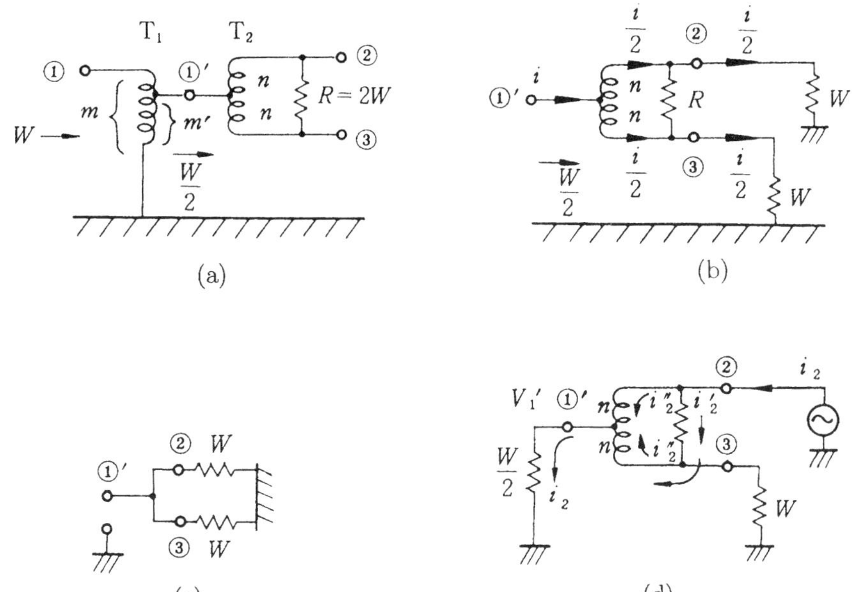

Figure 33. Explanation of the principle of two-way lumped element power divider: (a) construction; (b) the current flow when the signal is supplied to port 1′; (c) the impedance viewed from port 1′ toward the loads; (d) the current flow when the signal is supplied to port 2.

On the other hand, the voltage must be $(W/2)i_2$ as understood by Fig. 33d, which results in the following relation:

$$R = 2W \tag{19}$$

The impedance viewed from port 2 toward the input side can be obtained from

$$\frac{1}{2}\frac{i_2R}{i_2} = W$$

This means that port 2 matches the load impedance W.

To realize the principle mentioned above in the wide frequency range, T_1 and T_2 must be the wideband ideal transformers, which require a large inductance and tight coupling between coils in the transformers. To get a large impedance at the lower frequency, the number of the turns of the coil should be increased. It, however, increases the floating capacity, which deteriorates the performance of the higher frequency. To avoid such a problem, ferrite material with a high permeability at the lower frequency is used for the cores of T_1 and T_2. For example, ferrite materials with initial permeabilities greater than 1000 are used.

The general frequency performance of the relative complex permeability of the ferrite, $\dot{\mu} = \mu' - j\mu''$, is shown in Fig. 34. As understood from Fig. 34, μ' and μ'' decrease as the frequency increases in the region above the natural resonant frequency. When the inductance without ferrite, L_0, is loaded by the ferrite, the impedance Z_f is

$$Z_f = j\omega\mu_0\dot{\mu}_r = j\omega\mu_0\mu_r' L_0 + \omega\mu_0\mu_r'' L_0 \tag{20}$$

The first and second terms of the right hand side of Eq. (20) is the reactance and the resistance of Z_f, respectively, and their values can be maintained almost constant because μ_r' and μ_r'' decrease for the increase in ω. To verify this, the measured values of the impedance of the wire surrounded by a ferrite cylinder is shown in Fig. 35.

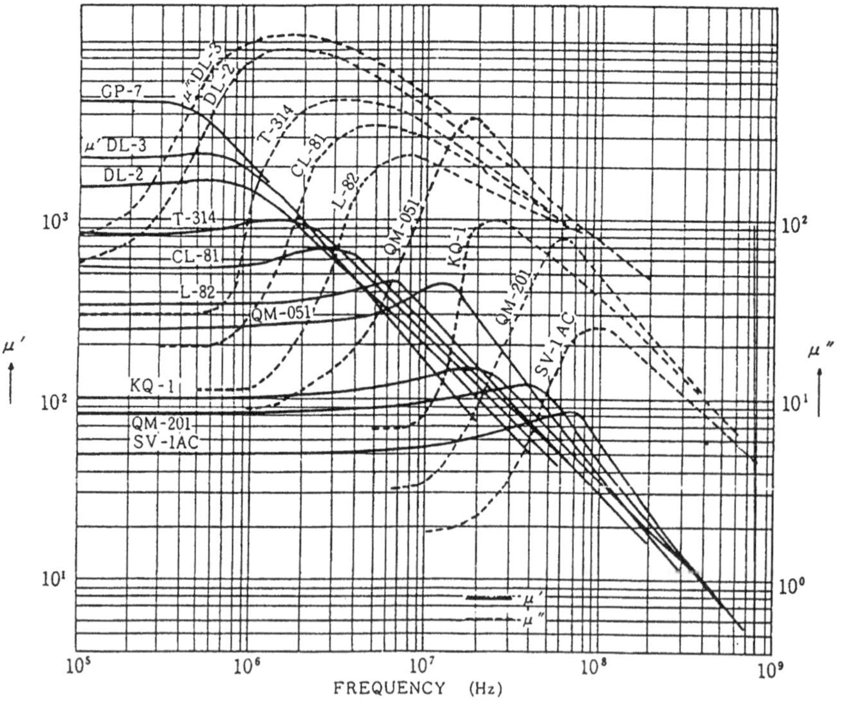

Figure 34. Examples of the complex relative permeability of ferrite. (Courtesy of TDK.)

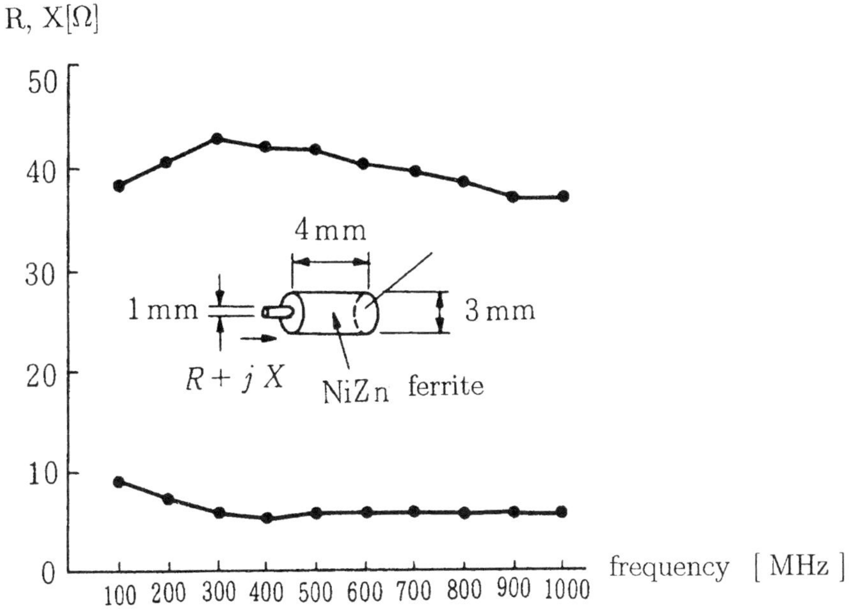

Figure 35. Impedance of wire loaded by ferrite.

When we use tightly coupled wires surrounded by a ferrite cylinder as shown in Fig. 36a, we can obtain the two-way power divider under the principle described as follows. The current supplied at port 1 in Fig. 36a is divided equally between ports 2′ and 3′ and they appear at ports 2 and 3. Because the two wires are tightly coupled, the magnetic flux does not take place in the ferrite, which results in the same potential at ports 1, 2, and 3, as shown in Fig. 36b.

Next, when the signal is supplied to port 2 as shown in Fig. 36c, the current flows from port 2 through port 1′ and should not flow through the pass between ports 3′ and 3 because the voltage of port 3 should be zero, which means the no current in the resistance is connected at port 3. In this case, we must have

$$V_{3-0} = V_{1-0} + V_{3-3'} = iR - iZ_f \tag{21}$$

where V_{i-j} means the voltage at the ith point viewed from the jth point of Fig. 36c. In the case $\mu_r' \ll \mu_r''$, $Z_f \simeq \omega\mu_0\mu_r''L_0$ in Eq. (20).

Taking into account $R = W/2$,

$$\omega\mu_0\mu_r''L_0 = \frac{W}{2} \tag{22}$$

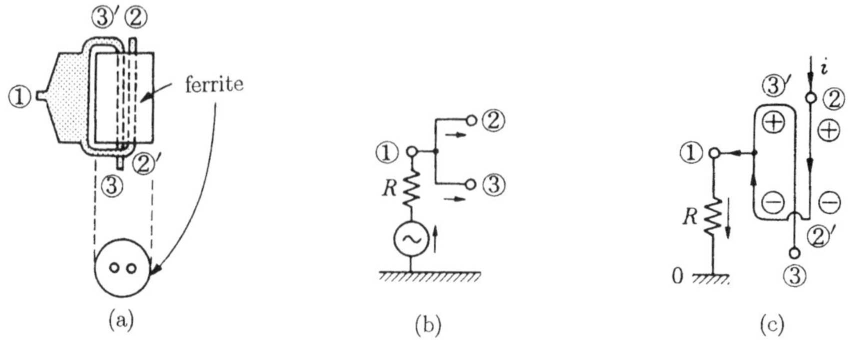

Figure 36. Explanation of the two-way power divider with two tightly coupled wires surrounded by a cylindrical ferrite: (a) construction; (b) current flow when the signal is supplied at port 1; (c) current flow when the signal is supplied at port 2.

is the condition of the isolation between ports 2 and 3. Because $\omega\mu_0\mu_r''L_0$ is almost constant in the wide frequency range as shown in Fig. 35 and $\omega\mu_0\mu_r'L_0 \ll \omega\mu_0\mu_r''L_0$, we can almost satisfy the condition of Eq. (22). Because ferrite has a Curie point, the performances of the isolation and the input VSWR are affected by the change of temperature. To decrease such an effect, an additional resistance R' is usually connected between ports 2 and 3, where the proper values of R' are designed together with consideration of the complex impedance caused by the complex permeability of ferrite.

Three-Way Power Divider

The construction is shown in Fig. 37a; the signal supplied at port 1 appears at P and P′ via T_2 the operation of which is that of the two-way power divider mentioned earlier. The signals are divided again through T_3 and T_4, and they appear at ports 2, 3, and 4, where the output ports of T_2 and T_3 are connected to each other to make port 3.

Since potentials 2, 3, and 4 must be the same, the currents should not flow through R_2. Therefore, denoting the current flowing in W connected at ports 2, 3, and 4, the current flowing in T_2 and T_3 must take the values as shown in Fig. 37b. Under the condition of the ideal transformer of T_3 and T_4, the number of turns of the coil wound in T_3 or T_4, n' and n, respectively must satisfy

$$n'i - n\frac{i}{2} = 0$$

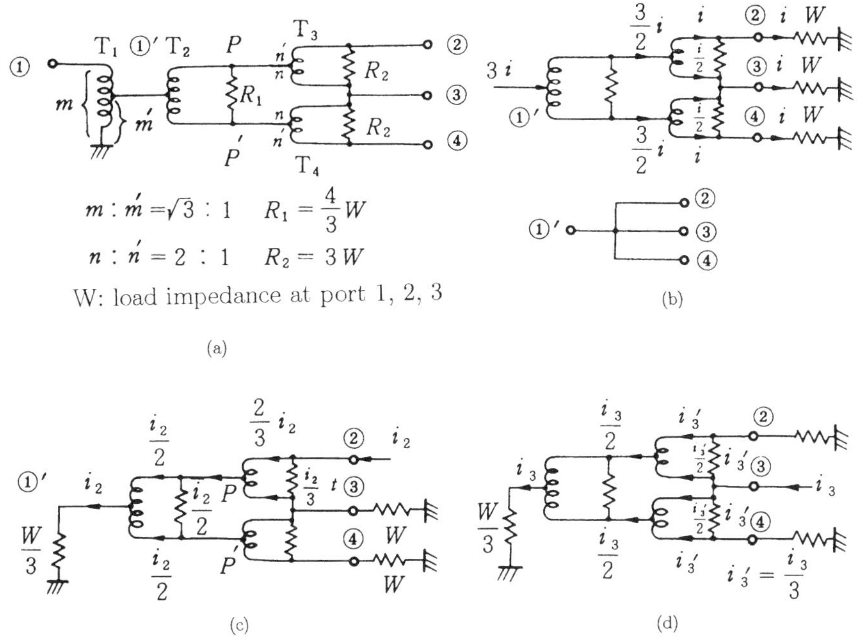

Figure 37. The construction and the explanation of three-way power dividers: (a) construction; (b) current flow when the signal is supplied to port 1′; (c) current flow when the signal is supplied to port 2; (d) current flow when the signal is supplied to port 3.

that is,

$$\frac{n}{n'} = 2 \tag{23}$$

Because the potential at points P and P′ of Fig. 37a are the same, we obtain the equivalent network shown in the bottom part of Fig. 37b. Therefore, the impedance viewed from port 1′ takes the values $W/3$, which requires the matching transformer T_1 with the ratio of the number of turns of $m{:}m' = \sqrt{3}{:}1$.

Next, we describe the isolation among ports 2, 3, and 4. When the signal is supplied to port 2 with current i_2 as shown in Fig. 37c, the potential of ports 3 and 4 should be zero. Therefore, the current does not flow in R_2 connected between ports 3 and 4, which results in no current in T_4. In the results, the current i_2 must be divided between T_3 and R_2 connected between 2 and 3.

From the condition of the ideal transformer of T_3, the current flowing into T_3 and R_2 should be $2i_2/3$ and $i_2/3$, respectively, as shown in Fig. 37c. The current flowing into point P, is also divided between T_2 and R_1 with the same values of $i_2/2$, because T_2 is also the ideal transformer and the current flowing into T_4 via point P′ must be zero.

Next, we will obtain the values of R_1 and R_2. In Fig. 37c, the voltage of port 4, V_4, must be zero. However, because

$$V_4 = V_{1'} - \frac{1}{2}(V_p - V_{p'}) = \frac{W}{3} i_2 - \frac{R_1}{4} i_2 = 0$$

we get

$$R_1 = \frac{4}{3} W \tag{24a}$$

Because the voltage at port 3, V_3 should also be zero, we obtain

$$V_3 = V_{1'} + \frac{1}{2}(V_p - V_{p'}) - \frac{2}{3}(V_2 - V_3)$$

$$= \frac{W}{3} i_2 + \frac{R_1}{4} i_2 - \frac{2R_2}{9} i_2 = \left(\frac{W}{3} + \frac{W}{3} - \frac{2R_2}{9}\right) i_2 = 0$$

Then

$$R_3 = 3W \tag{24b}$$

When the signal is supplied to port 4, it should not appear at ports 2 and 3. Under the conditions, we can get the same results of Eqs. (24a) and (24b) from the symmetry of the circuit.

Next, we will consider the case when the signal is supplied to port 3. Considering the symmetry of the circuit, the current flows as shown in Fig. 37d. In this case, the voltage of port 2, V_2, can be expressed by

$$V_2 = V_{1'} - \frac{1}{3}(V_3 - V_2) = \frac{W}{3} i_3 - \frac{R_3}{9} i_3$$

Applying Eq. (24b), we get

$$V = 0$$

From the symmetry of the circuit, we also get

$$V_4 = 0$$

Next, we will discuss the output impedance viewed from port 2. The voltage at port 2 in Fig. 37c, V_2, can be obtained by

$$V_2 = V_{1'} + \frac{1}{2}(V_p - V_{p'}) + \frac{1}{3}(V_2 - V_3) = \left(\frac{W}{3} + \frac{R_1}{4} + \frac{R_2}{9}\right) i_2$$

Substituting Eqs. (24a) and (24b) into the above equation, we get

$$\frac{V_2}{i_2} = W$$

In the same way, the voltage at port 3 of Fig. 37d, V_3, can be obtained by

$$V_3 = V_{1'} + \frac{2}{3}(V_3 - V_4) = \left(\frac{W}{3} - \frac{2R_3}{9}\right) i_3 = W i_3$$

This means that port 3 is matched with the load impedance W.

The descriptions presented above are concerned with the principle of the ideal network. When we obtain the network of Fig. 37a, we can also use the tightly coupled wires surrounded by ferrite material as in the case of the two-way power divider.

In the higher frequency, the isolations between output ports and the VSWR of the input port are deteriorated for the following reasons:

1. The leakage inductances existing at wires and the region between port $1'$ and the input point of T_2.
2. The floating capacitances.
3. The coupled lines should be considered as distributed lines rather than lumped elements, when the current flowing in two wires cannot be cancelled out by the electrical angle. In this case, T_2 and T_3 are no longer the ideal transformers.

To overcome the third reason, the length of the coupled line should be sufficiently short. We, however, encounter another problem in such an approach—the impedance of the wire at a low frequency becomes small, which results in the deterioration of the lower frequency. Then, we need a ferrite material with higher permeability.

The analysis of the power divider with distributed constant lines rather than a lumped element, is made in connection with the ferrite characteristics such as the initial permeability and the natural resonant frequency [16].

On the other hand, in the lower-frequency region, the impedance of the wires with ferrite becomes almost maximum at the natural resonant frequency, f_r. In the frequency range below f_r, the impedance of the transformer approach to short impedance, when the power supplied to the one of the output ports will be divided between an input port and an output port. It results in an isolation of 6 dB [17]. The frequency performance, therefore, becomes as shown in Fig. 38 [17].

The complex relative permeability of the ferrite, $\dot{\mu}_r$, can be shown by the empirical equation [18].

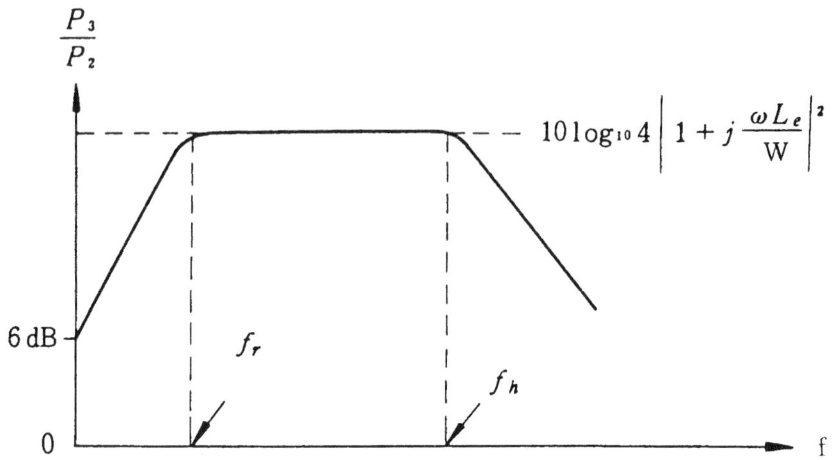

Figure 38. Frequency performance of isolation. (From Ref. 16.)

$$\dot{\mu}_r(f) \simeq 1 + \frac{K_r}{1 + j(f/f_r)} \tag{25}$$

in the frequency range $f >> f_r$. In the ordinary ferrite materials, the measured results show

$$K_r f_r \leq 8 \text{ GHz} \tag{26}$$

To construct a wideband power divider, large values of $K_r f_r$ and small values of f_r are preferred.

3 DIRECTIONAL COUPLER AND BRIDGE

3.1 Principal of Directional Coupler

Definition of Directional Coupler

As shown in Fig. 39a, the electromagnetic wave transmits from a transmitter to an antenna; sometimes, the reflected wave occurs when the antenna is mismatched with the feeder. To detect the component of the incident wave and the reflected wave, a four-port network, the so-called directional coupler, is used, as shown in Fig. 39b. When the incident power, P_i, is supplied to port 1, the power appears at ports 2 and 3, with the quantities of P_i' and P_i'', respectively, and not at port 4.

In the same way, when the reflected power, P_r, is supplied to port 2, the power appears at ports 1 and 4 with the quantities of P_r' and P_r'', respectively, and not at port 3. In this case, the quantity C defined as

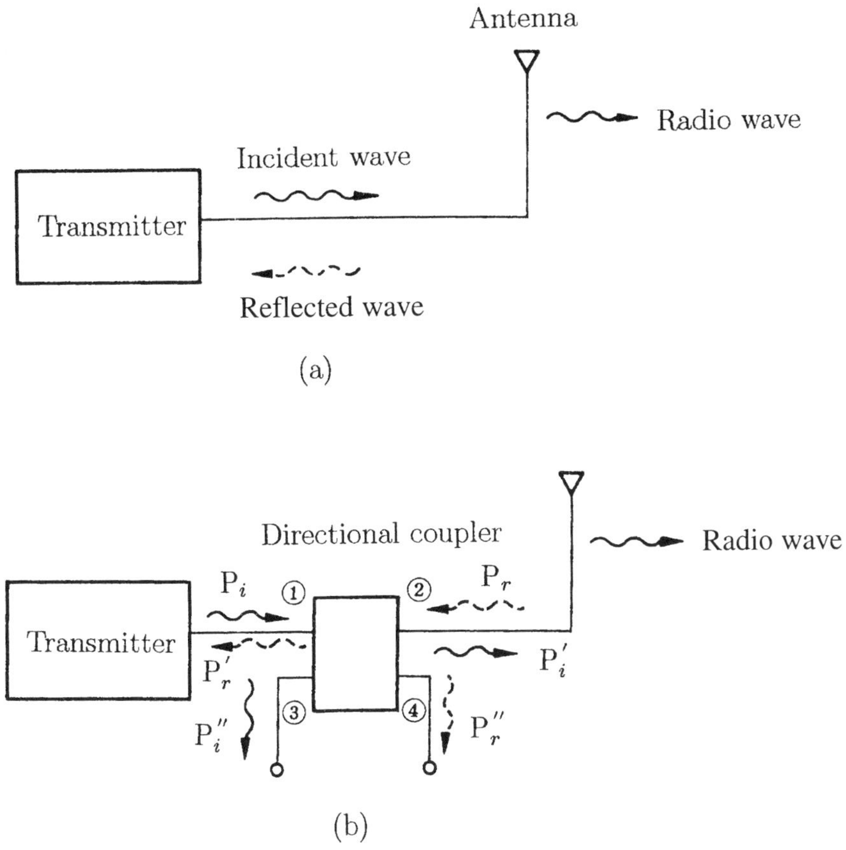

Figure 39. Example of applications of directional coupler: (a) reflective wave caused by mismatching in antenna; (b) directional coupler to detect incident wave at port 3 and reflection wave at port 4.

$$C = 10 \log_{10} \left(\frac{P_i}{P''_i} \right) = 10 \log_{10} \left(\frac{P_r}{P''_r} \right) \tag{27}$$

is called the coupling of the directional coupler, where C is the attenuated quantity from port 1 to port 3.

The description presented above is the ideal case, where P_i does not appear at port 4. In the practical case, however, a very small amount of the power P'''_i appears at port 4. In this case, the values of D defined as

$$D = 10 \log_{10} \left(\frac{P''_i}{P'''_i} \right) = 10 \log_{10} \left(\frac{P''_r}{P'''_r} \right) \tag{28}$$

is called the directivity of the directional coupler. In the ordinal devices, D is 20–30 dB.

Principle of Operation

Method of Using the Difference of the Relative Direction of E and H Between the Incident and the Reflected Waves As shown in Fig. 40, the relative directions of $\boldsymbol{E}$ and $\boldsymbol{H}$ are opposite to the incident and the reflected waves. The characteristics can be used to construct the directional coupler.

When the incident power, P_i, is supplied to port 1 in Fig. 41a, the circuit is designed such that the quantity A proportional to the $\boldsymbol{E}_i$ of P_i appear at ports 3 and 4 in the same phase and the quantity B proportional to the $\boldsymbol{H}_i$ of P_i appear at ports 3 and 4 in the opposite phase. If the design is made such that $A = B$, the incident wave does not appear at port 4.

Next, when the reflected wave is supplied to port 2, the phase of the magnetic field becomes opposite for the same electric field. Therefore, the reflected wave appears at port 4 and does not appear at port 3, as shown in Fig. 41b. An example of the principle is the loop-coupled directional coupler described later.

Different Ports Connected by a Multipass When the electric angle between ports 1 and 2 or 3 is $\pi/2$, the power of the incident wave supplied to port 1, P_i, is added in the same phase and appears at port 3, as shown in Fig. 42a. On the other hand, passes I′ and II′ have a difference of π radian in electric angle, which results in the disappearance of P_i at port 4, as shown in Fig. 42b. In the same way, the reflected power, P_r, supplied to port 2

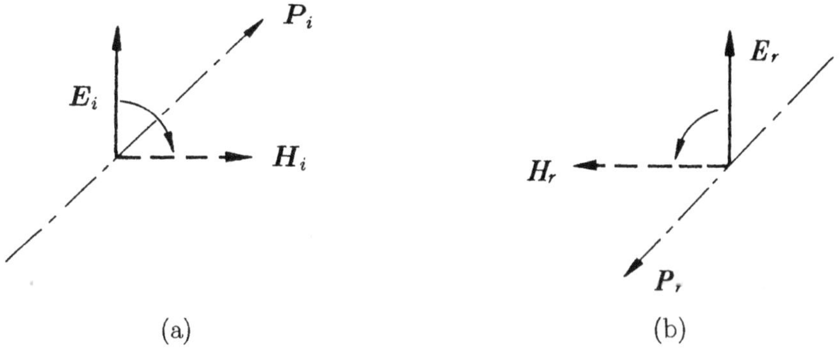

Figure 40. Direction of $\boldsymbol{E}_i$ (electric field) and $\boldsymbol{H}_i$ (magnetic field) corresponding to (a) the incident wave and (b) the reflected wave. The arrow from $\boldsymbol{E}_i$ to $\boldsymbol{H}_i$ is the direction of rotation for the right screw. The arrows of P_i and P_r show the directions of the wave propagation.

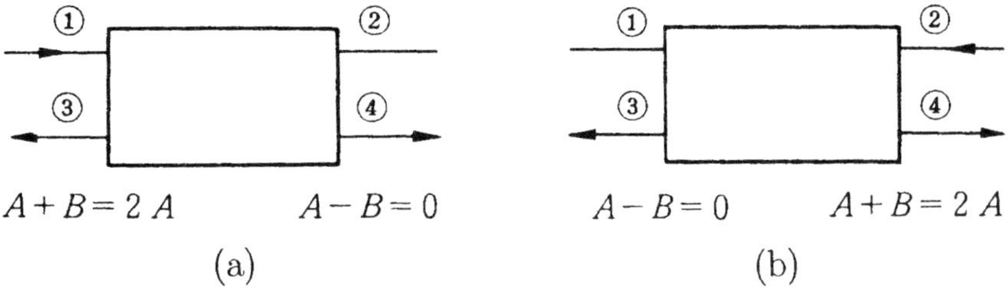

Figure 41. Explanation of the directional coupler by using the characteristics of the difference in the relative directions of ***E*** and ***H*** between the incident and the reflected waves: (a) direction of the incident wave and (b) direction of the reflected wave.

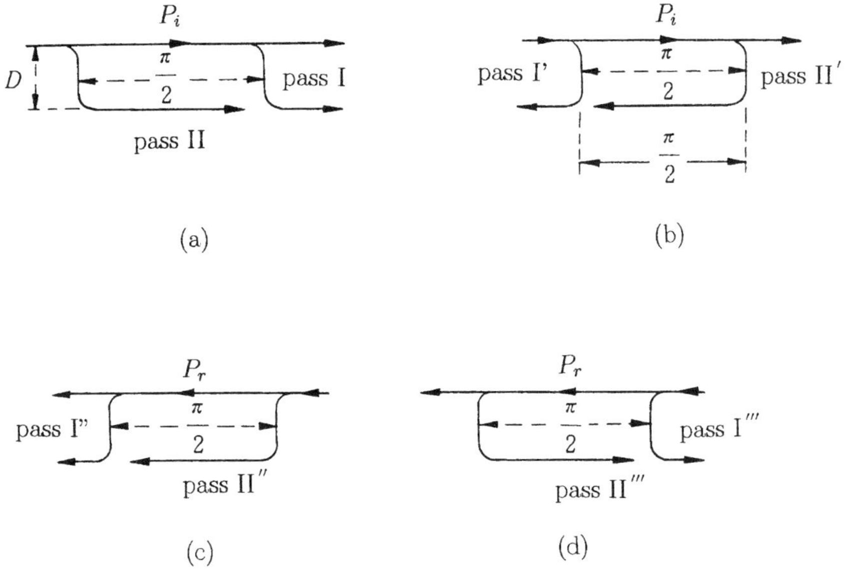

Figure 42. Different ports are connected by two passes and their phase difference is zero and π corresponding to the incident and the reflected wave, respectively: (a) P_i appear at port 3; (b) P_i disappears at port 4; (c) P_r appears at port 4; (d) P_r disappears at port 3.

appears at port 4, as shown in Fig. 39c, and does not appear at port 3, as shown in Fig. 42d.

An example of the principle is a hybrid ring (described later).

3.2 Practical Examples

Loop-Coupled Directional Coupler

This is an example of the case in Section 3.1. A loop is coupled to a coaxial line and a $TE^{\square}_{10}$ waveguide with the construction shown in Figs. 43a and 43b, respectively.

The displacement currents caused by the electric fields, $j\omega KE$, are divided between ports 3 and 4 with the same phase, as shown in Fig. 43c. The

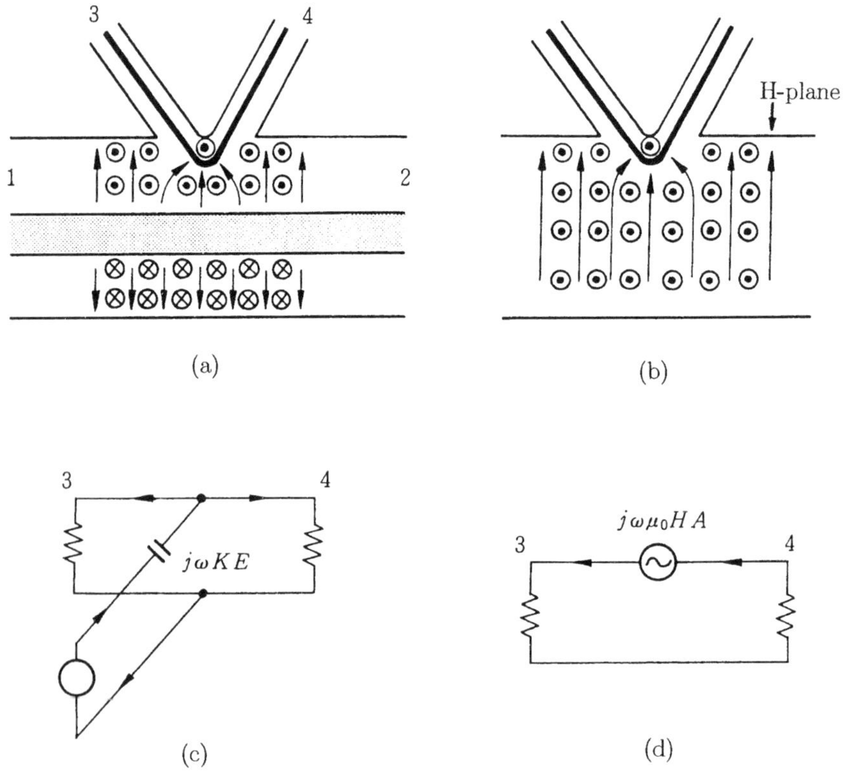

Figure 43. Explanation of a loop-coupled directional coupler: (a) electromagnetic field in the case of loop coupling to a coaxial line; (b) electromagnetic field in the case of loop coupling to a $TE^{\square}_{10}$ waveguide; (c) current flowing in a loop caused by electric fields; (d) current flowing in a loop caused by the magnetic fields.

electromotive force caused by the magnetic fields crossing in a loop, $j\omega\mu_0 HA$, produces the current as shown in Fig. 43d. To cancel out the two kinds of currents at port 4, the relation

$$\frac{j\omega KE}{2} = \frac{j\omega\mu_0 HA}{2R}$$

is necessary. However, because we have the relation $E = ZH$ (Z = wave impedance), the above relation becomes

$$KZ = \frac{\mu_0 A}{R} \tag{29}$$

The directional coupler, therefore, can be obtained by adjusting K or A because Z and R are defined.

On the other hand, H of the reflected wave takes the opposite phase from that of the incident wave under the same phase of E. Therefore, the reflected wave appears at port 4 and not port 3.

To get the high directivity, sometimes the loop is rotated to adjust the quantity of the crossed magnetic fields.

As in the constructions of Figs. 43a and 43b, the short subline parallel to the main line (coaxial of waveguide) is sometimes used, as shown in Fig. 44a. The sectional view and the equivalent network are as shown in Figs. 44a and 44c, respectively.

The displacement currents I_d can be expressed by using the notations of Fig. 44b as follows:

$$I_d = j\omega C_{12}(V_1 - V_2)$$

where V_1 and V_2 are the voltages of the main line and sublines, respectively. For $V_1 >> V_2$, we get

$$I_d \simeq j\omega C_{12}$$

As understood from the explanation in Fig. 43, we have

$$\frac{j\omega C_{12}}{2} V_1 = \frac{j\omega M I_1}{2R} = \frac{j\omega M}{2R} \frac{V_1}{Z_c},$$

where I_1 is the incident wave and Z_c is the characteristic impedance of the coaxial line. We, therefore, get

$$\frac{M}{C_{12}} = Z_c R \tag{30}$$

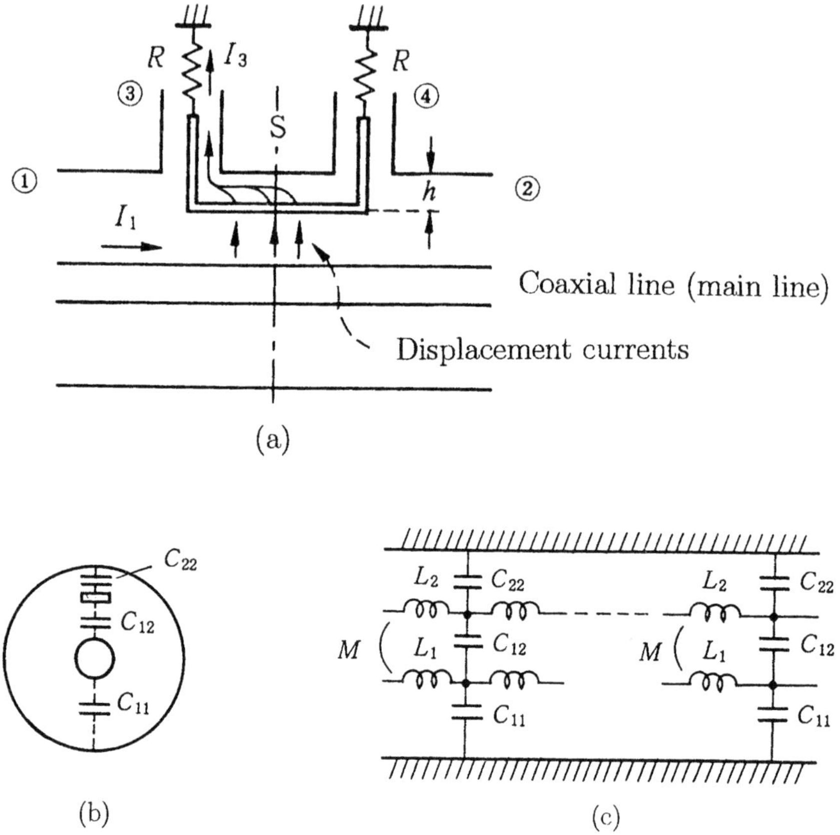

Figure 44. Loop of Fig. 43a is replaced by a short distributed constant line: (a) construction and displacement currents; (b) sectional view and capacities; (c) equivalent network of coupled distributed lines expressed by L and C.

In the case when the main line and subline exist in a same medium, such as air, the TEM (transverse electric magnetic) wave can propagate. In this case, we have

$$\frac{M}{\sqrt{L_1 L_2}} = \frac{C_{12}}{\sqrt{C_1 C_2}}, \qquad C_1 = C_{11} + C_{12}, \qquad C_2 = C_{22} + C_{12} \tag{31a}$$

then

$$\frac{M}{C_{12}} = \sqrt{\frac{L_1}{C_1}}\sqrt{\frac{L_2}{C_2}} \tag{31b}$$

When the coupling between the main and sublines is weak, we get

$$\sqrt{\frac{L_1}{C_1}} \simeq \sqrt{\frac{L_1}{C_{11}}} = Z_c \tag{31c}$$

for $C_{11} >> C_{12}$.

From Eqs. (30), (31b), and (31c), we get

$$\sqrt{\frac{L_2}{C_2}} = R \tag{31d}$$

This means that the characteristic impedance of a subdistributed constant line has the same load impedance. When we want to change the coupling, the distance h in Fig. 44a should be changed under the same characteristics.

Distributed-Coupling Directional Coupler

Because this can be considered as the extension of the loop-coupling directional coupler (which will be explained later), it is also an example of the case in Chapter 6, Section 3.1.

We will consider the construction in Fig. 45, where the length of the subdistributed constant line is l.

The displacement currents flow from the main distributed constant line to the subdistributed constant line everywhere, as shown in Fig. 45. When the displacement current density at $z = 0$ and $t = 0$ takes the values of $I(0)$, it decreases as z increases and it becomes zero at $z = \lambda/4$, as shown in Fig. 45. The currents corresponding to z can be shown by the vectors on a complex plane, as in Fig. 46. Such currents flow together into port 3. When l increases, the $\lambda/4$ exists in the second quadrant and the current cancels $I(0)$ at $z = 0$. Therefore, $I(0)$ decreases and approaches zero in the case of $l = \lambda/2$. The relation between the currents of port 3 and l is shown in Fig. 47.

As understood from Fig. 47, the loop-coupled directional coupler is the region $l << \lambda_g/4$. It is clear from Fig. 45 that the directivity, D is infinite at any frequency.

Although the above description is concerned with a weak coupling, the symmetrical structure as shown in Fig. 48a is usually used in the case of a strong coupling. The symmetrical structure shown in Fig. 48 is widely used as a distributed-coupling directional coupler not only for the strong coupling but also the weak coupling.

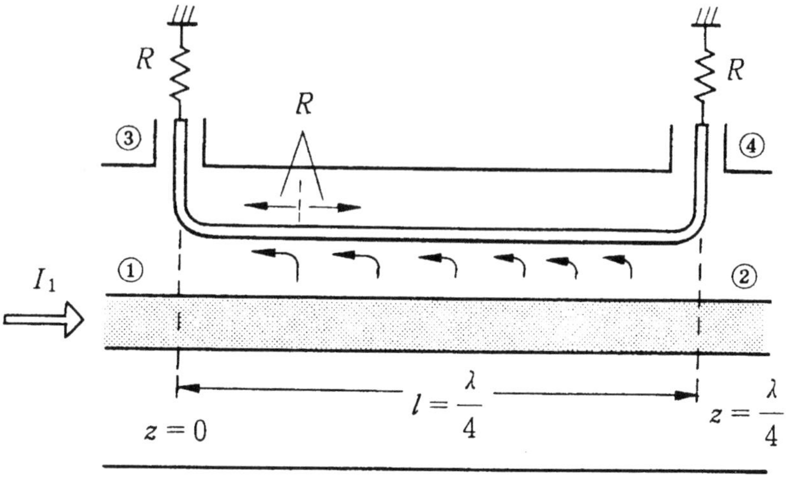

Figure 45. Construction of distributed-coupling directional coupler.

In the case of such a symmetrical structure, the coupling of the directional coupler, C (dB), is obtained from (see Appendix 6)

$$C\ (\text{dB}) = 10 \log_{10} |\beta|^2$$

$$|\beta| = \left(\frac{Z_e - Z_o}{Z_e + Z_o} \left\{ 1 + \frac{Z^2}{(Z_e + Z_o)^2} \left(\cot \frac{\theta}{2} - \tan \frac{\theta}{2} \right)^2 \right\}^{-1/2} \right) \tag{32a}$$

$$Z_e Z_o = Z^2$$

In Eqs. (32), Z_e and Z_o are the characteristics impedances of parallel lines for even and odd modes, respectively. For example, the electromagnetic fields in Fig. 48b for even and odd modes are shown in Fig. 49a and 49b,

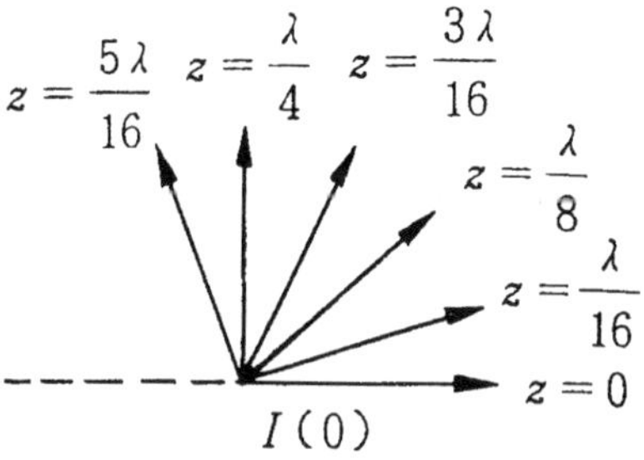

Figure 46. Current component at Z.

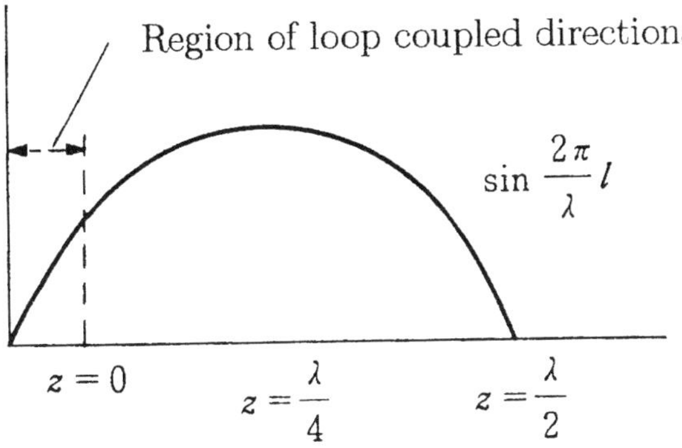

Figure 47. Relation between current at ports 3 and 1 in Fig. 46.

respectively, where the solid lines and dotted lines show electric fields and magnetic fields, respectively.

For $\theta = \pi/2$, we get the maximum values of $|\beta|$, $|\beta|_{max}$,

$$|\beta|_{max} = \frac{Z_e - Z_o}{Z_e + Z_o} \tag{32b}$$

and

$$Z_e = Z \sqrt{\frac{1 + |\beta|_{max}}{1 - |\beta|_{max}}}, \qquad Z_o = Z \sqrt{\frac{1 - |\beta|_{max}}{1 + |\beta|_{max}}} \tag{32c}$$

Therefore, when the coupling is weak, we get

$$Z_e \simeq Z_o \simeq Z_c \tag{33a}$$

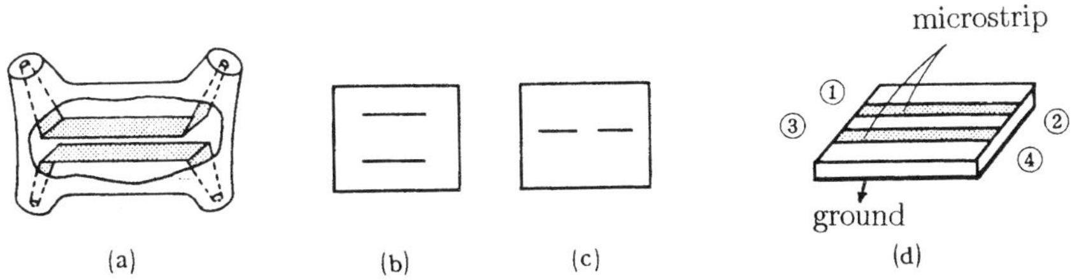

Figure 48. Distributed-coupling directional coupler: (a) broadside coupling; (b) sectional view of (a); (c) narrow-side coupling; (d) microstrip coupling.

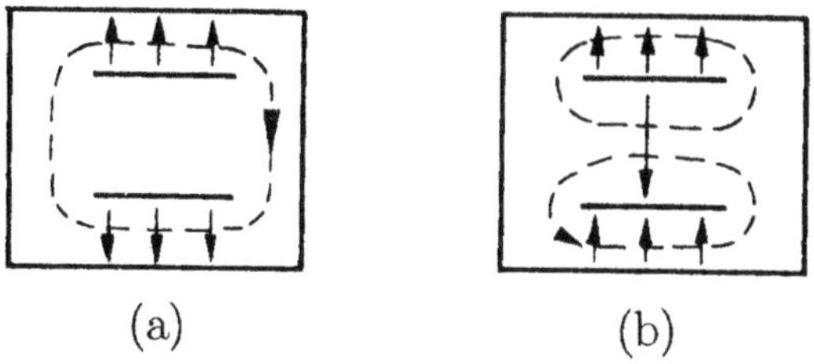

Figure 49. Electromagnetic field distributions: (a) even mode; (b) odd mode.

and

$$C\ (\mathrm{dB}) \simeq 10 \log_{10} \left(\frac{Z_e - Z_o}{Z1_e + Z_o} \sin\theta \right)^2 \tag{33b}$$

Therefore, the current flowing into port 1 appears at port 3 in a quantity proportional to sin θ, the result of which coincides with the explanation in Fig. 47.

In the case in Fig. 48d, the electric fields of even and odd modes is presented in Figs. 50a and 50b, respectively.

For the even mode, the symmetrical plane becomes the magnetic wall and almost all the electric fields exist in the dielectric substrate. On the other hand, for the odd mode, the symmetrical plane becomes the electric wall and the electric fields exist not only in the dielectric substrate but also in the air, which results in a lower effective dielectric constant. Therefore, we have the relation $\theta_e > \theta_o$, which deteriorates the directivity. To avoid this problem, sometimes the construction with the sectional view in Fig. 51 is used and it is called the overlay structure.

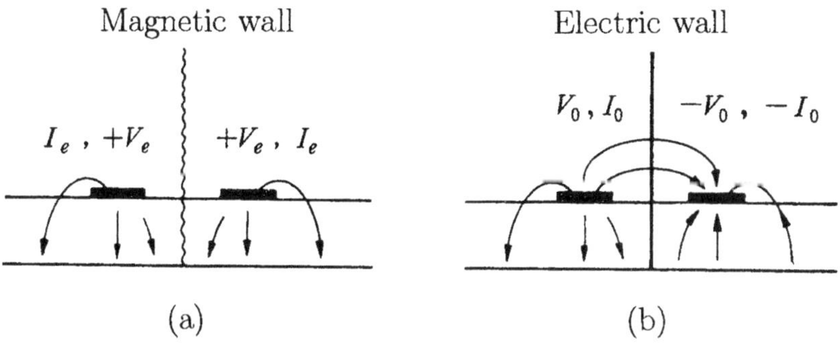

Figure 50. Electric fields, voltage, and current for even (a) and odd (b) modes.

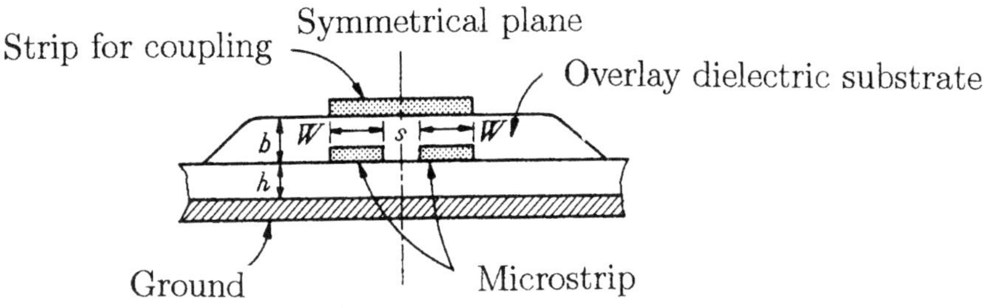

Figure 51. Distributed-coupling directional coupler.

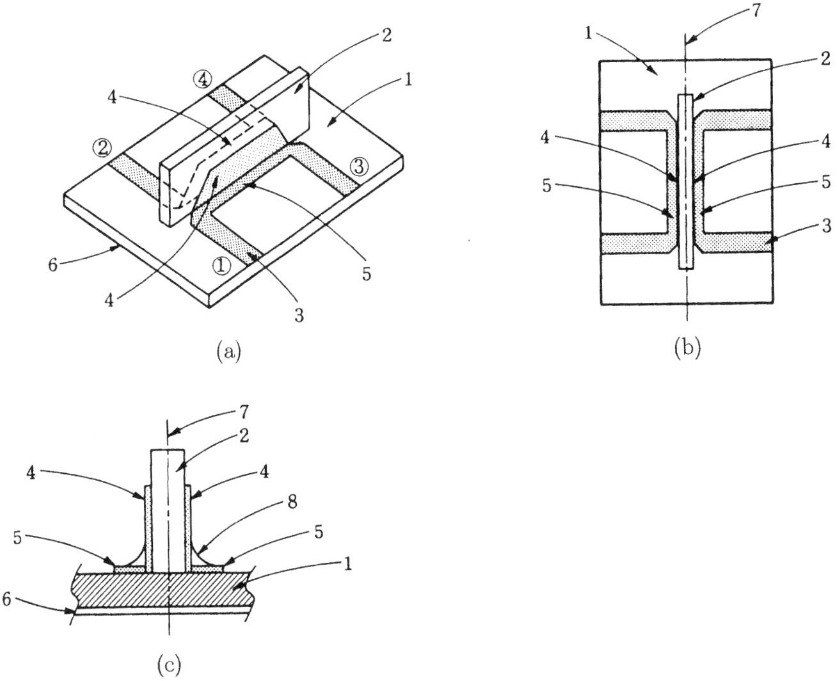

Figure 52. Vertically installed planar circuit directional coupler: (a) construction; (b) top view; (c) sectional view. ①, ②, ③, ④: port of directional coupler; 1: first dielectric substrate; 2: second dielectric substrate; 3: microstrip line; 4: parallel strip made on ②; 5: microstrip contacted to ④; 6: ground; 7: symmetrical plane; 8: soldering.

To obtain a strong coupling of such a 3-dB coupler, we can use the construction of Fig. 52, which is simply made on a substrate [18,19]. It is important to realize for the distributed constant-line-type directional coupler that the phase velocity of the even and odd mode are the same; that is, we have

$$\sqrt{L_e C_e} = \sqrt{L_o C_o} \tag{34a}$$

where L_e and C_e, and L_o and C_o are the inductances and capacitances per unit length for the even and odd modes, respectively.

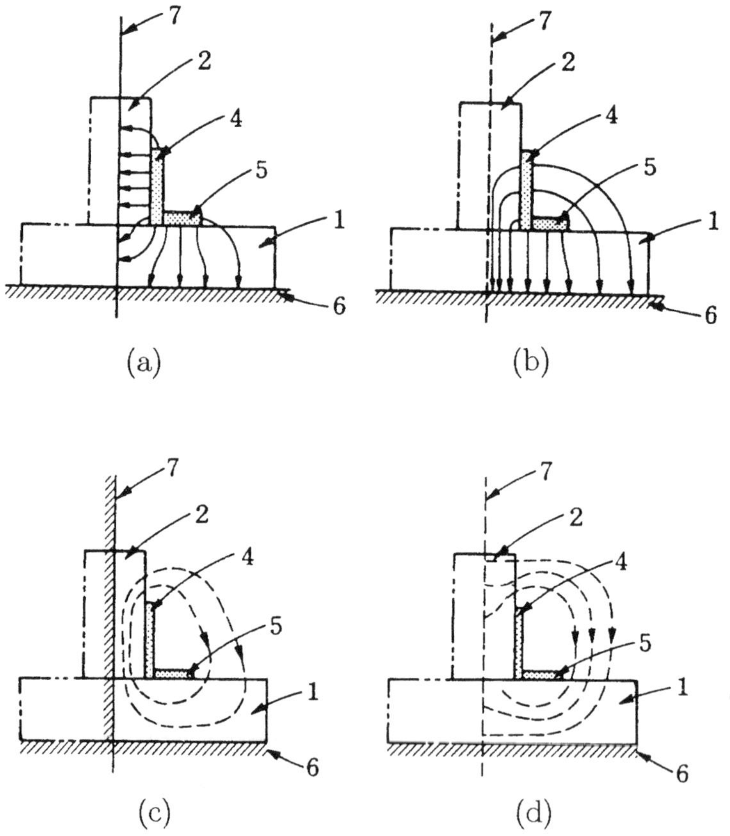

Figure 53. Electromagnetic field distributions; (a) odd-mode electric fields; (b) even-mode electric fields; (c) odd-mode magnetic fields; (d) even-mode magnetic fields. See legend of Fig. 52 for meaning of numbers.

On the other hand, because we have the relationships in Eq. (32c), $Z_e > Z_o$, that is,

$$\sqrt{\frac{L_o}{C_o}} < \sqrt{\frac{L_e}{C_e}} \tag{34b}$$

is required. From Eqs. (34a) and (34b), we get

$$C_o > C_e \tag{34c}$$

$$L_o < L_e \tag{34d}$$

We, therefore, examine the above conditions by the electromagnetic distributions in Figs. 53a–53d. Comparing the electric fields in Figs. 53c and 53b, the fields passing the air region is more in the even mode than the odd mode, which results in Eq. (34c).

In the case of the odd mode, because the currents on both lines flow in opposite directions, the magnetic fluxes partly cancelled out each other, which decreases the inductance. On the other hand, the magnetic flux for the even mode is increased by the coupling of the other line, because the current flows in the same direction. This results in the condition of Eq. (34d). Therefore, we can obtain Eqs. (34a) and (34b) if the proper sectional size is chosen together with the proper dielectric constant of the second substrate used for the coupled distributed lines [19].

Bethe Hole Coupler

As shown in Fig. 54, the sub-$TE_{10}^{\square}$ waveguide is piled on the main $TE_{10}^{\square}$ waveguide as they contact each other in the *H* plane, and they are coupled through a hole.

The electrical fields of the incident wave traveling from port 1 toward port 2 in the main $TE_{10}^{\square}$ waveguide become as shown in Fig. 54b, which is coupled to the sub-$TE_{10}^{\square}$ waveguide. The electrical fields are equivalent to those generated from an electrical dipole, P, as shown in Fig. 54c; the electromagnetic fields are shown in Fig. 54d.

On the other hand, the magnetic fields take the distribution as shown in Fig. 54e. The direction of the magnetic fields corresponding to the electric fields can be understood from the direction of the Poynting power toward port 2 from port 1 in the main waveguide. The magnetic fields are equivalent to those generated from a magnetic dipole, M, as shown in Fig. 54f. The electromagnetic fields radiated from M are shown in Fig. 54g. Comparing the electric fields of Fig. 54d with that of Fig. 54g, the electric fields are opposite in port 4 and are added in the same phase in port 3.

As shown in Fig. 54a, when the relative angle between the both waveguide, θ, is adjusted, the magnetic fields by M change according to cos θ

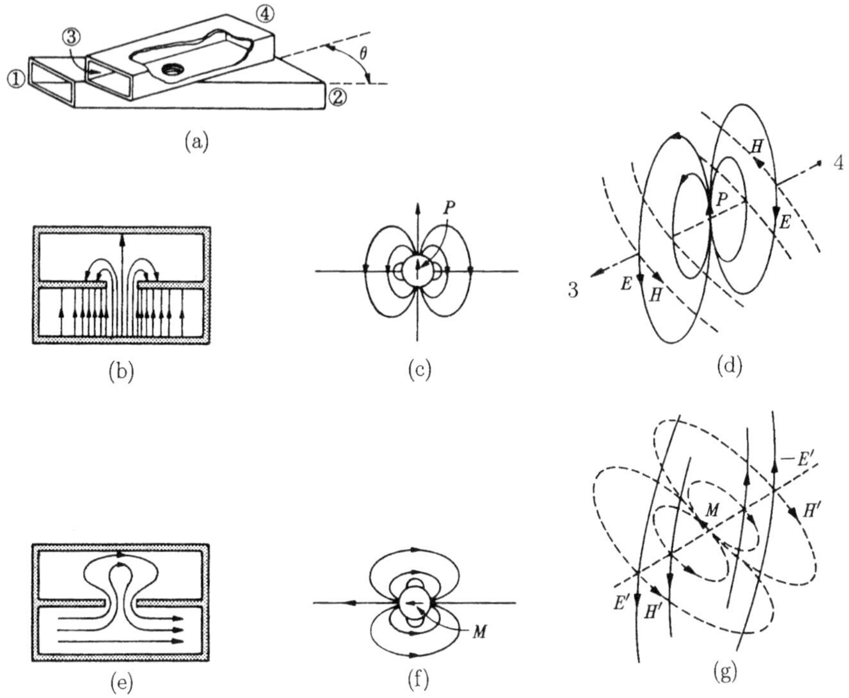

Figure 54. Construction and explanation of the Bethe hole coupler: (a) construction; (b) electrical coupling through hole; (c) equivalent electric dipole P vertical to the hole and the electric field; (d) radiation from P; (e) magnetic coupling through hole; (f) equivalent magnetic dipole M parallel to the hole and the magnetic field; (g) radiation from M.

even if the electric fields by P do not change. Therefore, the wave toward port 4 can be adjusted to zero by changing θ. In this case, the coupling of the directional coupler, C (dB), is obtained from Eq. (35) [20]:

$$C = 20 \log_{10} \left(\frac{2\pi d^3}{3ab\lambda_g} \right) \cos \theta \quad \text{(dB)} \tag{35}$$

On the other hand, in the case of the reflected wave traveling from port 2 toward port 3, the magnetic fields are in reverse direction from that of Figs. 54e–54g as understood from the Poynting theorem under the same electric fields. Therefore, the reflected wave appears in port 4 and not at port 3.

As understood from the above-mentioned principle, the Bethe hole coupler belongs to the principle mentioned in Section 3.1.

Cross-Type Directional Coupler

The electromagnetic fields of $TE_{10}^{\square}$ waveguides are shown in Fig. 55. When the wave propagates from the left to the right, the magnetic fields existing between the *E* plane and the vertical plane parallel to the *E* plane at the center of the *H* plane take the circularly rotating fields, which is seen from the magnetic fields of Fig. 55c traveling toward the right. The cross-type directional coupler, however, is made from the main waveguide and the subwaveguide being tiled by contacting their *H* planes and they are coupled through the hole with a cross shape shown in Fig. 56. In Fig. 56, the magnetic fields at hole A rotate clockwise in both waveguides as shown in Fig. 55c. Therefore, the wave traveling from port 1 to port 2 in the main waveguide is coupled to the wave traveling from port 3 to port 4. In the practical case, an additional hole, B, is made (as in Fig. 56) to increase the coupling power.

In Fig. 56, the phase delay through the pass of 1 → A → 4, θ_A, is obtained from

$$\theta_A = \theta_{1A} + \theta_{A4} + \frac{\pi}{2} \tag{36a}$$

where θ_{1A} and θ_{A4} are the delays of passes 1 → A and A→ 4, and the delay at hole A is $\pi/2$, because both waveguides are twisted 90°.

In the case of the pass 1 → B → 4, the total phase delay, θ_B, is

$$\theta_B = \theta_{1B} + \theta_{BA} - \frac{\pi}{2} \tag{36b}$$

where θ_{1B} and θ_{BA} are the delays of passes 1 → B and B → A, and $-\pi/2$ is the delay at hole B, because the magnetic fields at hole B rotate counterclockwise. As understood from Fig. 56, $\theta_{1B} = \theta_{1A} + \pi/2$ and $\theta_{BA} = \theta_{A4} + \pi/2$. Substituting the relations into Eq. (36b), we get

$$\theta_B = \theta_{1A} + \frac{\pi}{2} + \theta_{A4} + \frac{\pi}{2} - \frac{\pi}{2} = \theta_A \tag{36c}$$

Therefore, the wave through holes A and B are added in the same phase and travel to port 4.

As is clear from the principle mentioned above, if there are some electrical coupling, the directivity becomes worse. To avoid this, the hole is made in a cross shape, which makes it possible to give the magnetic coupling only.

For the reflective wave traveling from 2 → 1, the coupled wave to the subwaveguide propagates to port 3, because the rotation at the holes is reversed, as is the case of the incident wave described before.

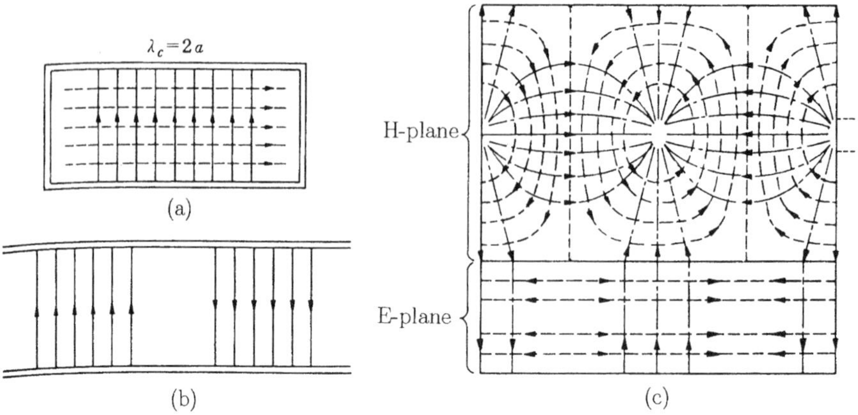

Figure 55. Electromagnetic field distributions of $TE^{\square}_{10}$ waveguide: (a) sectional view; (b) side view; (c) distribution on the H and E planes. —→: electrical fields; - - -→: magnetic fields; –·→: currents.

This directional coupler uses the rotational direction occurring in a waveguide and the direction is related to the electric field under the Poynting theorem. As such, the Bethe hole coupler belongs to the category of the principle of Section 3.1.

Multihole-Coupled-Type Directional Coupler

Two $TE^{\square}_{10}$ waveguides are arranged in such a way that one of their E planes is used in common, as shown in Fig. 57. On the common E plane, more

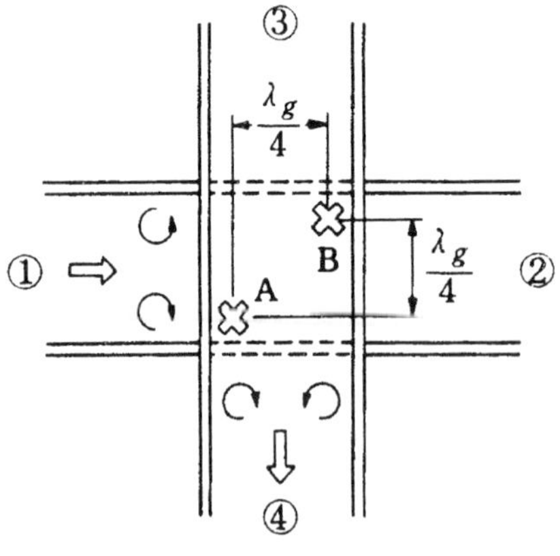

Figure 56. Cross-type directional coupler.

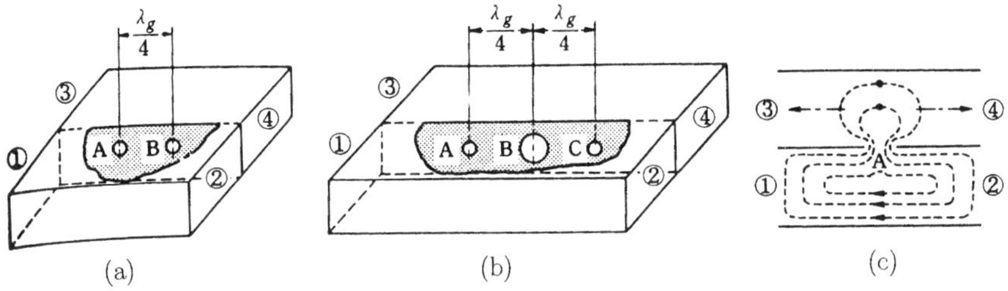

Figure 57. Multihole-type directional coupler: (a) two hole; (b) three hole; (c) magnetic fields distribution in a hole.

than two coupling holes are made with a distance of $\lambda_g/4$. The magnetic fields of the incident wave traveling from port 1 to port 2, are coupled through the hole and they are divided in the direction of ports 3 and 4 as shown in Fig. 57c. In Fig. 57a, the wave at port 4 is the sum of the waves through the passes 1 → A → 4 and 1 → B → 4 with the same phase.

On the other hand, the wave at port 3 is the sum of the waves through the passes 1 → A → 3 and 1 → B → 3 with the reverse phases, which results in the cancellation. This is the operation of the directional coupler. As understood from the explanation, this directional coupler belongs to the principle described in Section 3.1. The cancellation, however, takes place only at the frequency which satisfies the condition of $\lambda_g/4$. The bandwidth of the directivity, therefore, is very narrow. To improve the performance, more than three holes on the common E plane are made, as shown in Fig. 57b. When we denote the coupling coefficients of holes A, B, and C by C_1, C_2, and C_3, and we denote the wave through passes 1 → A → 3, 1 → B → 3, and 1 → C → 3 by A_1, A_2, and A_3 with the phase of each pass being θ_1, θ_2, and θ_3, the voltage appearing at port 3, V_3, can be expressed by

$$V_3 = (C_1 e^{j\theta_1} + C_2 e^{j\theta_2} + C_3 e^{j\theta_3})V_1 \tag{37a}$$

where V_1 is the applied voltage at port 1. When the frequency increases with Δf, θ_2 and θ_3 deviate $\Delta\theta$ from the center frequency, which can be expressed as

$$\theta_2 = \theta_1 + \pi + \Delta\theta$$

$$\theta_3 = \theta_1 + 2\pi + 2\Delta\theta_2$$

$$\Delta\theta = \pi \frac{\Delta f}{f_0}$$

Substituting these equations into Eq. (37a), we get

$$\left|\frac{V_3}{V_1}\right| = \left|C_1 e^{j\theta_1}\right| \cdot \left|1 - \frac{C_2}{C_1} e^{j\Delta\theta} + \frac{C_3}{C_1} e^{j2\Delta\theta}\right|$$

$$\simeq C_1 \left\{\left(1 - \frac{C_2}{C_1} + \frac{C_3}{C_1}\right) + \left(\frac{C_2 - 2C_3}{C_1}\right)^2 (\Delta\theta)^2\right\} \tag{37b}$$

For $V_3 = 0$, therefore, we must have

$$C_1 = C_3 = \frac{C_2}{2} \tag{38}$$

This requires that the center hole should have twice the coupling of end holes A and C.

By extending such a concept, for more than two holes, we can get the wideband performance by choosing the coupling of the *n*th hole, as shown in Fig. 58.

The coupling coefficient, C_n, through the hole with radius r_n, is calculated from [21]

$$C_n = -\frac{4}{3} r_n^3 \frac{(\pi/a)^2}{ab\beta} \tag{39}$$

N		C1	C2	C3	C4	C5
2	....	1	1			
	+		1	1		
3	....	1	2	1		
	+		1	2	1	
4	....	1	3	3	1	
	+		1	3	3	1
5	....	1	4	6	4	1
....						

Figure 58. Coupling coefficients of the *n*th hole, C_1, C_2, . . . , of *N* holes of a coupled-type directional coupler.

Multi-branch-line Directional Coupler, Ring-type Directional Coupler, and the Equivalent Lump Element Direction Coupler

The constructions using coaxial cables and microstrip lines are shown in Figs. 59a and 59b, respectively, where ports 1–4 are connected with a $\lambda_g/4$ distributed constant line.

The characteristic impedances of the distributed constant lines are Z_1 and Z_2, as shown in Fig. 59. The signal supplied at port 1 does not appear at port 4 because the electrical angles through passes 1 → 2 → 3 → 4 and 1 → 4 are different than π radian, which results in cancelling the waves at port 4. This is the brief concept. Therefore, this type belongs to the principle based on the multipass described in Section 3.1.

When we obtain the values of Z_1 and Z_2, assuming the voltage at port 4 to be zero, we can make the simple circuit in Fig. 60b, for the signal flow supplied at port 1. It is easily understood because the impedance viewed from ports 1 and 3 toward port 4 becomes infinite in the case of $V_4 = 0$.

When we want to design the coupling from port 1 to port 3, $C_{1\to3}$ (dB) of

$$C_{1\to3} = 10 \log_{10}|\beta|^2 \tag{40a}$$

the coupling from port 1 to port 2 is calculated from Eq. (40b) by the unitary condition in a lossless network:

$$C_{1\to2} = 10 \log_{10}|\alpha|^2$$
$$|\alpha|^2 = 1 - |\beta|^2 \tag{40b}$$

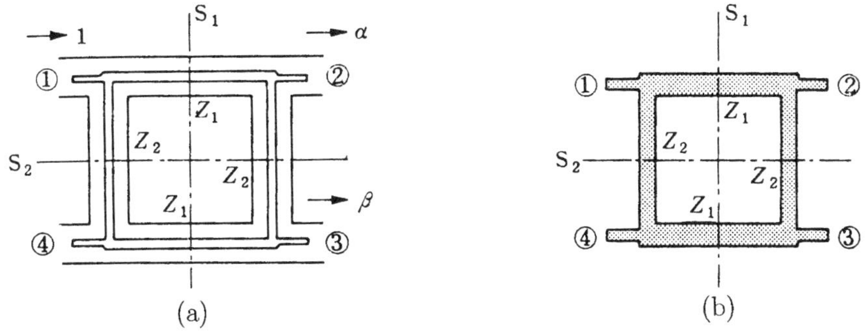

Figure 59. A two-branch-line directional coupler: (a) coaxial line; (b) microstrip line.

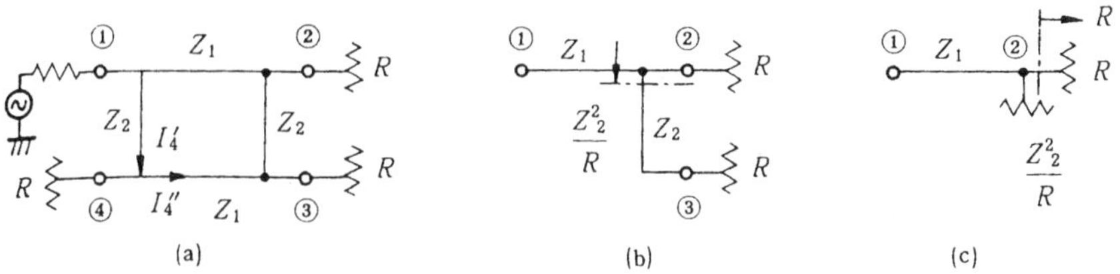

Figure 60. A two-branch line directional coupler: (a) construction; (b) equivalent network for the signal supplied at port 1; (c) impedance viewed from port 2.

However, because the impedance viewed from port 2 to port 3 becomes Z_2^2/R, we get

$$\frac{|\alpha|^2}{|\beta|^2} = \frac{Z_2^2}{R^2}$$

then

$$Z_2 = R\left|\frac{\alpha}{\beta}\right| = R\frac{\sqrt{1-|\beta|^2}}{|\beta|} \tag{41a}$$

On the other hand, under the matching condition at port 1, we get

$$R = Z_1^2\left(R\Big/\Big/\frac{Z_2^2}{R}\right)^{-1} = Z_2^2\left(\frac{Z_2^2}{R + Z_2^2/R}\right)^{-1} = \frac{RZ_1^2\,(1 + Z_2^2/R^2)}{Z_2^2}$$

Therefore, we obtain Z_1 from

$$Z_1 = Z_2\frac{1}{\sqrt{1+(Z_2/R)^2}} = \frac{R|\alpha/\beta|}{\sqrt{1+|\alpha/\beta|^2}} = R\sqrt{1-|\beta|^2} \tag{41b}$$

In conclusion, Z_1 and Z_2 can be obtained from Eqs. (41b) and (41a) if $|\alpha/\beta|$ or $|\beta|$ is given. For example, in the case

$$|\alpha| = |\beta| \tag{42a}$$

we get

$$Z_1 = \frac{R}{\sqrt{2}}, \qquad Z_2 = R \tag{42b}$$

and

$$C_{1\rightarrow 2}\ (\text{dB}) = 3\ (\text{dB}) \tag{42c}$$

In the above description, we assumed that $V_4 = 0$. In this case, we will examine the currents flowing into port 4 through the distributed lines connected to port 4.

In conclusion, the current flowing into port 4 through the distributed constant line connecting ports 1 and 4, i'_4, flows into the distributed constant line connecting between ports 4 and 3 keeping the same values, that is, $i'_4 = i''_4$ in Fig. 60a. This is easily verified by the characteristics of the $\lambda_g/4$ distributed constant line assuming V_1 and V_3, which is defined from β. Therefore, the current of R connected at port 4 becomes zero and the voltage V_4 also becomes zero.

The explanations mentioned above are only concerned with the center frequency meeting the condition $\lambda_g/4$ for each distributed constant line. When the center angular frequency, ω_0, deviates by $\Delta\omega$, the length of the distributed constant line does not meet $\lambda_g/4$, which results in the deterioration of D.

The quantity D corresponding to $\Delta\omega/\omega_0$ can be calculated from (see Appendix 6)

$$D = 20 \log_{10}\left(\frac{1}{|\gamma|}\right) \quad \text{(dB)} \tag{43a}$$

$$|\gamma| \simeq \sqrt{\frac{1 + |\beta|}{1 - |\beta|}}\,\frac{\pi}{4}\,\frac{\Delta\omega}{\omega} \tag{43b}$$

Therefore, the bandwidth ratio to satisfy the specified D, w, can be expressed by

$$w = \frac{2\Delta\omega}{\omega} = \frac{8}{\pi}\sqrt{\frac{1 - |\beta|}{1 + |\beta|}}\,|\gamma| \tag{44}$$

As as example, for w to have directivity of 20 dB in the case of a 3-dB coupler can be obtained as follows.

Because $|\beta| = 1/\sqrt{2}$ in the case of a 3-dB coupler, $|\gamma| = 0.1$ from Eq. (43a); substituting these values into Eq. (44), we get $w = 0.1055$.

To make small size the construction of Fig. 59, it is realized by lumped element circuits as shown in Fig. 61 under the consideration of the equivalent lumped element circuit of $\lambda_g/4$ distributed constant line.

The $\lambda_g/4$ distributed constant line with the characteristics impedance of $Z_{1,2}$ can be expressed by the L,C,π network at the center frequency; the values of L and C are

$$L = \frac{R}{\omega_0}, \qquad C = \frac{2}{\omega_0 R} \tag{45}$$

Therefore, replacing each $\lambda_g/4$ line of Fig. 59 by the L,C,π network, we get the network of Fig. 61, where L_1, L_2, and C_0 are calculated from

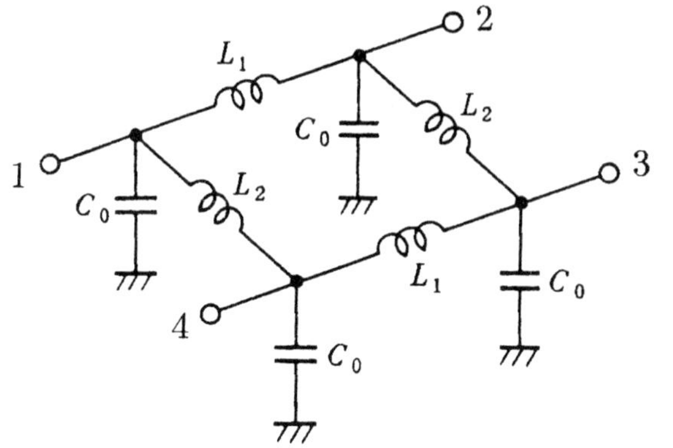

Figure 61. Realization by lumped element circuit of Fig. 57.

$$L_1 = \frac{Z_1}{\omega_0}$$

$$L_2 = \frac{Z_2}{\omega_0}$$

$$C_0 = \frac{2}{\omega_0}\left(\frac{1}{Z_1} + \frac{1}{Z_2}\right) \tag{46}$$

Because the equivalent network of $\lambda_g/4$ constructed by an L,C network is almost available in the bandwidth of the required directivity of greater than 20 dB, the bandwidth ratio of the isolation and input VSWR almost coincide with that of Fig. 59. The insertion loss, however, resulted from the no-load Q values of L_i ($i = 1, 2$) and C_0, and is slightly larger than that of the construction in Fig. 59. For example, the insertion loss of the construction in Fig. 61 is about 0.3–0.4 dB at 100 MHz.

To make a wide band, the cascade connection of the construction in Fig. 59 can be used as shown in Fig. 62. Such an idea is possible by considering the equivalent network of $\lambda_g/4$ as shown in Fig. 25 of Chapter 3, and the susceptances of parallel tuned networks connected at both ends of the ideal $\pi/2$ radian can be cancelled out by cascade connections.

For example, by using the values of the characteristic impedances of

$$Z_1 = Z_3 = 2.1R$$

$$Z_2 = 0.84R$$

$$Z_4 = Z_5 = 0.7R$$

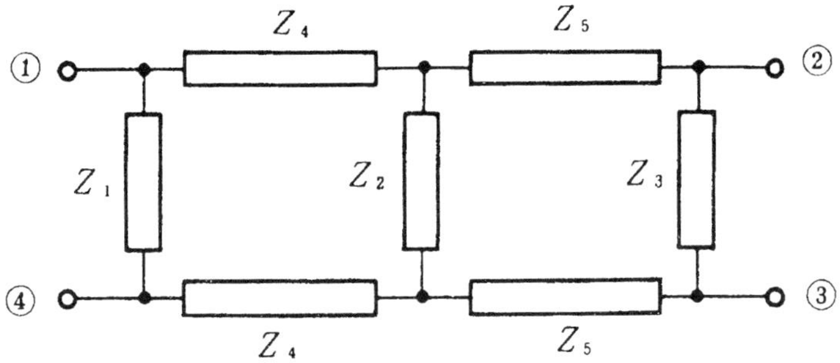

Figure 62. Three-branch-line directional coupler.

the directivity of more than 20 dB between ports 1 and 4 or ports 2 and 3 is achieved with the bandwidth ratio of 36% [22].

From the square constructions of Fig. 59, we can use the ring construction. As an example, the pattern as shown in Fig. 63 is used for a microstrip 3-dB coupler, which is called a hybrid ring.

In the constructions of Figs. 59 and 63, the values of Z_1 and Z_2 should take different values. We, however, can achieve the 3-dB coupler with the construction of Fig. 64, where the characteristic impedance of the line connecting each port, Z, takes the values

$$Z = \sqrt{2}R \tag{47}$$

This is clear because the signal supplied to port 1 arrives at port 4 through passes 1 → 2 → 4 and 1 → 3 → 4, and the signal is cancelled out at port 4. Therefore, the impedance looking from ports 2 and 3 to port 4 becomes infinite. The signals appears at ports 2 and 3. For the matching at port 1,

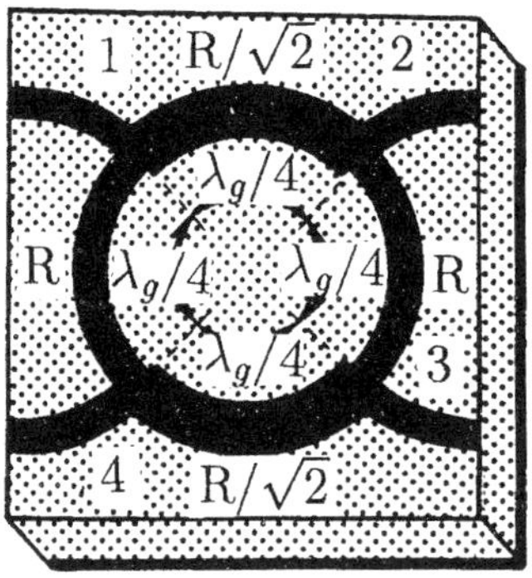

Figure 63. Microstrip hybrid ring.

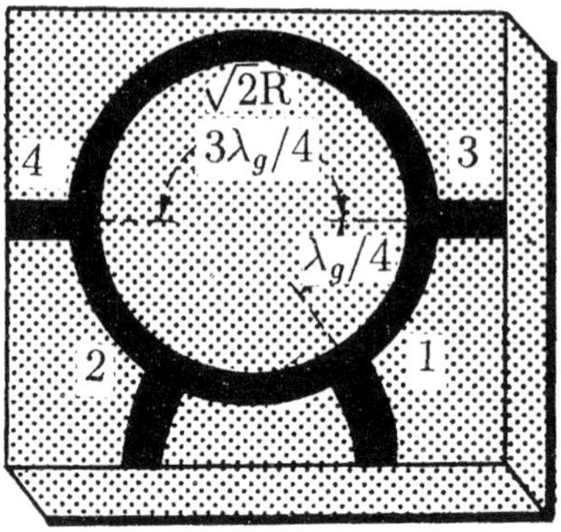

Figure 64. Microstrip ratrace hybrid.

therefore, we need the impedance viewed from port 1 to ports 2 and 3 with the values $2R$, which results in the characteristic impedance in Eq. (47). The construction is called a ratrace hybrid.

Tightly-Coupled-Coil-Type Directional Coupler

The directional couplers described in the previous two subsections use magnetic and electric couplings. We can, therefore, consider the construction in which the magnetic coupling is brought together at the center coupled coils and the electric coupling is brought together at both end capacitors, as shown in Fig. 65. In this construction, a signal of 1 V supplied to port 1 appears at port 2, with $|\alpha|$ V and $|\beta|$ V at port 4 assumed by comparing the constructions of Fig. 45 and 65.

By analyzing the network, we get the following results (see Appendix 6.4):

(i) The coupling of the coil should be complete.
(ii) The relation

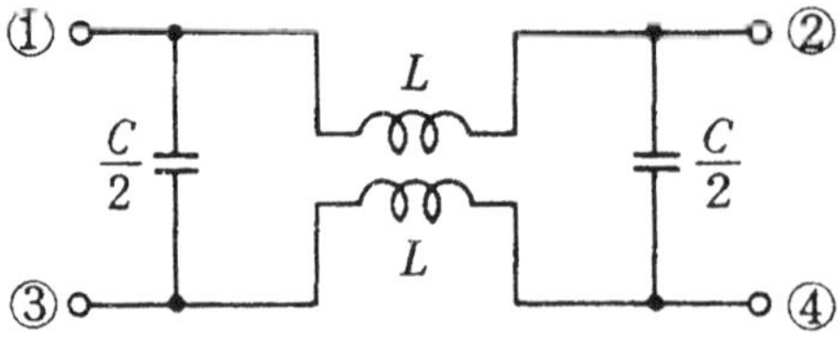

Figure 65. Tightly-coupled-coil-type directional coupler.

$$\sqrt{\frac{L}{C}} = R \tag{48}$$

is required.

(iii) α and β take the following values:

$$\alpha = \frac{1}{1 + j\omega L/R} = \frac{1}{\sqrt{1 + (\omega L/R)^2}} e^{-j\theta_\alpha} \tag{49a}$$
$$\theta_\alpha = \tan^{-1}\left(\frac{\omega L}{R}\right)$$

$$\beta = \frac{1}{1 - jR/\omega L} = \frac{1}{\sqrt{1 + (R/\omega L)^2}} e^{-j\theta_\beta} \tag{49b}$$
$$\theta_\beta = \tan^{-1}\left(\frac{R}{\omega L}\right)$$

(iv) The difference in phases between ports 3 and 2 is π/2 radian for the whole frequency range; that is,

$$\theta_\beta - \theta_\alpha = \frac{\pi}{2}$$

which is easily obtained from Eqs. (49a) and (49b). This relation is certainly understood from the general theory for the directional coupler with symmetrical structure, as explained in Appendix 6, Eq. (9).

Calculated results from Eqs. (48a) and (49b) are shown in Figs. 66a and 66b. It is found that the amplitudes of α and β are

$$|\alpha| = |\beta| = \frac{1}{\sqrt{2}} \tag{50a}$$

$$\omega_0 = \frac{R}{L} \tag{50b}$$

This means that the directional coupler becomes a 3-dB coupler at ω_0.

Ferrite-Loaded Wideband Lumped Element Directional Coupler

This is constructed by two transformers. When the signal is supplied to port 1′, the signal appears at load impedances W connected at ports 2 and 3 and does not appear at port 4. The operation is as follows.

As shown in Fig. 67b, when the signal current i_1 supplied at port 1′, we can assume that the current i_1 flows through the primary coil of T_1 to port 2, because the current should not flow through T_2 because of the zero voltage

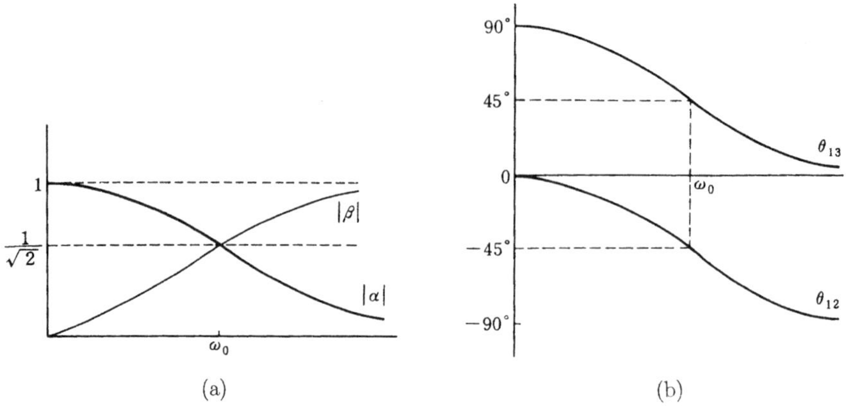

Figure 66. The values of S_{12} and S_{13} of the directional coupler in Fig. 65: (a) $|S_{12}| = |\alpha|$ and $|S_{13}| = \beta$; (b) θ_{12} and θ_{13} ($S_{12} = |\alpha|e^{j\theta_{12}}$, $S_{13} = |\beta|e^{j\theta_{13}}$).

at port 4. Because T_1 is the ideal transformer, the current i_1/n should be flowing in the secondary coil of T_1 and it must be flowing into port 3 because there is no signal at port 4. The voltage at port 3, V_3, therefore, takes the values i_1W/n.

On the other hand, because $V_4 = 0$, we obtain

$$V_3 = \frac{i_1 W}{n} = \frac{V_1'}{m} \tag{51a}$$

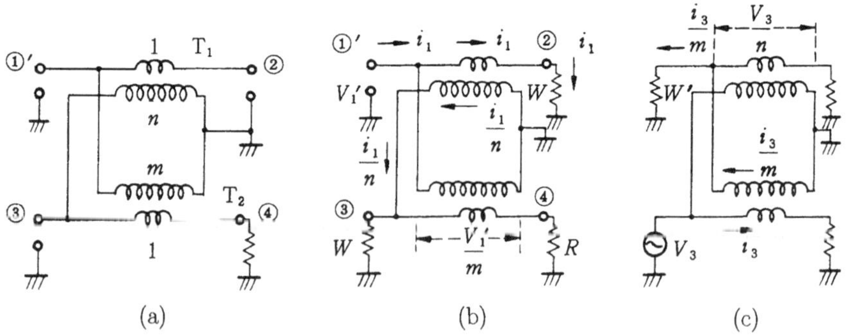

Figure 67. Construction and principle of ferrite-loaded wideband lumped element directional coupler: (a) construction; (b) currents flow when signal is supplied to port 1′; (c) currents flow when signal is supplied to port 3.

which is clear from Fig. 67b. Because the voltage at port 2, V_2, is i_1/W, we get

$$V_1' = i_1 W + \frac{V_3}{n} = i_1 W \left(1 + \frac{1}{n^2}\right) \tag{51b}$$

Substituting Eq. (51b) into Eq. (51a), we get

$$m = n \left(1 + \frac{1}{n^2}\right) \tag{51c}$$

The impedance at port 1′ viewed toward the network can be obtained from

$$\frac{V_1'}{i_1} = W \left(1 + \frac{1}{n^2}\right) \tag{51d}$$

Therefore, we need the transformer at port 1′ with ratio

$$1 : \frac{n}{\sqrt{1 + n^2}} \tag{51e}$$

3.4 Hybrid Coupler or Bridge Circuit

When we consider the hybrid coil shown in Fig. 68, the signal applied to port 1 appears at ports 2 and 3 and does not appear at port 4. In the case when every port is matched by choosing the proper values of n and n' of an ideal transformer, the scattering matrix $[S]$ is

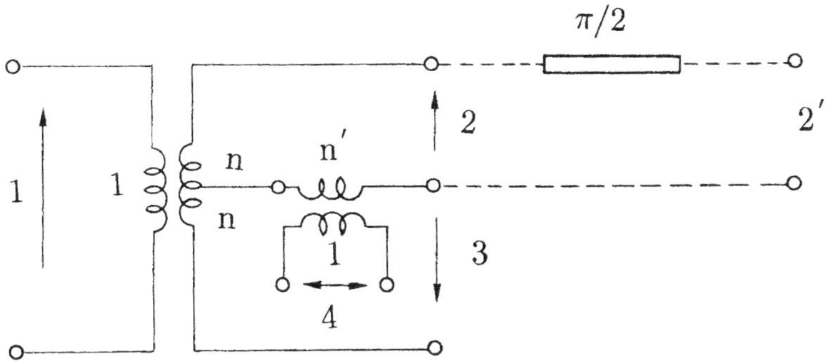

Figure 68. Hybrid coil.

$$[S] = \begin{bmatrix} 0 & \frac{1}{\sqrt{2}} & -\frac{1}{\sqrt{2}} & 0 \\ \frac{1}{\sqrt{2}} & 0 & 0 & \frac{1}{\sqrt{2}} \\ -\frac{1}{\sqrt{2}} & 0 & 0 & \frac{1}{\sqrt{2}} \\ 0 & \frac{1}{\sqrt{2}} & \frac{1}{\sqrt{2}} & 0 \end{bmatrix} \tag{52a}$$

Next, when we connect the $\lambda_g/4$ line to port 2 as shown in Fig. 68 by the dotted line, the scattering matrix $[S']$ of the circuit of ports 1, 2′, 3, and 4 is calculated from Eq. (52b), where the sign takes the upper part.

$$[S'] = \begin{bmatrix} 0 & -j\frac{1}{\sqrt{2}} & -\frac{1}{\sqrt{2}} & 0 \\ -j\frac{1}{\sqrt{2}} & 0 & 0 & -j\frac{1}{\sqrt{2}} \\ -\frac{1}{\sqrt{2}} & 0 & 0 & \frac{1}{\sqrt{2}} \\ 0 & -j\frac{1}{\sqrt{2}} & \frac{1}{\sqrt{2}} & 0 \end{bmatrix} \tag{52b}$$

As shown in Eqs. (52a) and (52b), there is the infinite directivity (i.e., the isolation between ports 1 and 4), and the absolute values of α and β are both $1/\sqrt{2}$. The circuits with such characteristics are called a hybrid circuit or a bridge circuit. Since the bridge circuits realized by a 3-dB directional coupler have been described, we will describe the other types of bridge as follows.

Magic T

As shown in Fig. 69a, the *E*-plane T junction and *H*-plane T junction are combined in such a way points A, B, C, and D of the figure are placed in a symmetrical plane. The constructions in Fig. 70, where they take a T shape viewed from the E plane and H plane of the $TE_{10}^{\square}$ waveguide are called *E*-plane and *H*-plane T junctions, respectively.

When the signal is supplied at port 1 of Fig. 69a, the electric fields become as shown in Fig. 69b, and the signal appears at ports 2 and 3 with opposite phases. In this case, port 4 is mainly excited by the TE_{2m} mode as seen by the electrical fields in Fig. 69b. Because the TE_{2m} mode is a cutoff

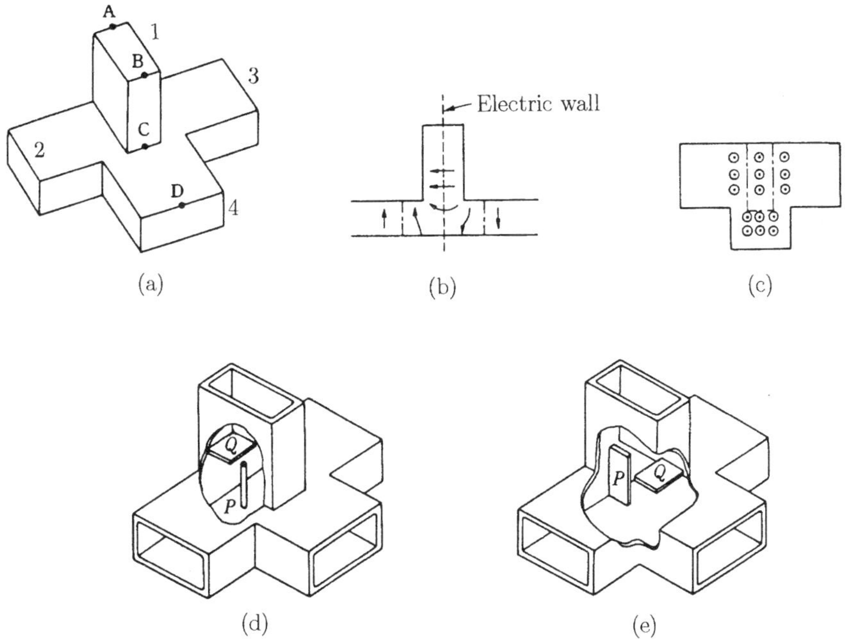

Figure 69. Magic T: (a) construction; (b) electric fields parallel to *E* plane when ports 1 and 2 are excited by the odd mode; (c) electrical fields distribution vertical to the *H* plane when ports 1 and 2 are excited by the even mode; (d) example of matching at ports 3 and 4; (e) another example of matching at ports 3 and 4.

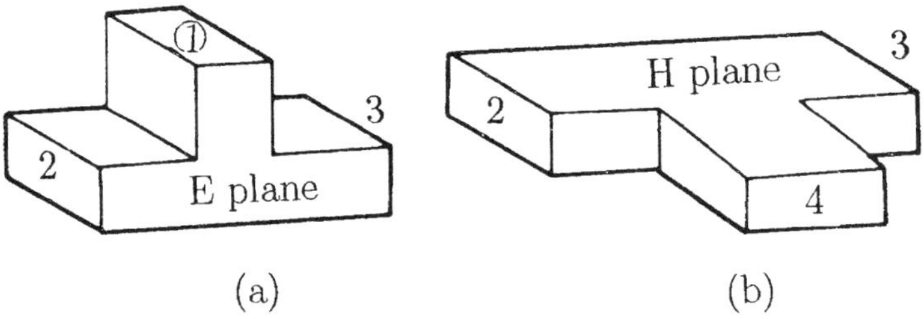

Figure 70. $\mathrm{TE}_{10}^{\square}$ waveguide T junction: (a) *E* plane T junction; (b) *H* plane T junction.

mode, the signal does not appear at port 4. Under this condition, port 1 can be matched by placing a window Q and a post P, keeping the symmetry about the symmetrical plane. In this situation, ports 2 and 3 are loaded in series to port 1, which is equivalent to the hybrid coil shown in Fig. 68.

Next, the signal supplied to port 4 appear at ports 2 and 3 in phase, as in Fig. 69c. As understood from Fig. 69c, port 1 is excited by the *E*-wave cutoff mode, and the signal does not appear at port 1. In this case, port 3 has to match in the same way, keeping the symmetry. In this situation, ports 2 and 3 are connected to port 3 in parallel, and the equivalent network also becomes that of Fig. 68.

In the practical case, the post or the plate P in Fig. 69d or 69e is only effective for the matching of port 4, because P is placed in the zero potential for the excitation from port 1. On the other hand, Q in Fig. 69d or 69e is only effective for the matching of port 1, because Q exists in a cutoff region for the excited signal at port 4.

Slotted Coaxial Bridge [23]

The bridge is usually used for the constant-impedance notch diplexer and the vestigial sideband filter of a TV transmitter.

As shown in Fig. 71, the outer conductor of the coaxial line A connected by port 1 is split in half by slots over the length of $\lambda/4$, and the end of the center conductor is connected to the outer conductor of the coaxial at point p. The point p and the counterpoint q on the other side of the outer conductor of the coaxial line A, are connected to the center conductor of the coaxial line of ports 2 and 3, respectively. On the coaxial line A, a concentric shielding conductor is placed and it makes another coaxial line B together

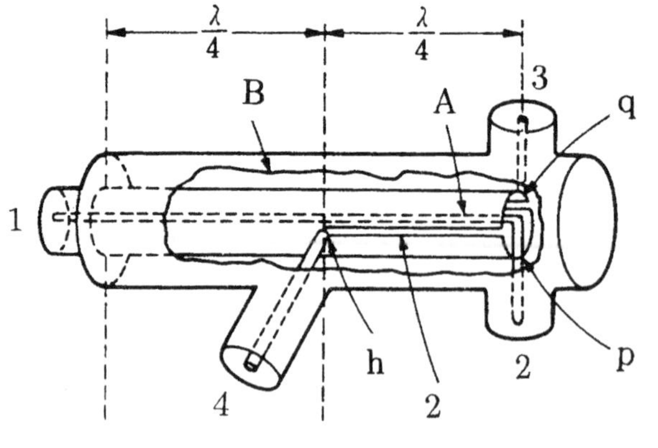

Figure 71. Construction of the slotted coaxial bridge.

with the outer conductor of the coaxial line A. We will explain the operation below.

When we supply the power at port 1, the electric field in the sectional plane including ports 2 and 3 becomes as shown in Fig. 72a, where the electric fields distribute in the upper half of coaxial line A. The fields distribution is decomposed as shown in Figs. 72b and 72c.

The mode in Fig. 72b almost coincides with the TEM mode of a coaxial line, and that of Fig. 72c is the balance mode or odd mode existing in two conductors and have the components of the TE_{11} mode of a coaxial line.

We denote the electric fields of the TEM mode and the balance mode by solid lines and dotted lines, respectively; they are shown in Fig. 72d. The currents corresponding to the TEM and the balance modes are denoted by solid lines and dotted lines, respectively, they are shown in Fig. 72e.

When the length of the slot is $\lambda/4$, the impedance looking from p and q to the left in Fig. 72e becomes infinite and it results in $I' = 0$. In such a case, the current is as shown in Fig. 72f.

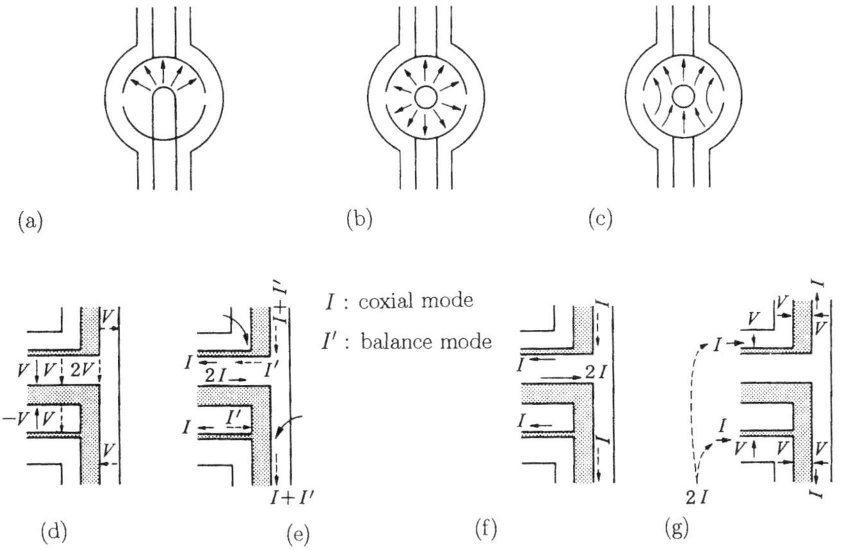

Figure 72. Principle of the slotted coaxial bridge: (a) electric fields at sectional plane including ports 2 and 3; (b) electric fields of the coaxial mode; (c) electric fields of the coaxial balance ($TE_{11}^{\square}$) mode; (d) voltages are divided into coaxial and balance mode voltages; (e) currents are divided into coaxial and balance mode currents; (f) currents when balance mode is zero; (g) voltages and currents when port 4 excited by a signal.

The current flowing in the center conductor of the coaxial line A with $2I$, therefore, is divided to the outer conductor of the coaxial line A and port 2 with the same values of I. The current at port 3 flows into another outer conductor of the coaxial line A. When the load impedance connected at ports 2 and 3 is W (Ω), we have $W = V/I$ as seen in Fig. 72g.

On the other hand, the impedance viewed from the coaxial line A toward the right at the reference plane including p and q is $W/2$ (Ω), because the voltage and the current of the coaxial line A are V and $2I$ as understood from Figs. 72d and 72f. Therefore, matching $W/2$ (Ω) to the source impedance $W/2$ (Ω) at port 1, a matching circuit such as the $\lambda/4$ transformer with a characteristic impedance of $W/\sqrt{2}$ (Ω) is required.

Next, when the signal power is supplied to port 4, the coaxial line B is excited and the currents and the voltages at the sectional plane at p and q become as shown in Fig. 72g, and the currents flow out to ports 2 and 3 with the same phase, which results in an impedance of $W/2$ of looking to the right at the sectional plat at p and q of the coaxial line B. To match the impedance to the source impedance of port 4, W, we need the following two items: (i) Because the impedance looking to the left from port 4 should be infinite, the length of the coaxial line B between the shorten portion at the left end of the coaxial line B and port 4 should be $\lambda/4$. (ii) The characteristic impedance of the coaxial line B between port 4 and the point B should be $W/\sqrt{2}$ (Ω).

As mentioned above, the signal supplied at port 1 flows to ports 2 and 3 in series, and the signal supplied at port 4 flows to ports 2 and 3 in parallel. This operation coincides with that of the hybrid coil, as shown in Fig. 68, and it becomes a bridge circuit.

In the above description, we neglect the effect of the coupling through the slots which are made on the outer conductor of the coaxial A. However, considering the effect of the slot, we can get the directivity, D (dB), as [23].

$$D = 20 \log_{10} \left(0.105 \frac{(1/2W_1 + 1/\sqrt{2})\theta^2}{\ln(r_c/r_b)} \right)^{-1}$$

$$W_1 = \frac{1}{2v_0 C_{12}\ (\text{F/m})}\ (\Omega) \tag{53}$$

where v_0 is the velocity of light (m/s).

In Eq. (53), we find that the directivity is increased as the slot angle θ shown in Fig. 73a decreases. As a practical example, in the case $W_4 = 50/\sqrt{2}\ \Omega$ and $W_1 = 41.7\ \Omega$, the calculated values of D (dB) is shown in Fig. 73b, where W_4 and W_1 are the characteristic impedance required for a one-stage $\lambda_g/4$ transformer to match at port 4 and a two-stage transformer to give the wideband matching in port 1.

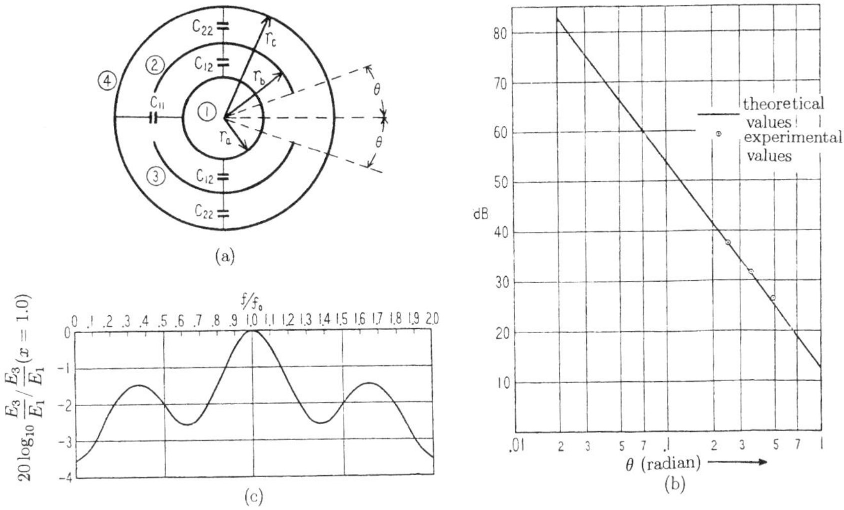

Figure 73. Relationship between the directivity, D (dB) and the sectional construction of the portion of slots: (a) sectional view at the portion of slots; (b) D (dB) in the case $W_4 \simeq 60 \ln(r_c/r_b) = 50/\sqrt{2}$ (Ω) and $W_1 \doteqdot 60 \ln(r_b/r_a) = 41.7$ (Ω); (c) frequency performance of the directivity of the slotted coaxial bridge.

As understood from Fig. 73b, the slot angle, 2θ, should be less than 0.4 radian to keep the directivity more than 40 dB. The directivity is kept in a very wide frequency band as shown in Fig. 73c [23], where the deviation of the quantities is less than 3–4 dB in a full-frequency region.

4 CIRCULAR POLARIZER

When the electric fields with the x direction, E_x, and the y direction, E_y, have a phase difference of 90° and the same amplitude, the resultant vector of the electric field becomes a circularly rotating field, that is, circular polarized wave. In the case when E_y is delayed by 90° with respect to E_x, as shown in Figs. 74b and 74c, the resultant electric field rotates to the right, against the propagating direction of the z-axis, which is called the positive circularly polarized wave. Therefore, when E_x is delayed by 90° with respect to E_y, the resultant wave becomes a positive polarized wave.

Therefore, if we want to make a circularly polarized wave from a linearly polarized wave, the method shown in Fig. 75a is used. As shown in Fig. 75a, a thin dielectric plate is inserted in a TE_{11}° waveguide. The linearly polarized wave, where the angle between the direction of the electric field

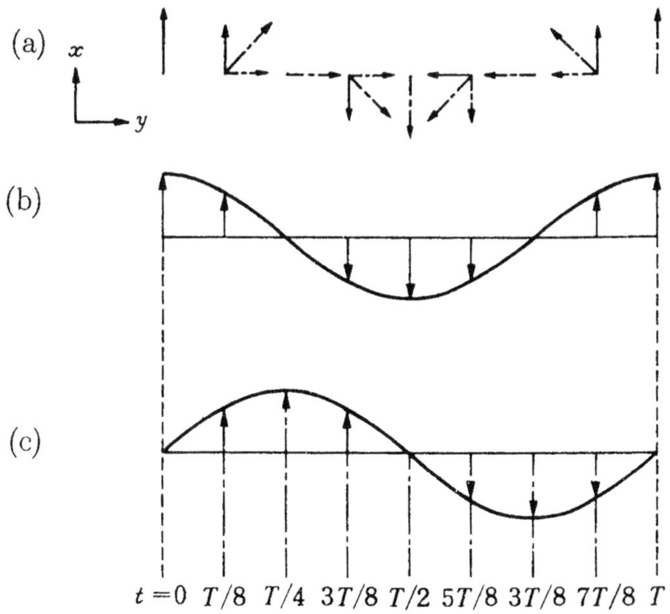

Figure 74. Generation of circularly polarized wave: (a) resultant electrical field; (b) E_x; (c) E_y.

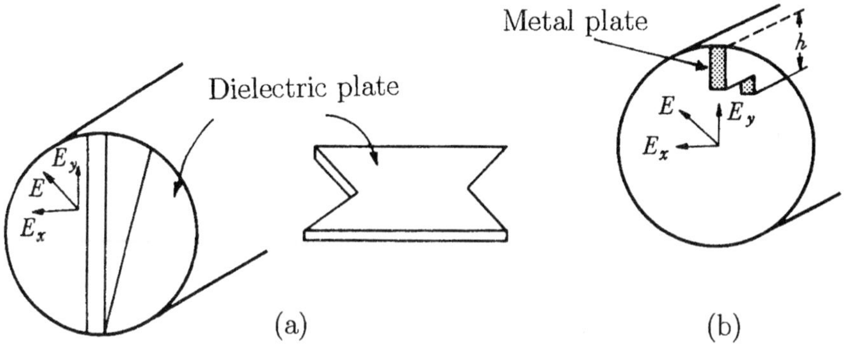

Figure 75. Construction of polarizer where the circularly polarized wave is produced from the linearly polarized wave: (a) method using a dielectric plate; (b) method using a metal plate.

and the normal vector to the surface of dielectric plate is 45°, is supplied to a TE_{11}° waveguide. The electric field can be separated into E_x and E_y as shown in Fig. 75a. Here, E_x and E_y are the normal and the parallel components to the surface of the dielectric plate, respectively. Even if the wave with the E_x component is not affected by the dielectric plate, the phase of the TE_{11}° wave with the E_y component is affected by the dielectric plate; it is delayed because the dielectric plate is placed at the maximum values of E_y. Therefore, E_y can be delayed by 90° by using the proper length of the dielectric plate, which results in producing the positive circularly polarized wave. To keep the effect of the discontinuity small, the edge of the plate is made with a tapered shape. In the opposite case, the positive circularly wave is supplied from the output, and we get the linearly polarized wave with $\boldsymbol{E}$ at the input port.

As a second method, we can use the construction with the metal plate as shown in Fig. 75b. The wave with the electric field parallel to E_y in Fig. 75b, the phase velocity is slowed by the ridge construction because the cutoff frequency decreases.

On the other hand, the wave with the E_x component is only slightly affected because the electric field of E_x is very small at the metal plate. Therefore, the linearly polarized wave with $\boldsymbol{E}$ as shown in Fig. 75b can be transformed to the positive circularly polarized wave by using the proper length of the metal.

5 CIRCULATOR AND ISOLATOR

5.1 Introduction

In the n-port network with ports 1, 2, . . . , n connected by a load impedance, we will consider the case when the signal supplied to port 1 appears only at port 2. Next, the signal supplied to port 2 appears only at port 3. When the signal circulates through $1 \rightarrow 2 \rightarrow \cdots n \rightarrow 1$, the network is called a circulator.

In the two-port network, we consider the case when the signal supplied to port 1 appears at port 2 and the signal supplied to port 2 does not appear at port 1. Such a two-port network is called an isolator. Obviously, we can make an isolator with a circulator by connecting a load to port 3 when the isolator operates as $1 \rightarrow 2$. The circulator and isolator are the nonreciprocal circuit because the signal flows as $1 \rightarrow 2$ but not $2 \rightarrow 1$. Such a nonreciprocal circuit can be made by including ferrite in the dc magnetic field.

As another nonreciprocal circuit, there is the nonreciprocal phase shifter, where the phase delay through the pass $1 \rightarrow 2$ is not same to that through the pass $2 \rightarrow 1$. Using the nonreciprocal phase shifter together with the two

3-dB couplers mentioned in Section 3, we can make a circulator as shown in Fig. 76a. We have also another nonreciprocal circuit, the so-called Faraday rotator, where the signal with the same polarization supplied to port 1 changes the direction of the polarization at port 2. By using the Faraday rotator, we can make a circulator as shown in Fig. 76b.

There is another phenomenon: The RF (radio frequency) magnetic field rotating to the left toward the dc magnetic field H_0 does not attenuate in ferrite, although the RF magnetic field rotating to the right toward H_0 attenuates at the same dc magnetic field, the so-called resonant magnetic field. Putting ferrite at the point where the rotating magnetic field exists in a waveguide, we have an isolator, which is called the magnetic-resonant-type isolator. For rotating magnetic RF fields, there exists permeability with values μ_+ and μ_- corresponding to the different rotating directions.

When ferrite is put in a waveguide, the distribution of the electrical fields are different, corresponding to the incident wave or the reflected wave. Putting the resistive absorber at the place where the electric field is maximum for a reflected wave, the incident wave does not attenuate even if the reflected wave is attenuated by the absorber. By this principle, we have an isolator, the so-called field-displacement-type isolator as shown in Fig. 76c.

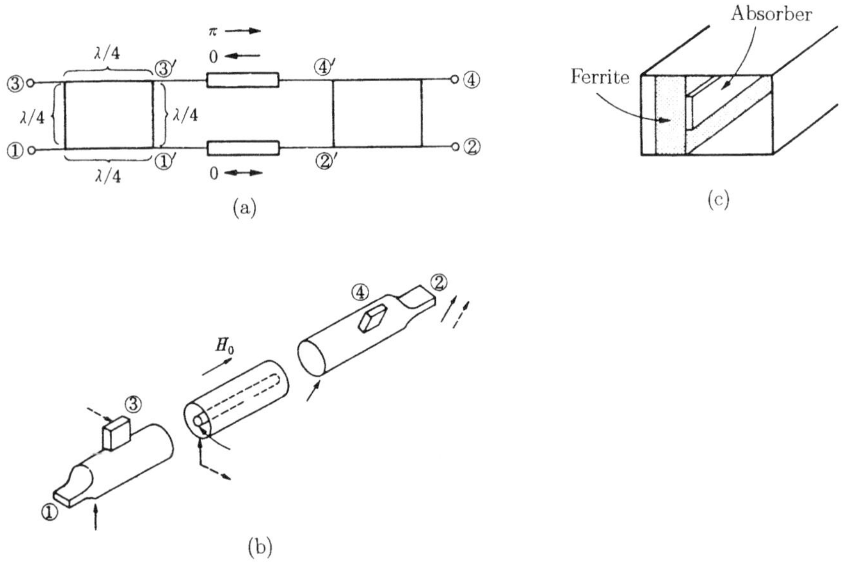

Figure 76. Example of circulator and isolator: (a) circulator with nonreciprocal phase shifter and two 3-dB couple; (b) circulator with a Faraday rotator; (c) field-displacement-type isolator.

By using the characteristics of the difference between μ_+ and μ_-, we can split the degenerated eigenvalues of the rotationally symmetrical three-port network, which make it possible to realize a three-port circulator such as lumped element and strip line circulators.

There are many kinds of circulator and isolator, and they are used to meet requirements such as size, power, loss, and others. Examples are shown in Fig. 77.

In this book, junction-type circulators and isolators, especially the lumped element type and the strip line type used for lower frequency such as the VHF to UHF band are explained.

In Section 4, the fundamental principle of the nonreciprocal circuit in connection with a circulator are explained together with the ferrite performance under a dc magnetic field. Junction-type circulators are also explained in Section 4.

5.2 Principle of Nonreciprocal Circuits and Related Basic Matters

As described in Chapter 2, Section 3, when the elements of the impedance or admittance matrixes are

$$Z_{ij} \neq Z_{ji} \tag{54a}$$

$$Y_{ij} \neq Y_{ji} \tag{54b}$$

the circuits are called nonreciprocal circuits. Because the elements of the scattering matrix are related by Eqs. (7h) and (7i) of Chapter 2, we also have

$$S_{ij} \neq S_{ji} \tag{54c}$$

As we learned in Chapter 2, Section 3, the nonreciprocal circuits should include the medium with tensor permeability $\hat{\mu}$ or the tensor dielectric constant $\hat{\varepsilon}$ to satisfy the unsymmetrical matrix; that is,

$$\hat{\mu} \neq \tilde{\hat{\mu}} \quad \text{or} \quad \hat{\varepsilon} \neq \tilde{\hat{\varepsilon}} \tag{54d}$$

Such a medium is obtained in ferrite in dc magnetic fields as follows. As shown in Fig. 78, the dc magnetic field is applied to ferrite in the direction of the z-axis. When the high-frequency magnetic fields h_x and h_y exist in the ferrite, the magnetic fluxes b_x and b_y is calculated as

$$\begin{aligned} b_x &= \mu_0(\mu h_x - j\kappa h_y) \\ b_y &= \mu_0(j\kappa h_x + \mu h_y) \end{aligned} \tag{55a}$$

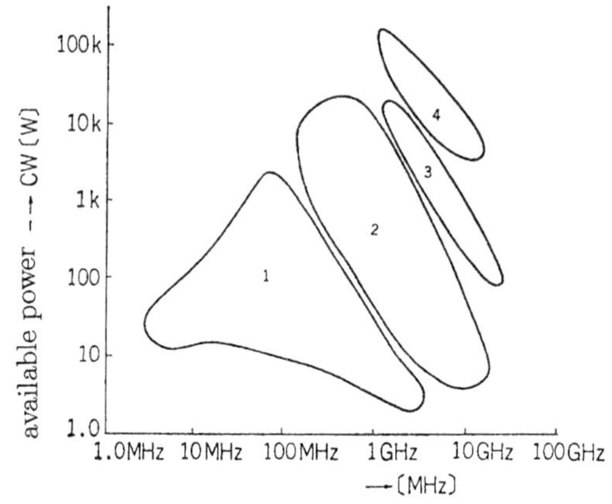

Figure 77. Frequency and available power used for types of circulator.

which is also expressed as

$$\begin{bmatrix} b_x \\ b_y \end{bmatrix} = \mu_0 \hat{\mu} \begin{bmatrix} h_x \\ h_y \end{bmatrix} \tag{55b}$$

$$\hat{\mu} = \begin{bmatrix} \mu & -j\kappa \\ j\kappa & \mu \end{bmatrix} \tag{55c}$$

μ and κ are obtained from

$$\mu = 1 + \frac{(\omega_0 + j\omega\alpha)\omega_M}{(\omega_0 + j\omega\alpha)^2 - \omega^2}$$

$$\kappa = \frac{-\omega\omega_M}{(\omega_0 + j\omega\alpha)^2 - \omega^2}$$

$$\omega_0 = -\gamma H_0, \qquad \omega_M = -\gamma M_0 \tag{55d}$$

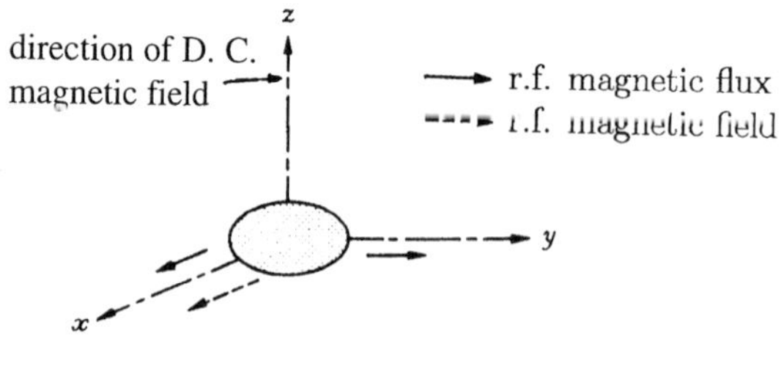

Figure 78. Ferrite in the dc magnetic field.

where γ is called the gyromagnetic coefficient, M_0 is the saturation magnetization, and α is the so-called damping parameter, which shows the loss terms of μ and κ. As understood from Eq. (55c) $\hat{\mu} \neq \tilde{\hat{\mu}}$.

Setting $h_y = 0$ in Eq. (55a), we get $b_x = \mu_0\mu h_x$ and $b_y = j\mu_0\kappa h_x$. Therefore, not only x component but also y component of the magnetic flux are generated corresponding to only the x component of the magnetic field; this is shown in Fig. 78 by the solid arrows of RF magnetic fluxes and the dashed arrows of the magnetic field.

Thus far, we mentioned that the tensor permeability or tensor dielectric constant as in Eq. (54d) are necessary to obtain the nonreciprocal circuits, and ferrite in the dc magnetic field is available.

In the opposite case, we can show a practical example to obtain the nonreciprocal circuit by using such a ferrite as follows (see Fig. 79). When the current i_1 is flowing in coil 1 and coil 2 is open, we get only the magnetic field with the x component, h_x, proportional to i_1, which produces b_y:

$$b_y = -j\mu_0\kappa h_x$$

By Faraday's law, therefore, we get the voltage, v_2, at port 2:

$$v_2 = j\omega(-j\kappa)Ai_1 = \omega A\kappa i_1$$

where A is the proportional constant with a positive real number. Therefore, we get

$$Z_{21} = \omega A\kappa$$

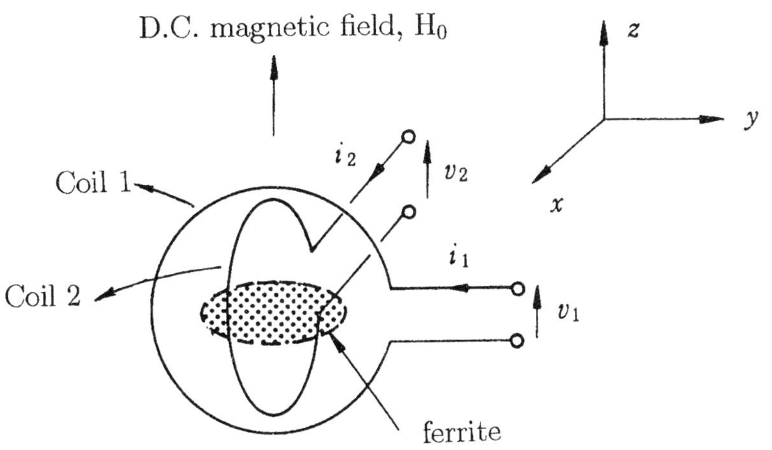

Figure 79. Two-port nonreciprocal circuit constructed from two coils including ferrite under a dc magnetic field.

Next, when the current i_2 flows and coil 1 is open, the voltage v_1 produced at coil 1 is

$$v_1 = j\omega(j\kappa)Aj_2 = -\omega A\kappa i_2$$

which results in

$$Z_{12} = -\omega A\kappa$$

Then, we get

$$Z_{12} \neq Z_{21}$$

Next, we will consider the construction with three recoils which cross each other making 120° angles as shown in Fig. 80a.

When the current flows in coil 1, the magnetic field h_x induced by coil 1 and the magnetic fluxes b_x and b_y are produced in ferrite. However, b_x and b_y have components that cross coils 2 and 3 as shown in Fig. 80b. As understood from Fig. 80b, b_x and b_y take opposite directions at port 2 and take the same direction at port 3. Therefore, the voltage generated at ports 2 and 3 have different values, as expressed by

$$v_2 = \alpha i_1, \qquad v_3 = \beta i_1, \quad \alpha \neq \beta$$

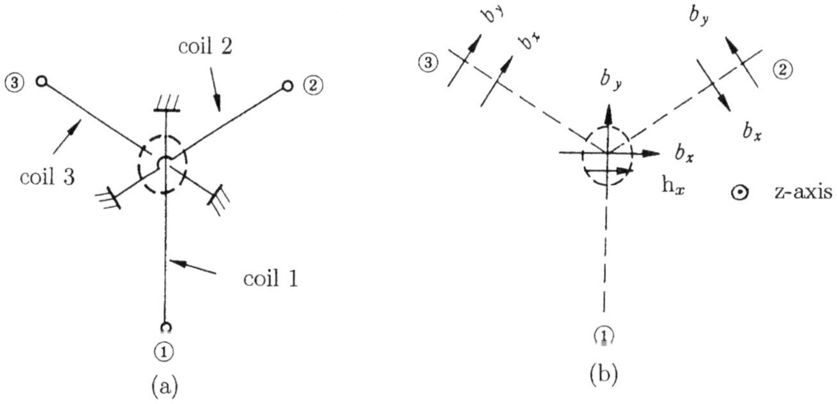

Figure 80. Three-port circuit constructed by three crossed coils making 120° angles, where ferrite is placed in the center of the cross point: (a) construction of coils; (b) magnetic fluxes and their components crossing each coil produced by current i flowing in port 1.

When we represent the impedance matrix Z of Fig. 80a by

$$Z = \begin{bmatrix} Z_{11} & Z_{12} & Z_{13} \\ Z_{21} & Z_{22} & Z_{23} \\ Z_{31} & Z_{32} & Z_{33} \end{bmatrix} \tag{56a}$$

we get

$$Z_{21} = \alpha, \qquad Z_{31} = \beta \tag{56b}$$

In the same way, considering the voltages appearing at ports 3 and 1 caused by i_3 at port 2, we get

$$Z_{32} = \alpha, \qquad Z_{12} = \beta \tag{56c}$$

In the same manner, from the voltages of ports 1 and 2 caused by i_3, we get

$$Z_{13} = \alpha, \qquad Z_{23} = \beta \tag{56d}$$

From Eqs. (56b)–(56d), we get

$$Z_{21} \neq Z_{12}, \qquad Z_{31} \neq Z_{13}, \qquad Z_{32} \neq Z_{23} \tag{56e}$$

When the magnetic fluxes crossing coil 2 are cancelled out, we have $\alpha = 0$. In this situation, when port 1 is matched as the power supplied at port 1 travels to the load at port 3, the circuit becomes a so-called circulator.

However, because the construction of Fig. 80 is rotationally symmetric about the z-axis, the eigenvectors $\boldsymbol{u}^{(1)}$, $\boldsymbol{u}^{(2)}$, and $\boldsymbol{u}^{(3)}$ are calculated from the values of Eq. (17k) in Chapter 2, that is,

$$\boldsymbol{u}^{(1)} = \frac{1}{3}\begin{bmatrix} 1 \\ 1 \\ 1 \end{bmatrix}, \qquad \boldsymbol{u}^{(2)} = \frac{1}{3}\begin{bmatrix} 1 \\ \alpha^2 \\ \alpha \end{bmatrix}, \qquad \boldsymbol{u}^{(3)} = \frac{1}{3}\begin{bmatrix} 1 \\ \alpha \\ \alpha^2 \end{bmatrix}$$

$$\alpha = e^{j2\pi/3} \tag{57}$$

$\boldsymbol{u}^{(1)}$ is the same-phase excitation where there is no magnetic field at the ferrite placed at the center of crossed coils, because the magnetic fields caused by each current flowing in coils 1, 2, and 3 are canceled.

$\boldsymbol{u}^{(2)}$ is the circularly rotated excitation about the z-axis which has the direction perpendicular to the surface of the paper. In the case of $\boldsymbol{u}^{(2)}$, the magnetic fields rotating counterclockwise, that is, toward the right hand about the z-axis, are generated at the ferrite. On the other hand, in the case of $\boldsymbol{u}^{(3)}$, the magnetic fields rotated clockwise, that is, toward the left about the z-axis, are generated at the ferrite.

However, because the network parameters can be determined by the eigenvalues of the corresponding eigenvectors as explained in chapter 2, section 2, the design and evaluation of a circulator with rotational symmetry construction can be made on the basis of the eigenvalues. It is, therefore, important to examine the permeability of the rotational magnetic fields about the z-axis, that is, in the direction of the dc magnetic field.

When the rotational fields toward the right and the left are denoted by $\boldsymbol{h}^+$ and $\boldsymbol{h}^-$, respectively, we can express them by h_x and h_y as follows:

$$\boldsymbol{h}^+ = h_x - jh_y$$
$$\boldsymbol{h}^- = h_x + jh_y \qquad (58a)$$

In the same way,

$$\boldsymbol{b}^+ = b_x - jb_y$$
$$\boldsymbol{b}^- = b_x + jb_y \qquad (58b)$$

Substituting Eq. (55a) into Eq. (58b) and using Eq. (58a), we get

$$\boldsymbol{b}^+ = \mu_0\mu_+\boldsymbol{h}^+$$
$$\boldsymbol{b}^- = \mu_0\mu_-\boldsymbol{h}^-$$
$$\mu_+ = \mu + \kappa$$
$$\mu_- = \mu - \kappa \qquad (58c)$$

As understood from Eq. (58c), the rotational magnetic fields $\boldsymbol{h}^+$ and $\boldsymbol{h}^-$ have scalar relative permeabilities μ_+ and μ_-, respectively. Therefore, the eigenvalues of a three-port circulator with a rotationally symmetric construction can be evaluated by using the scalar permeabilities μ_+ and μ_-, which are calculated from

$$\mu_+ = 1 + \frac{p}{(\sigma \mp 1) + j\alpha}$$
$$p = \frac{\omega_M}{\omega} = \frac{|\gamma|M}{\omega}, \qquad \sigma = \frac{|\gamma|H}{\omega} \qquad (58d)$$

Because $\mu_\pm$ are functions of the dc magnetic field $\boldsymbol{H}$, we can show the values corresponding to H_0 as shown in Fig. 81. μ''_+ takes maximum values at $\boldsymbol{H} = \omega/|\gamma|$, which is called the resonant field. The values of $\mu''_+ > \mu''_+/2$ exist in the very small range of the interval magnetic dc field as shown in Fig. 81. ΔH is called the linewidth.

As understood from Fig. 81, μ_+ and μ_- have quite different values. This difference contributes to the nonreciprocity, which is explained later.

5.3 Junction Circulator

Lumped Element Y Circulator

For each coil of Fig. 80, two parallel lines, i_a and i_b (i = 1,2,3) are used; the sectional view is as shown in Fig. 82c. The end of the parallel lines are

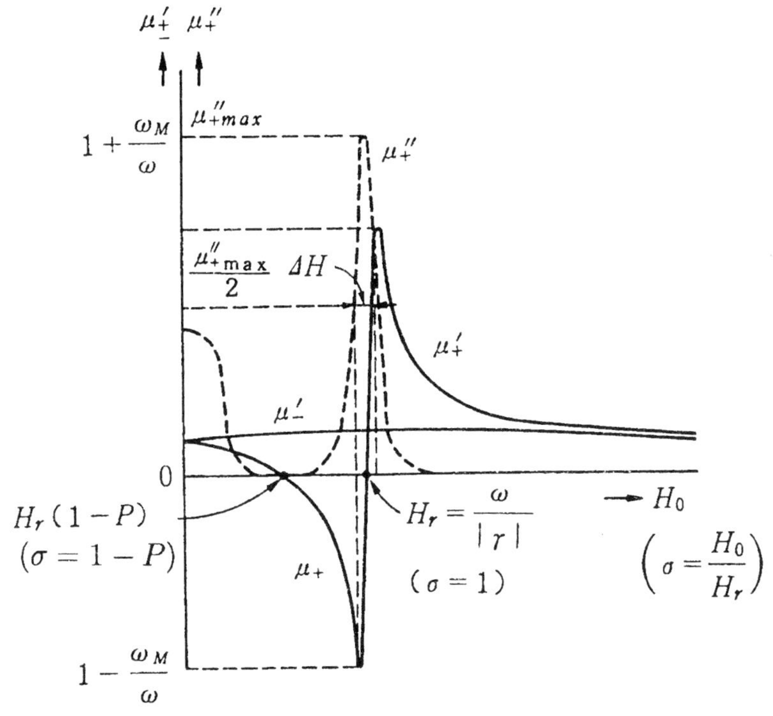

Figure 81. $\mu_\pm = \mu'_\pm - j\mu''_\pm$ corresponding to H_0.

connected to the ith port and the other end is grounded to the shielding case, as shown in Fig. 82a.

The current flows into the ith port and is divided into two parallel lines, which makes a magnetic field as shown in Fig. 82c by the dotted lines; shown to be almost parallel to the shielding metal plate placed above and below. When such conductors are constructed with 120° angles to each other, as in Fig. 82d, we can make a mesh over the surface of the ferrite. By using such a construction, the area for maintaining the circularly rotated fields for the excitation $\boldsymbol{u}^{(2)}$ and $\boldsymbol{u}^{(3)}$ becomes wider than for the case of one line, which results in an increase in the nonreciprocal filling factor and making a wide band [24].

In the case of the excitation $\boldsymbol{u}^{(1)}$, the impedance looking into the mesh from each port becomes zero, as mentioned earlier, and the admittance y_1 is shown on a Smith chart (Fig. 82e).

Next, the admittance looking from each port in the case of $\boldsymbol{u}^{(2)}$ and $\boldsymbol{u}^{(3)}$ takes the values y'_l when $\mu_\pm = 1$ is assumed, as shown in Fig. 82e; y'_l =

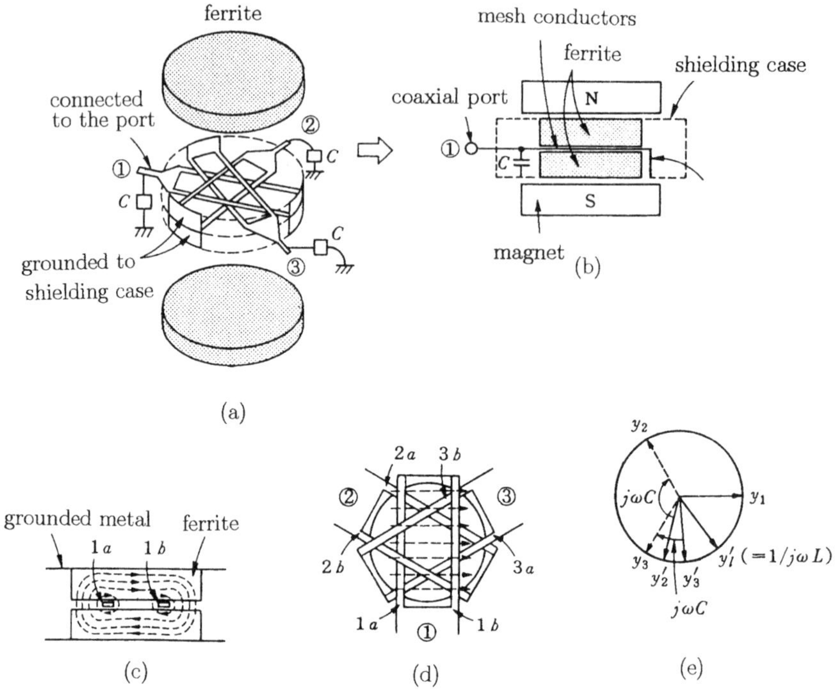

Figure 82. Lumped element circulator: (a) construction inside the shielding case; (b) sectional view including port 1 and the center of the disks: (c) RF magnetic field distribution; (d) top view of (c); (e) eigenvalues of the admittance matrix of the circulator on an admittance Smith chart.

$1/j\omega\xi$. The inductance ξ is 3/2 times the inductance connected to each port, which can be easily introduced by taking account of the coupling from the other two coils.

Considering the ideal case when the permeabilities of ferrite are μ_+ and μ_- over the whole region corresponding to the excitations $\boldsymbol{u}^{(2)}$ and $\boldsymbol{u}^{(3)}$, respectively, the admittances viewed from each port to the mesh y_2' and y_3', become $y_2' = 1/j\omega\mu_+\xi$ and $y_3' = 1/j\omega\mu_-\xi$, which are shown in Fig. 82e.

Next, connecting the capacity at each port in parallel, the admittances corresponding to $\boldsymbol{u}^{(2)}$ and $\boldsymbol{u}^{(3)}$ become y_2 and y_3, respectively. By such a method, we can design y_1, y_2, and y_3 such that their positions on a Smith chart keeps the relative angle of $2\pi/3$ (rad) with respect to each other, as shown in Fig. 82e.

In this state, we can construct the circulator by the following reasoning: If we sum the incident waves of $\boldsymbol{u}^{(1)}$, $\boldsymbol{u}^{(2)}$, and $\boldsymbol{u}^{(3)}$, we get

$$\boldsymbol{u}^{(1)} + \boldsymbol{u}^{(2)} + \boldsymbol{u}^{(3)} = \begin{bmatrix} 1 \\ 0 \\ 0 \end{bmatrix} \tag{59a}$$

This means that the incident wave with 1 V is supplied at port 1. Denoting the eigenvalues of the scattering matrix by s_1, s_2, and s_3 and assuming s_2 and s_3 are

$$\begin{aligned} s_2 &= s_1 e^{j2\pi/3} \\ s_3 &= s_2 e^{-j2\pi/3} \end{aligned} \tag{59b}$$

the reflected wave from the circuit, $\boldsymbol{b}$, is

$$\begin{aligned} \boldsymbol{b} &= s_1\boldsymbol{u}^{(1)} + s_2\boldsymbol{u}^{(2)} + s_3\boldsymbol{u}^{(3)} \\ &= s_1(\boldsymbol{u}^{(1)} + e^{j2\pi/3}\boldsymbol{u}^{(2)} + e^{-j2\pi/3}\boldsymbol{u}^{(3)} \\ &= s_1 \begin{bmatrix} 0 \\ \frac{1}{3}(1 + e^{j2\pi/3}e^{-j2\pi/3} + e^{-j2\pi/3}e^{j2\pi/3}) \\ \frac{1}{3}(1 + e^{j2\pi/3}e^{j2\pi/3} + e^{-j2\pi/3}e^{-j2\pi/3}) \end{bmatrix} \\ &= s_1 \begin{bmatrix} 0 \\ 1 \\ 0 \end{bmatrix} \end{aligned} \tag{59c}$$

We see from Eq. (59c) that the incident wave supplied to port 1 appears at port 2 with s_1, where $|s_1| = 1$. By reason of rotational symmetry, the signal supplied to port 2 appears at port 3. Therefore, the circuit is the circulator $1 \rightarrow 2 \rightarrow 3 \rightarrow 1$. In the case of Fig. 82, as $y_1 = \infty$ and $s_1 = -1$, the signal supplied to port 1 appears at port 2 with π radian.

To design the circulator, therefore, we use the following formulas:

$$\begin{aligned} y_1 &= \infty \\ y_2 &= j\omega C + \frac{1}{j\omega L_+}, \qquad L_+ = \mu_+\xi \\ y_3 &= j\omega C + \frac{1}{j\omega L_-}, \qquad L_- = \mu_-\xi \\ \omega C &= \frac{1}{2}\left(\frac{1}{\omega L_+} + \frac{1}{\omega L_-}\right) \\ \frac{\omega L}{R} &= \frac{\sqrt{3}}{2}\left(\frac{1}{\mu_+} - \frac{1}{\mu_-}\right) \end{aligned} \tag{60}$$

The explanation mentioned above is only appropriate for the center frequency. For a greater frequency deviation, s_2 and s_3 move clockwise, even if s_1 does not move because of no reactive energy for the excitation of $\boldsymbol{u}^{(1)}$. This is understandable because the variation of the eigenvalues is proportional to the total reactive energies, as in Eq. (20d) of Chapter 2. Therefore, the input matching and the attenuation in the reverse direction is sometimes called the isolation deteriorate.

Such a bandwidth ratio to meet the specified isolation I (dB) can be obtained as follows [24];

$$w = \frac{\Delta f}{f_0} = \frac{2\sqrt{3}|\eta||s''|}{\sqrt{1 + (3/4)\eta^2}}$$

$$\eta = \frac{\mu_+ - \mu_-}{\mu_+ + \mu_-}$$

$$I\ (\text{dB}) = 20 \log_{10}\left(\frac{1}{|s''|}\right) \tag{61}$$

To improve the wideband performance of the circulator, we have to add the network to increase the reactive energy only for the same-phase excitation of $\boldsymbol{u}^{(1)}$. One way is to add the series resonant circuit at each port, as shown in Fig. 83a. In this case, because the impedance at each port for $\boldsymbol{u}^{(2)}$ and $\boldsymbol{u}^{(3)}$ is high as understood from Fig. 82e, the current flowing in the series resonant is small. Then the series resonant circuits do not give the effect for the deviation of y_2 and y_3 by the frequency deviation.

On the other hand, the impedance for $\boldsymbol{u}^{(1)}$ at each port is zero; thus, the currents flowing in the series resonant circuits increases, which results in an increase in the reactive energy, contributing to the variation of y_1 for $\boldsymbol{u}^{(1)}$, keeping the relative angle of $2\pi/3$ (rad) with respect to y_2 and y_3 and creating a wideband circulator.

This can also be realized by using only one series resonant circuit connecting the shielding metal plate and the newly made ground, as shown in Fig. 83b [25]. Figure 83c shows the frequency performance of the wideband circulator. Although this explanation was made on the basis of reactive energies, it can be explained from the view point of an equivalent network.

The practical circulator, however, can be expressed by the ideal circulator together with the parallel tuned circuits connected at each port, as shown in Fig. 84 [26]. Therefore, the additional resonant networks provide the wideband by the network theory.

Strip-Line Circulator

As shown in Figs. 85a and 85b, the strip lines connected to ports 1, 2, and 3 join the center conductor and the ferrite disks are installed on both sides of the conductors.

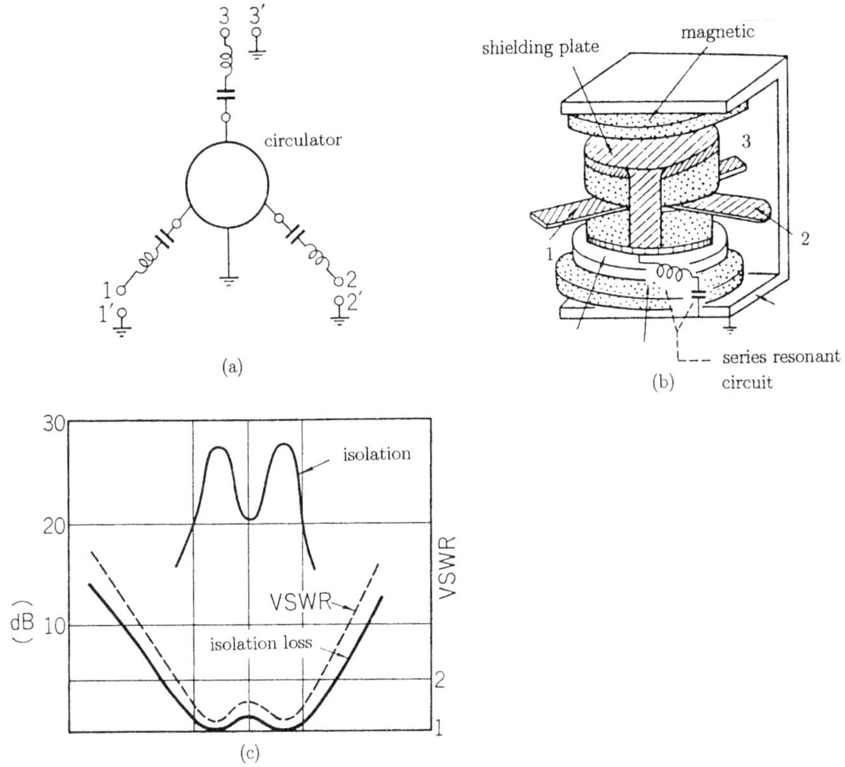

Figure 83. Wideband circulator: (a) series resonant circuits connected at each port, (b) one series resonant circuit connected between the shielding metal plate and ground; (c) measured performance.

After shielding, the dc magnetic field is supplied vertically to the surface of the ferrite. The simple concept is as follows. When the wave is supplied to port 1, the RF magnetic field is rotated in the direction shown in Fig. 85c by Faraday's effect (see Appendix 8). Therefore, the RF magnetic field is coupled to port 2 but not port 3.

The operation from the standpoint of the eigenvalues is as follows. The excitations of $\boldsymbol{u}^{(1)}$, $\boldsymbol{u}^{(2)}$, and $\boldsymbol{u}^{(3)}$ produce the electromagnetic fields shown in Figs. 86a and 86b.

The admittance corresponding the excitation is obtained as in Figs. 86c and 86d.

The frequency performances are shown in Fig. 87. When we choose as $y_1 = \infty$, as in the case of Fig. 82e, y_2' and y_3' become inductive. By adding

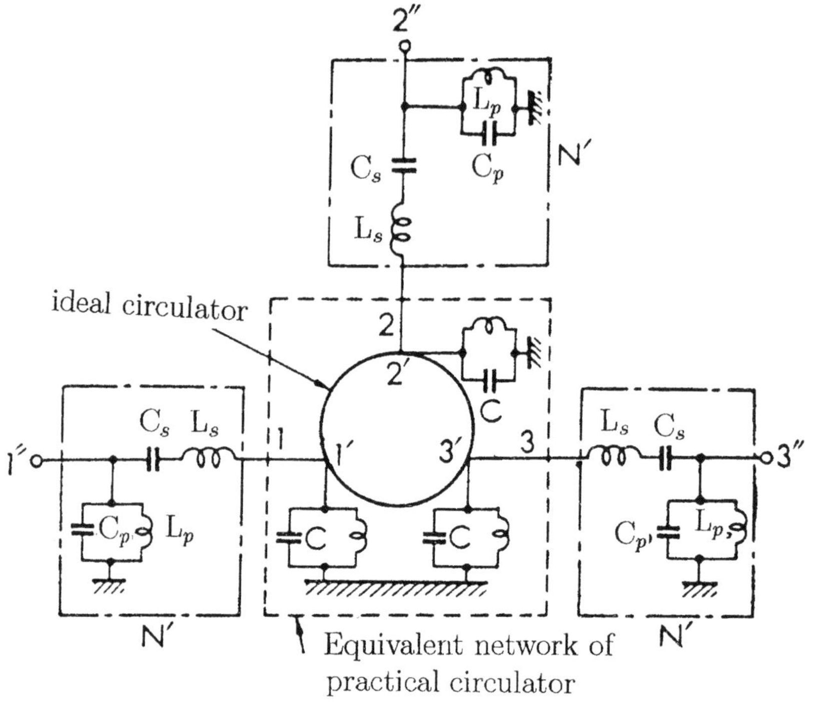

Figure 84. Equivalent network of a practical circulator and the wideband network. N' is the circuit used for making a wideband.

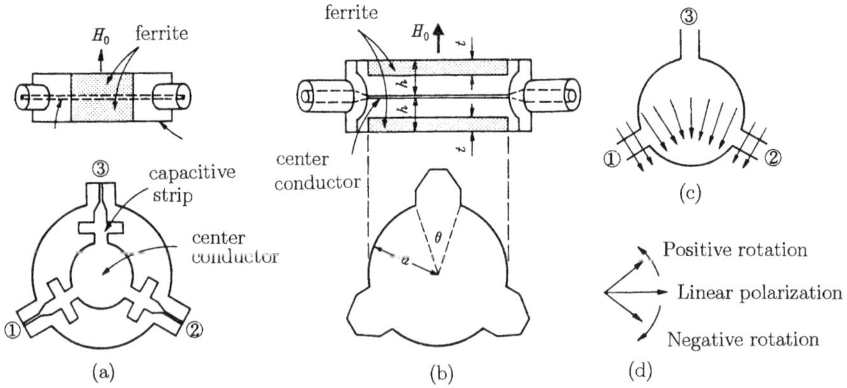

Figure 85. Construction of strip-line circulator: (a) ordinal circulator; (b) middle power circulator; (c) change of polarization; (d) decomposition of linear polarization.

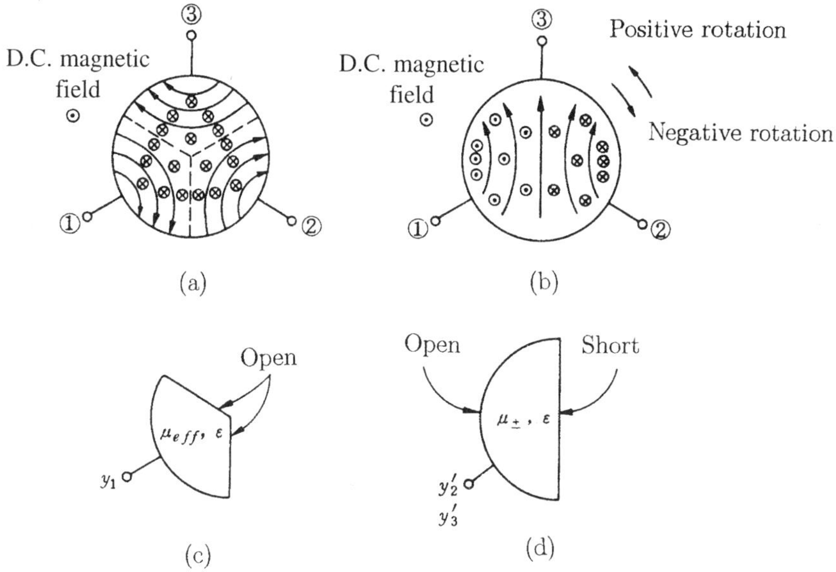

Figure 86. Electromagnetic fields for $\boldsymbol{u}^{(1)}$, $\boldsymbol{u}^{(2)}$, and $\boldsymbol{u}^{(3)}$, and the admittance: (a) $\boldsymbol{u}^{(1)}$; (b) $\boldsymbol{u}^{(2)}$ and $\boldsymbol{u}^{(3)}$; (c) equivalent network of (a) for the admittance; (d) equivalent network of (b) for the admittance.

the parallel capacity, we can get y_2 and y_3 to meet the condition of a circulator. The capacities are made by the strip lines as shown in Fig. 85a.

On the other hand, we can design y_2' and y_3' to take the values of y_2 and y_3 without any capacities. In this case, we design the size of the ferrite disks to resonate using the TM_{11} planar resonator described at Fig. 38 of Chapter 4, where the relative permeability of ferrite, $\tilde{\mu}_r$, is assumed to be

$$\tilde{\mu}_r = \frac{\mu_+ + \mu_-}{2} \tag{62}$$

Therefore, the diameter of the ferrite disk, D, is taken as

$$D \simeq \frac{17.7}{f\sqrt{\varepsilon_r \tilde{\mu}_r}} \text{ (cm)} \tag{63}$$

where ε_r is the relative dielectric constant of ferrite and $\tilde{\mu}_r = (\mu_+ + \mu_-)/2$.

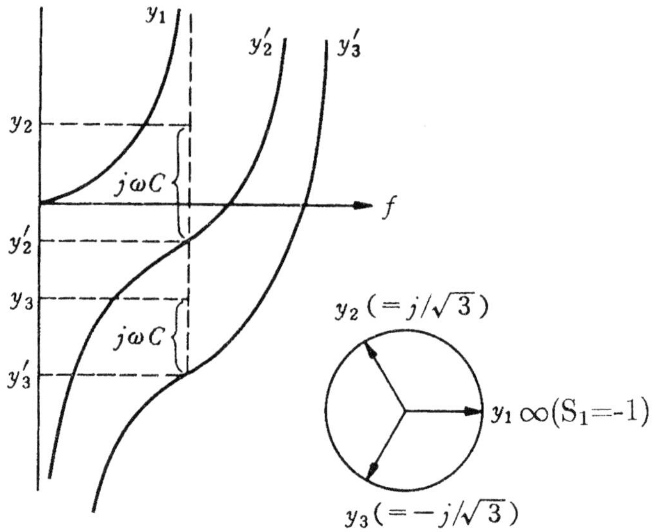

Figure 87. Frequency performance of admittances for excitations $\boldsymbol{u}^{(11)}$, $\boldsymbol{u}^{(2)}$, and $\boldsymbol{u}^{(3)}$ and the plot on a Smith chart. (This Smith chart is rotated, as $y > 0$ corresponds to the upper half of a circle.)

Equation (63) shows that the diameter of the ferrite disk decreases for a higher frequency and the available power decreases. Therefore, by increasing the air gap between the ferrite disk and the center conductor the diameter of the ferrite disk increases and the available power increases up to 10-fold.

The radius and the thickness of ferrite, a and t, are obtained from Reference [27].

6 MAGNETOSTATIC MODE RESONATOR AND FILTER

6.1 Mode Consideration

When an ellipsoidal ferrite sample is placed in a dc magnetic field toward the z-axis, as in Fig. 88, the ferrite sample takes resonant frequencies in the region $\mu_+ < 0$.

We consider the closed surface, S, surrounding the ferrite sample. (See Fig. 89) The complex pointing power toward inside, $\dot{P}$, must be zero when the sample is resonant. From Eq. (18c) of Chapter 2, $\dot{P}$ takes in the lossless network

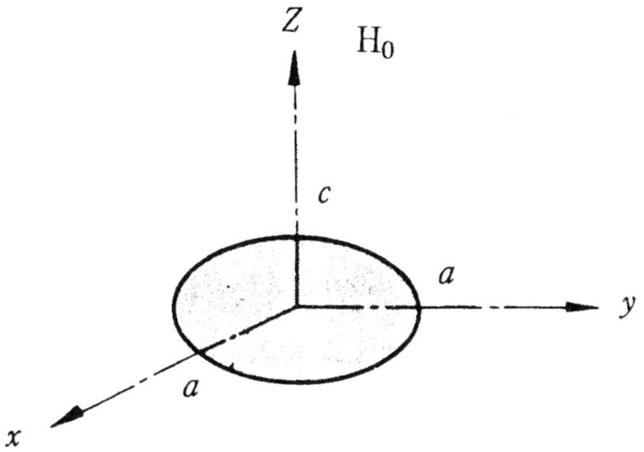

Figure 88. Ferrite with ellispsoidal shape placed in the dc magnetic field toward the z direction.

$$\begin{aligned}\dot{P} &= j\omega \iiint_\tau \{\mu_0 \boldsymbol{H}^* \cdot \hat{\mu}\boldsymbol{H} - \varepsilon_0 \boldsymbol{E} \cdot \hat{\varepsilon}^* \boldsymbol{E}^*\}\, d\tau \\ &= j\omega\mu_0 \iiint_\tau \boldsymbol{H}^* \cdot \hat{\mu}\boldsymbol{H}\, d\tau - \tilde{W}_e \\ &= j\omega\mu_0 \iiint_\tau \{\mu_+ |a^+|^2 |\boldsymbol{H}^+|^2 + \mu_- |a^-|^2 |\boldsymbol{H}^-|^2\}\, d\tau - W_e \qquad (64a)\end{aligned}$$

(see Note 2). Considering $\mu_+ < 0$ and $\mu_- > 0$, there exists the feasibility that

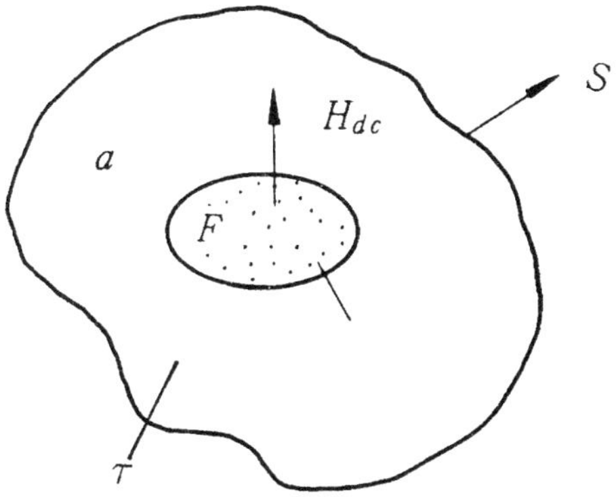

Figure 89. A ferrite sample surrounded by a closed surface S.

$$W_e = 0$$

$$\boldsymbol{E} = 0 \tag{64b}$$

From Maxwell's equation,

$$\nabla \times \boldsymbol{H} = j\omega \boldsymbol{E} = 0 \tag{64c}$$

This is the static magnetic field, which exists for $\omega = 0$. The mode resonating under the condition of Eq. (64c), is therefore called the magnetostatic mode.

By the above reason, the magnetostatic mode exists in the region $\mu_+ < 0$ because $\mu_- > 0$ always.

The simplest mode has magnetic fields and magnetic flux as shown in Fig. 90; this mode is the uniform mode obtained by Kittel.

Even if the explanation mentioned above started from the feasibility of magnetostatic mode, another magnetostatic mode can be obtained under the assumption of Eq. (64c). Walker [28] made calculations and obtained several modes about the ellipsoidal ferrite sample of Fig. 88. Here, the main results are introduced. We use the expressions

$$\Omega = \frac{\omega}{\omega_m}, \qquad \Omega_H = \frac{|\gamma| H_{\text{in}}}{\mu_m}, \qquad \alpha = \frac{c}{a} \tag{65}$$

where $\omega_m = |\gamma| 4\pi M_s$, H_{in} is the dc magnetic field in ferrite, M_s is the saturation magnetization, $|\gamma|$ is the gyromagnetic ratio, and ω is the resonant angular frequency.

The characteristic equation to determine ω is expressed by the functions including P_n^m and Q_n^m, and they include Ω, Ω_H, and α. Therefore, Ω is obtained if m, n, α, and Ω_H are given. The number of solutions of Ω is generally the maximum integer less than $1 + \frac{1}{2}(n - |m|)$. Therefore, the Ω with maximum values is denoted by $(m, n, 0)$ and the next one is $(m, n, 1)$, $(m, n, 2)$ and so on. For example, in the case of $m = n$ and $n = m + 1$, only one solution of Ω exists for the given m, n, α, and Ω_H. Ω always exists in the following range [28]:

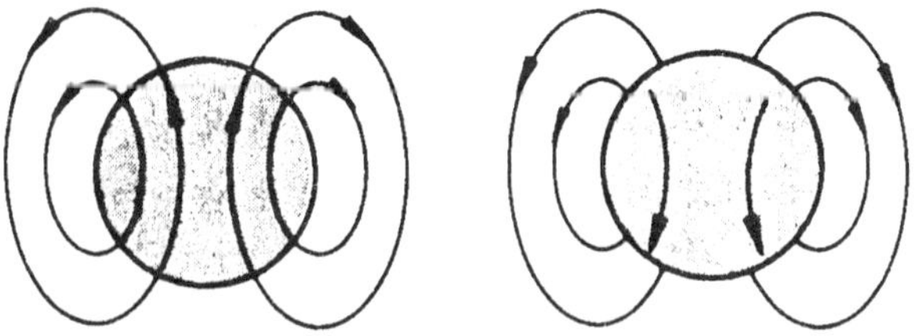

Figure 90. Magnetic flux and field of magnetostatic uniform mode.

$$0 < \Omega - \Omega_H < 0.5 \tag{66}$$

To satisfy Eq. (64c), however, $\boldsymbol{H}$ can be expressed by the scalar function ψ. It is also found that ψ takes the values

$$\psi \propto (x - jy)^m \quad \text{for the } (m, n, 0) \text{ mode}$$
$$\psi \propto z(x - jy)^m \quad \text{for the } (m + 1, m, 0) \text{ mode} \tag{67}$$

As a simple case, $m = n = 1$, that is, in the case of the (1, 1, 0) mode, we have

$$\boldsymbol{H} \propto (i_x - ji_y)$$

from Eq. (67), which shows the uniform mode or Kittel mode as shown in Fig. 90. This (1.1.0) mode is the most important in the practical use.

In this mode, Ω is given by

$$\Omega = \Omega_H + N_t = \Omega_{H0} - N_z + N_t \tag{68}$$

where $\Omega_{H0} = |\gamma|H_0/\omega_m$, H_0 is the external dc magnetic field, N_z is the demagneting coefficient in the z direction, and N_T is the demagneting coefficient in the transverse direction; where we have the relation

$$N_x + N_y + N_z = N_z + 2N_t = 1 \tag{69}$$

where N_x and N_y are the demagnetizing coefficients in the x and y directions, respectively. For example, we can obtain the resonant angular frequency of Eq. (70):

$$N_z = 0, \qquad N_t = \frac{1}{2}, \qquad \omega_r = |\gamma|H_0 + \frac{|\gamma|4\pi M_s}{2} \qquad \text{(fine post)}$$
$$N_z = \frac{1}{3}, \qquad N_t = \frac{1}{3}, \qquad \omega_r = |\gamma|H_0 \qquad \text{(sphere)}$$
$$N_z = 1, \qquad N_t = 0, \qquad \omega_r = |\gamma|H_0 - |\gamma|4\pi M_s \qquad \text{(thin disk)} \tag{70}$$

The values of the internal dc magnetic field, H_i, is obtained by substituting Eq. (70) into Eq. (68) to obtain

$$H_i = \frac{1}{|\gamma|}\left(\omega - \frac{4\pi M_s}{2}\right) \qquad \text{(fine post)}$$
$$H_i = \frac{1}{|\gamma|}\left(\omega - \frac{4\pi M_s}{3}\right) \qquad \text{(sphere)}$$
$$H_i = \frac{\omega}{|\gamma|} \qquad \text{(thin disk)} \tag{71}$$

H_i of Eq. (71) is shown in Fig. 91. As understood from Eq. (70), ω is

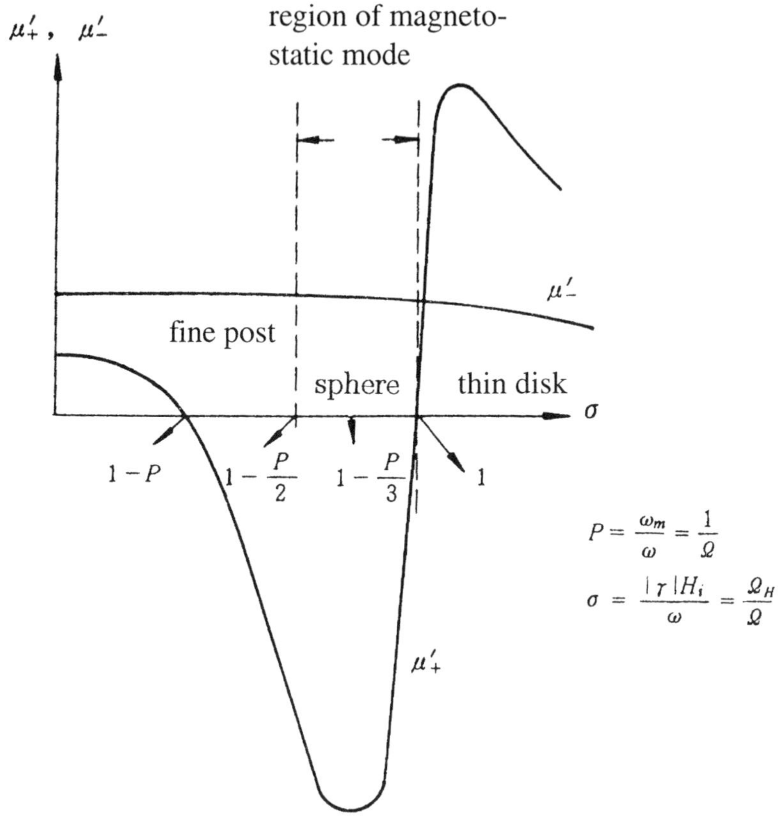

Figure 91. The dc magnetic field corresponding to the magnetostatic resonance with a typical shape.

proportional to H_0 and is not affected by M_s. Therefore, ω does not change with the change of the temperature.

Next, from Eq. (71), the internal dc magnetic field is strongest in the case of a thin disk. In the case of a lower frequency range such as VHF to UHF, H_i decreases; we then encounter the problem of low field loss caused by unsaturation. This deteriorates the Q values of the resonator. Then, the stronger internal dc magnetic field is preferable. For this reason, the thin disk is usually used for lower frequency.

It is also understood from Eq. (71) that the small values of M_s is preferable for obtaining a strong internal dc magnetic field for the lower frequency. In practical material, however, the Curie temperature decreases for

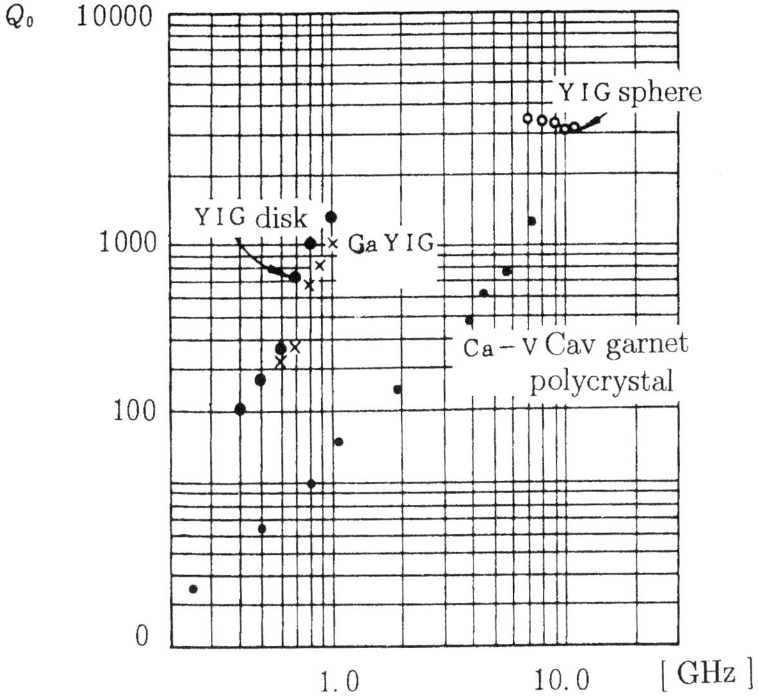

Figure 92. Q values of several magnetostatic resonators.

low values of M_s, which results in the deterioration of the temperature performance. Therefore, a compromise has to be made.

6.2 *Q* Values

The loss of the resonator is caused by μ''_+, which is the loss term of $\dot{\mu}_+$ in Eq. (58d). However, the loss term is released to the ΔH of Fig. 81 by

$$\alpha = \frac{|\gamma|\Delta H}{2\omega} \tag{72a}$$

(see Note 3). Q values of the magnetostatic mode can be expressed as

$$Q = \frac{\omega}{|\gamma|\Delta H} \tag{72b}$$

(see Note 4). Generally, the loss term μ''_+ does not take the Lorenzian distribution when H_i is separated from H_r. Therefore, ΔH should be calculated under the assumption of Lorenzian distribution. The values of ΔH obtained in such a way are denoted by ΔH_{eff}. Thus, Q values should be expressed by $\omega/|\gamma|\Delta H_{\text{eff}}$. Examples of Q values are shown in Fig. 92.

A summary of the feature of the magneto-static mode are as follows:

(i) The high Q with small size irrelevant to frequency.

- A sphere with a diameter less than 1 mm is used for the microwave band.
- A thin disk with a diameter of a few millimeters is used for the VHF–UHF band.

(ii) The resonant frequency can be controlled by the external dc magnetic field. A 3–26.5-GHz frequency can be changed by changing the current of the coil for the dc magnetic field from 45 mA to 360 mA.

The resonators are used for the bandpass filler (BPF) and the band-reflection filter (BRF), examples of which are shown in Fig. 93. When there is no ferrite sample, the input and the output ports are not coupled by their orthogonal construction in the case of BPF. When there is a ferrite sample,

Figure 93. Filters with magnetostatic modes: (a) principle of BPF with orthogonal coil and ferrite magnetostatic mode; (b) waveguide BPF; (c) strip-line BPF; (d) strip-line BRF; (e) three-stage BPF with half-turn coils. (From Ref. 29.)

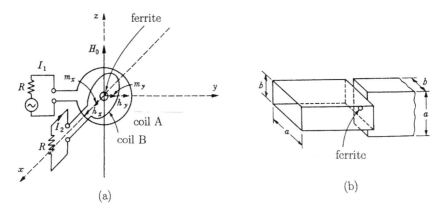

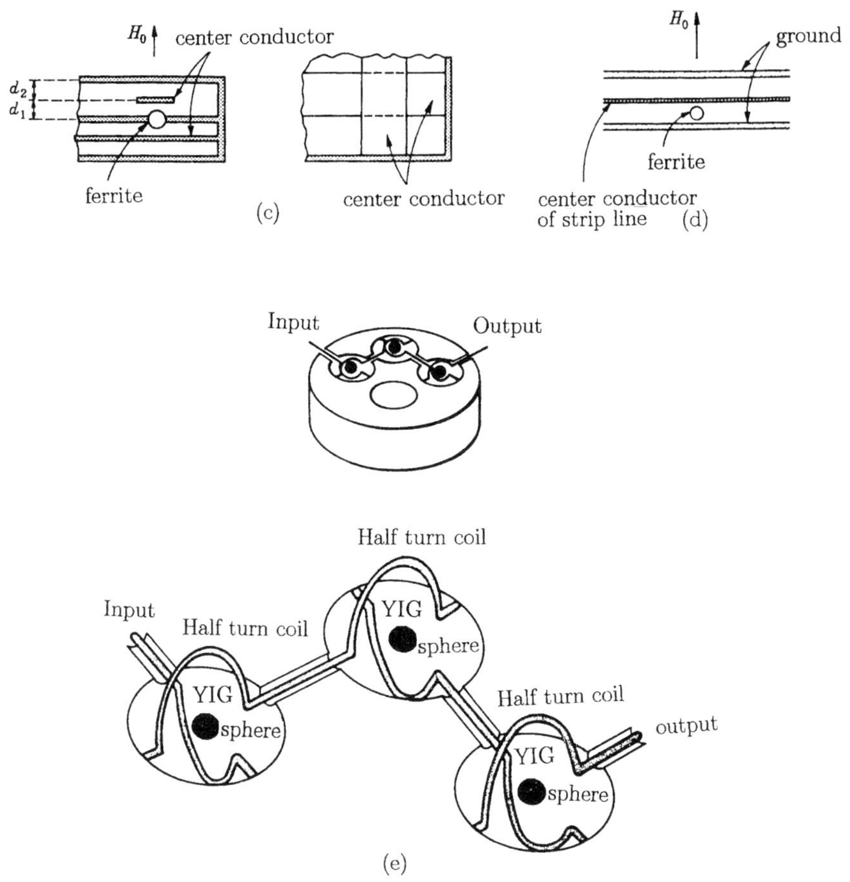

Figure 93. Continued.

the positive rotational magnetic fields take place in the sample; the input and the output ports are coupled with the ferrite with a phase delay of $\pi/2$ radian. It results in the PBFs shown in Figs. 93a–93c.

In the case of Fig. 93d, the magnetic coupling between the strip line and the ferrite sample, makes an equivalent parallel turned circuit connected to the strip line in series. It then works as a BRF.

The equivalent networks of a BPF and a BRF are shown in Fig. 94a and 94b, respectively [30]. In the case of the BPF, the magnetic coupling between port 1 and the ferrite is made by the nonreciprocal transformer; in the case of the BRF, the magnetic coupling is made by the reciprocal coupling.

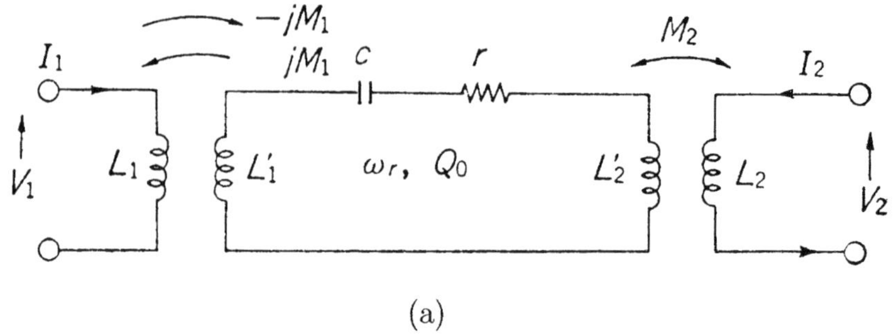

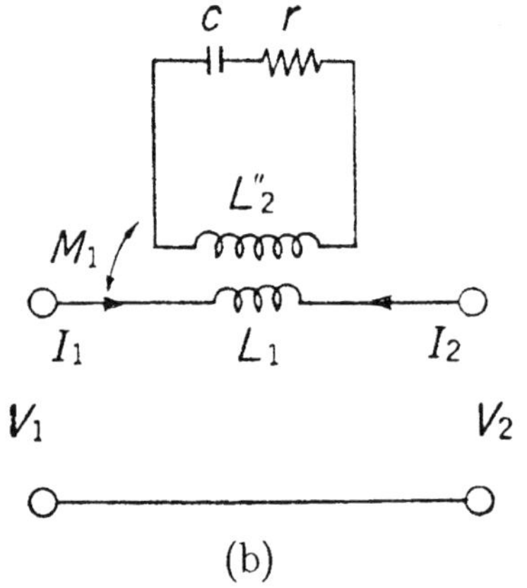

Figure 94. Equivalent networks of filters with a magnetostatic mode of a ferrite sample: (a) BPF; (b) BRF.

Thus far, we were concerned the uniform mode of (110). In the practical case, however, higher modes, explained earlier, give the unwanted spurious performance usually at the level 10–15 dB below than that of the main mode. To suppress them, the multistage filter is used. By this method, we cannot only suppress spurious performance but also get the maximum flat or Tchebycheff performances, as in the case of ordinal multistage filters. An example of a three-stage BPF is shown in Fig. 93e.

A multistage BRF can be obtained by placing plural ferrite samples coupled to the strip line with each distance of $\lambda_g/4$ between neighboring ferrite samples.

APPENDIX

Note 1

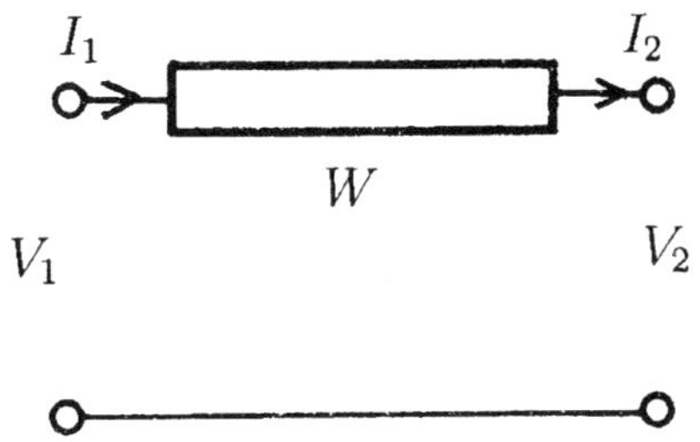

$$\begin{bmatrix} V_1 \\ I_1 \end{bmatrix} = \begin{bmatrix} 0 & jW \\ j\dfrac{1}{W} & 0 \end{bmatrix} \begin{bmatrix} V_2 \\ I_2 \end{bmatrix} \tag{1}$$

From Eq. (1), we get $I_1 = jV_2/W$ and $V_2 = WI_1/j$. We also get the results $V_1 = jWI_2$ and $I_2 = V_1/jW$.

Note 2

We can express $\boldsymbol{H}$ by rotational magnetic fields $\boldsymbol{H}^+$ and $\boldsymbol{H}^-$ as $\boldsymbol{H} = a^+\boldsymbol{H}^+ + a^-\boldsymbol{H}^-$ and $\boldsymbol{H} = a^+\boldsymbol{H}^+ + a^-\boldsymbol{H}^-$, where $\boldsymbol{H}^+ = \boldsymbol{i}_x - j\boldsymbol{i}_y$, $\boldsymbol{H}^- = \boldsymbol{i}_x + j\boldsymbol{i}_y$, $a^+ = (H_x + jH_y)/2$, and $a^- = (H_x - jH_y)/2$; $\boldsymbol{i}_x$ and $\boldsymbol{i}_y$ are the unit vectors toward x and y, respectively, and h_x and h_y are the x and y component $\boldsymbol{H}$, respectively.

Note 3

Considering the neighboring of H_r, we may set

$$\sigma = 1 + \frac{|\gamma|\delta H}{\omega} \tag{1}$$

Expressing μ'' by δH from Eq. (58d),

$$\mu'' = \frac{\alpha}{(|\gamma|\delta H/\omega)^2 + \alpha^2} \tag{2}$$

By the definition of ΔH of Fig. 81, we get

$$\alpha = \frac{|\gamma|\Delta H}{2\omega} \tag{3}$$

Note 4

From Eq. (68), we get

$$\delta\dot{\Omega} = \delta\dot{\Omega}_H \tag{1}$$

From Eq. (65),

$$\delta\dot{\omega} = |\gamma|\delta\dot{H} = \omega\delta\dot{\sigma} \tag{2}$$

Because

$$\delta\dot{\sigma} = j\alpha = \frac{j|\gamma|\delta H}{2\omega}, \qquad \delta\dot{\omega} = \frac{j|\gamma|\Delta H}{2} \tag{3}$$

The Q values are expressed by complex angular frequency as

$$\omega + \frac{\omega}{2Q} = \omega + \delta\dot{\omega} \tag{4}$$

Comparing Eqs. (3) and (4), we get

$$Q = \frac{\omega}{|\gamma|\Delta H} \tag{5}$$

REFERENCES

1. Watkins, J., Circular resonant structures in microstrip, *Electr. Lett.*, 524–525 (October 1969).
2. Terman, F. E., *Radio Engineer's Handbook*, McGraw-Hill, New York, 1943, p. 51, Eq. (26).
3. Caulton, M., Knight, S. P., Dary, D. A., Hybrid integrated lumped-element microwave amplifiers, *IEEE Trans., ED-15*(7), 459–466 (1968).
4. Terman, F. E., *Radio Engineer's Handbook*, McGraw-Hill, New York, 1943, pp. 35–36.
5. Terman, F. E., *Radio Engineer's Handbook*, McGraw-Hill, New York, 1943, p. 58, Eq. (45).
6. Dill, H. G., Designing inductors for thin film applications, *Electr. Design*, 52–59 (February 1964).
7. Standley, R. D., Frequency response of strip-line incident-wave directional filter, *IEEE Trans. Microwave Theory Tech., MTT-11, Electron. Devices*, 264 (July 1963).
8. Moreno, T., *Microwave Transmission Design Data*, McGraw-Hill, New York, 1984.
9. Marcuvitz, N., *Waveguide Handbook*, McGraw-Hill, New York, 1951.
10. Caulton, M., The lumped element approach to microwave integrated circuits, *Microwave J.*, *13*, 51–58 (May 1970).

11. Alley, C. D., Interdigital capacitors and their application to lumped element microwave integrated circuits, *IEEE Trans. Microwave Theory Tech., MTT-18*(12), 1028–1033 (1970).
12. Wilkinsn, E. J., An *N*-way hybrid power divider, *IEEE Trans. Microwave Theory Tech., MTT-8*(1), 116–118 (1960).
13. Saleh, A. A. M., Planar electrically symmetric *n*-way hybrid power dividers/combiners, *IEEE Trans. MTT-28*(6), 555–563 (1980).
14. Gonzalez, F. J. O., Martin, J. L. J., High-power amps blend low cost, high performance, *Microwaves RF*, 87–103 (1995).
15. Yee, H. Y., Chang, F. C., Audeh, N. F., *N*-way TEM-mode broad-band power dividers, *IEEE Trans. Microwave Theory Tech., MTT-18*, 682–688 (1970).
16. Konishi, Y., et al. Consideration of the frequency limitation of Wilkinson power divider with coupled wires in ferrite, *IEEE Trans. Broadcasting, B30*(4), 364–367 (1993).
17. Naito, Y., Formulation of frequency dispersion of ferrite permeability, *Trans. IEICE, 56-c*, 113 (February 1973).
18. Konishi, Y., et al. Newly proposed vertically installed planar circuit and its application, *IEEE Trans. Broadcasting, BC-33*, 1–7 (March 1987).
19. Konishi, Y., et al., A directional coupler of a vertically installed planar circuit structure, *IEEE Trans. Microwave Theory and Techniques, MTT-36*, 1057–1063 (June 1988).
20. Monttomery, C. G., *Technique of Microwave Measurements*, McGraw-Hill, New York, 1947.
21. Collin, R. E., ''Foundations for microwave engineering,'' McGraw-Hill, Inc., (1966).
22. Muraguch, M., Yukitakem, T., Naito, Y., Optimum design of 3-dB branch-line couplers using microstrip lines, *IEEE Trans. Microwave Theory and Tech., MTT-31*(8), 674–678 (1983).
23. Konishi, Y., Analysis of slotted bridge, *Tech. J. Japan Broadcasting Corp., 12*(2), 15–31 (1960).
24. Konishi, Y., Lumped element Y circulator, *IEEE Trans. Microwave Theory Tech., MTT-13*, 852–864 (1965).
25. Konishi, Y., Design of a new broad-band isolator, *IEEE Trans. Microwave Theory Tech., MTT-19*(3), 260–269 (1971).
26. Konishi, Y., VHF–UHF Y Circulators, *Tech. J. Japan Broadcasting Corp., 17*(2), 87–121 (1965).
27. Konishi, Y., A high power UHF circulator, *IEEE Trans. Microwave Theory Tech., MTT-15*(12), 700–708 (1967).
28. Walker, L. R., Magnetostatic modes in ferromagnetic resonance, *Phy. Rev., 105*, 390 (1957).
29. Helszain, J., *YIG resonators and filters*, Wiley–Interscience, New York.
30. Konishi, Y., *Recent Microwave Ferrite Circuit Technology*, The Institute of Electronics, Information and Communication Engineers, New York, 1972. pp. 143–148.

7

Practical Applications on Systems

1 SATELLITE BROADCASTING SYSTEMS

Figure 1 shows the block diagram of transponder of BS-3 for Japanese satellite broadcasting. The signal of 14 GHz transmitted from the ground station is received by the parabolic antenna, and it is introduced to a 14-GHz low-noise receiver through the multiplexer. The signal is converted to 12 GHz by the mixer and the local oscillator. Because the signal includes the A, B, and C channels, it is divided by the multiplexer MUP_i and amplified by each traveling wave tube. Amplified signals are combined again by the multiplexer and they are supplied to the antenna through the multiplexer. The multiplexer is generally constructed with a combiner and band-pass filters (BPF) of each channel.

In the practical case, there are one more set of the parts surrounded by the broken line, it is switched over at the time of an accident at point A. The output of the receiver is connected at the isolated port of a 3-dB coupler (or bridge) and the left two ports are connected to the two sets of transmitters as shown in Fig. 2. Each channel in the transmitter can be switched over by SW_o, as shown in Fig. 2.

In the home, the receiving system is used. (See Fig. 3) When the wave of a 12-GHz band is received by the antenna, it is converted to a 1-GHz band, which is fed by a coaxial cable and introduced to the broadcast system

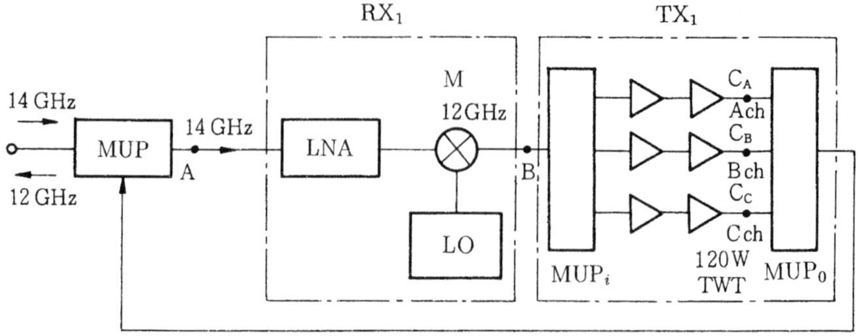

Figure 1. Block diagram of transponder of BS-3 for Japanese satellite broadcasting (in the practical case, a spare system is installed). MUP: multiplexer; LNA: low-noise amplifier; LO: local oscillator; MUP_i: multiplexer to divide each channel; MUP_0: multiplexer to combine each channel; TWT: traveling wave tube; RX_1: receiver; TX_1: transmitter. (From Ref. 1.)

(BS) tuner. In the BS tuner, a channel is selected and the video and the sound are supplied to the TV display.

In the BS converter, a local oscillator with a dielectric resonator as shown in Fig. 4 is mounted. The 12-GHz band is usually received at the offset parabolic antenna, as shown in Fig. 5. It is collected at a primary horn and introduced to the circular waveguide. In the circular waveguide, the circular polarized wave of the BS wave is transformed to the linear polarized wave through a polarizer.

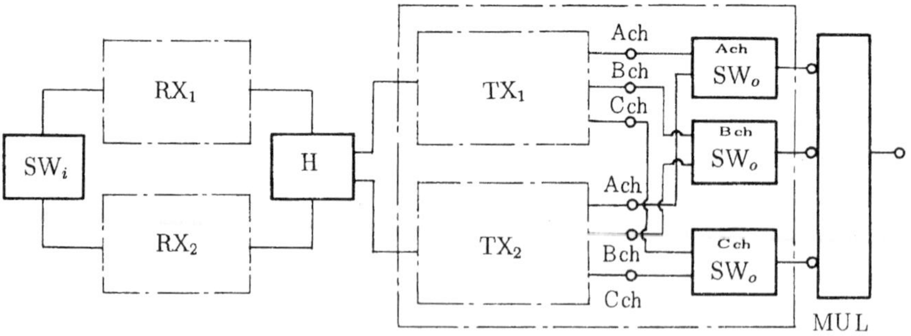

Figure 2. System of transponder with spare. SW_i: Input switch; SW_o: output switch; RX_1: receiver of Fig. 1; RX_2: spare receiver; H: 3-dB coupler (or hybrid); TX_1: transmitter of Fig. 1; TX_2: spare transmitter.

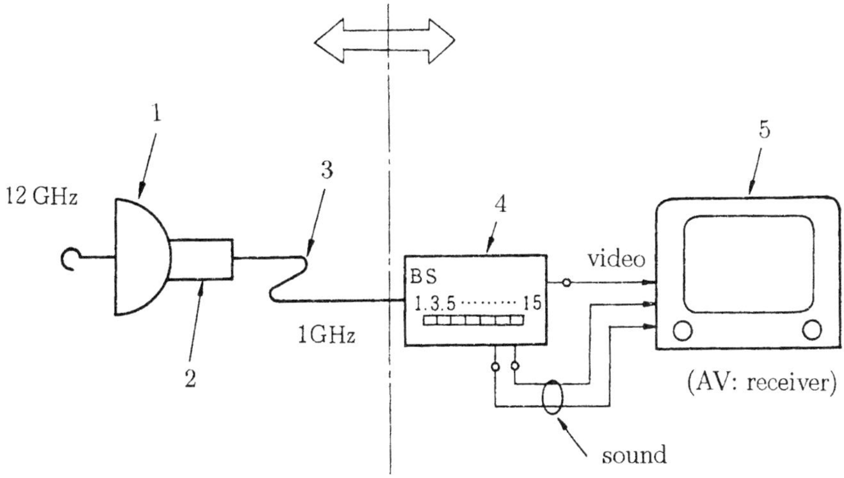

Figure 3. Home receiving system of satellite broadcasting. 1: Receiving antenna; 2: BS converter; 3: cable; 4: BS tuner; 5: TV display.

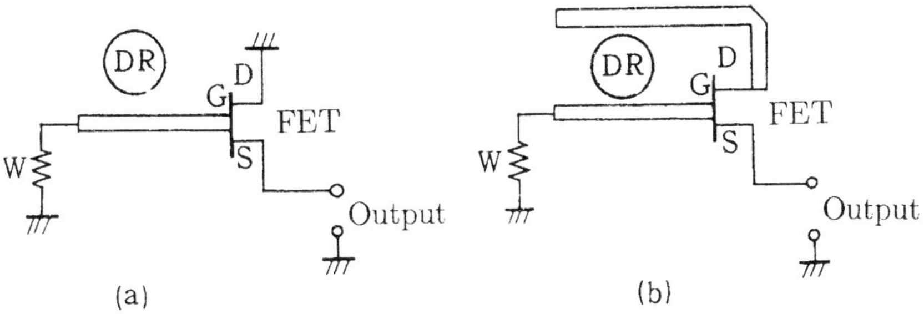

Figure 4. Dielectric resonator (DR) oscillator to convert 12 GHz to 1 GHz used for the local oscillator of a BS converter: (a) colpitz type; (b) feedback from drain to gate through the DR. DR: Dielectric resonator of 10.678 GHz; W: stabilized resistance; G, D, and S: gate, drain, and source of Fourier electron transmission (FET), respectively.

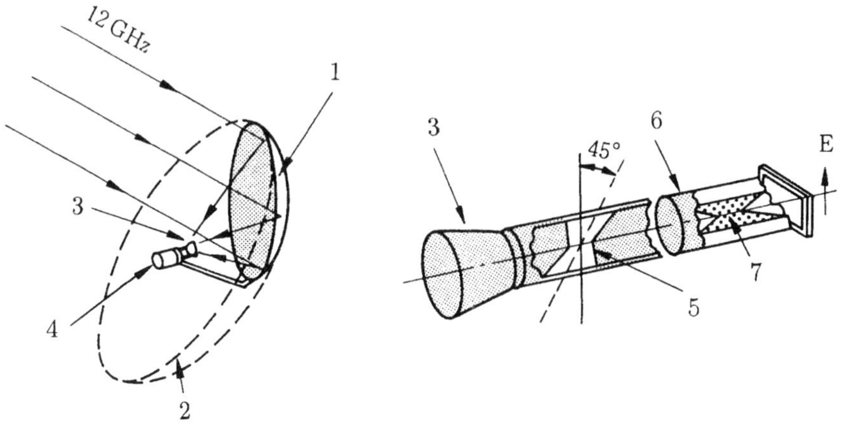

Figure 5. Offset parabolic antenna and polarizer in BS outdoor unit. 1: Reflector; 2: parabolic surface; 3: primary horn; 4: BS converter; 5: polarizer; 6: transformer from circular to rectangular waveguide; 7: absorber for polarizer.

2 MOVABLE COMMUNICATION SYSTEM

In a movable communication system, there are two kinds of systems: the duplex system and the simplex system. In the duplex system, we need a combiner and two BPFs corresponding to the receiving frequency band and the transmitting frequency band, as shown in Fig. 6. The duplexer should

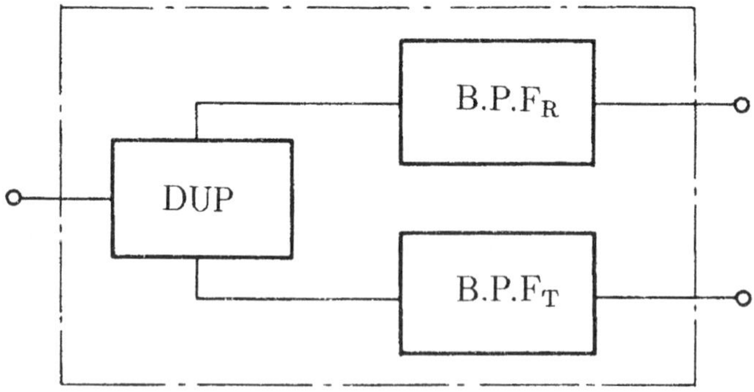

Figure 6. Components of a duplexer with two BPFs. DUP: Duplexer; BPF_R: BPF for receiving frequency; BPF_T: BPF transmitting frequency.

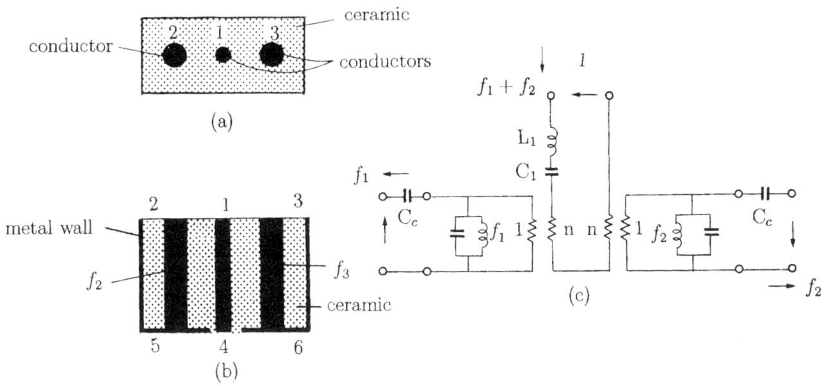

Figure 7. Duplexer with one-stage BPFs: (a) top view; (b) side view; (c) equivalent network.

be compact, lightweight, and have a low insertion loss. Therefore, ceramics with high dielectric constants and low-loss material are usually used. Examples are shown in Figs. 7 and 8.

Figure 7 shows a duplexer with one-stage BPFs; port 1 is used for the commonly used antenna, and ports 2 and 3 are connected to the receiver and transmitter. Port 1 is connected to the $\lambda_g/4$ line with an open end, which makes the $\lambda_g/4$ interdigital coupling to the BPFs. The equivalent network is shown in Fig. 7c.

To get a sharper cutoff performance at the output of the pass band, we can use the two-stage BPF to obtain the maximum flat or Tchebycheff performances. Figure 8 shows the duplexer with two-stage BPFs, where interdigital couplings are made; the equivalent network is shown in Fig. 8c.

The inductive coupling between resonators is caused by the air holes made between resonators, described in detail in Fig. 66 of Chapter 4.

3 IMAGE REJECTION RECEIVING SYSTEM

When one of the many signals with frequency $f_{s1}, \ldots, f_{sn}$ distributed in a wide frequency range is received, the signal must be converted to an intermediate frequency, f_{IF} by a mixer.

In the output of the mixer, f_{IF} corresponding to the f_s and f_m components shown in Fig. 9 appear in the case of a simple mixer. When the signal is

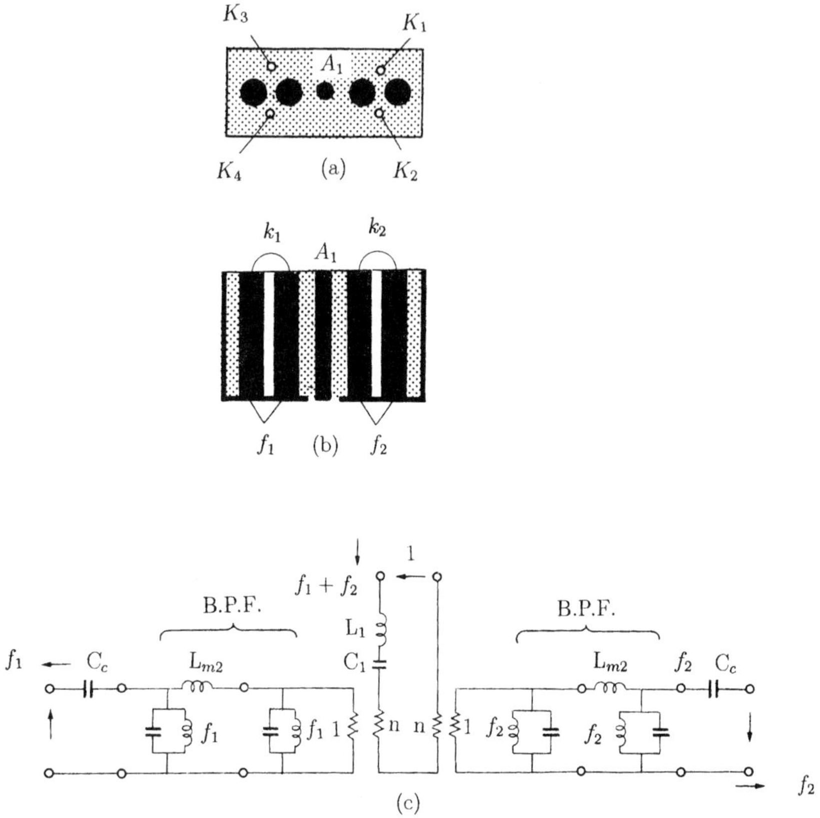

Figure 8. Duplexer with two-stage BPFs: (a) top view; (b) side view; (c) equivalent network.

distributed from f_1 through f_2 as in Fig. 9, the f_m component is not received. In such a case, an image rejection receiver is required. An example of the image rejection mixer used for the receiving system is shown in Fig. 10, which shows both the 90° and 0° Hybrids.

When the signal is supplied at port 1 of Fig. 11, it appears at ports 2 and 3 with a phase difference of $\pm\pi/2$; no signal appears at port 4. H_1 denotes the 90° Hybrid. A 3-dB coupler with two symmetrical planes as described in Chapter 6, Section 3 belongs to H_1; the equivalent network is in Fig. 66 of Chapter 6, where ports 1, 2′, and 4 are selected.

On the other hand, when the signal supplied at port 1′ appears at ports 2 and 3 with the same phase and no signal appears at port 4, it is called the

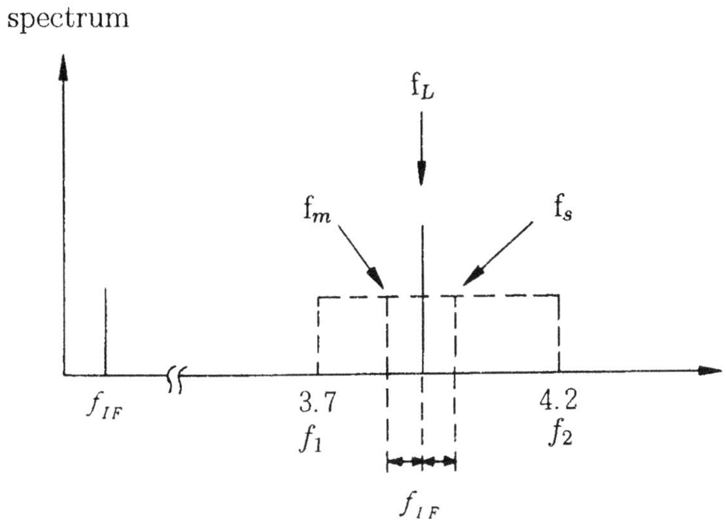

Figure 9. Frequency spectrum of a converter. f_s: Signal frequency; f_1: lowest signal frequency; f_m: image frequency; f_L: local frequency; f_2: highest signal frequency.

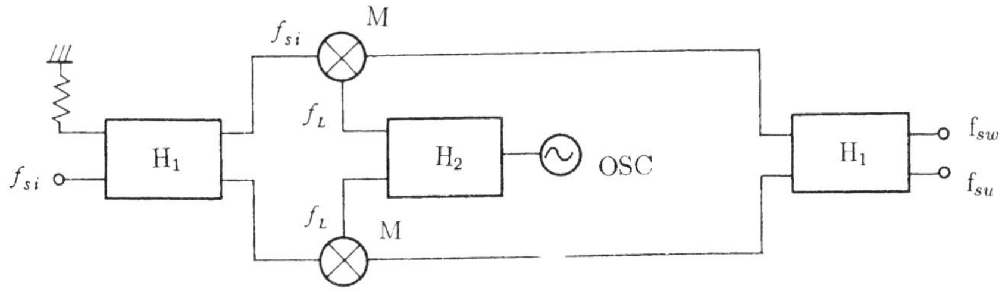

Figure 10. Components of an image rejection mixer. H_1: 90° Hybrid; H_2: 0° Hybrid; OSC: frequency changeable oscillator; f_{sw}: wanted signal frequency; f_{su}: unwanted signal frequency; M: mixer.

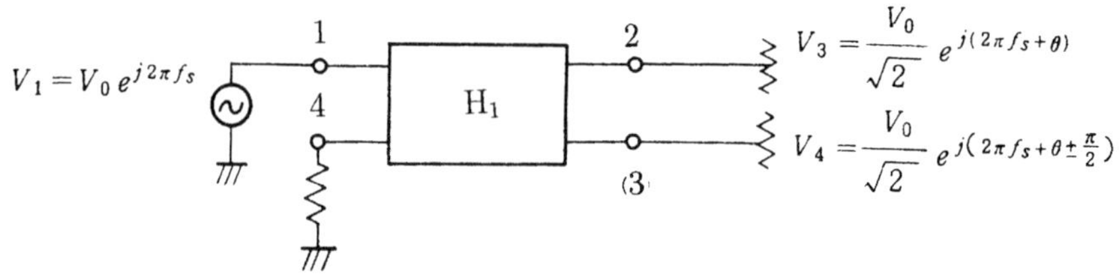

Figure 11. The 90° hybrid.

0° Hybrid. The equivalent network is that of Fig. 66 of Chapter 6, where ports 1, 2, 3, and 4 are selected.

Examples (from Chapter 6) are the magic T (Fig. 68), the slotted coaxial bridge (Fig. 69), and the ferrite-loaded lumped element directional coupler (Fig. 65); all are 0° Hybrids.

4 AMPLIFIER

As mentioned in Chapter 6, Section 2, power dividers are used at the input and the output of the amplifier, together with isolators.

The circuit of Fig. 12 is the balance-type amplifier. The input signal flows following the solid lines and it appears at the output port shown in Fig. 12. If reflection exists at each amplifier, the reflected waves flow following the dotted lines; they are added at the terminated resistance *R* in phase and do not appear at the input port. It makes the wideband matching in the input port. This circuit is used for the wideband amplifier and the multistage amplifier, where the input and output matchings are sometimes difficult.

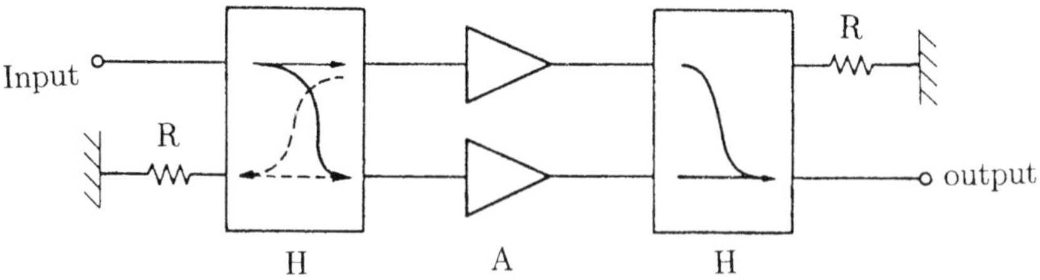

Figure 12. Balance-type amplifier. H: 90° Hybrid; R: terminated resistance; A: amplifier.

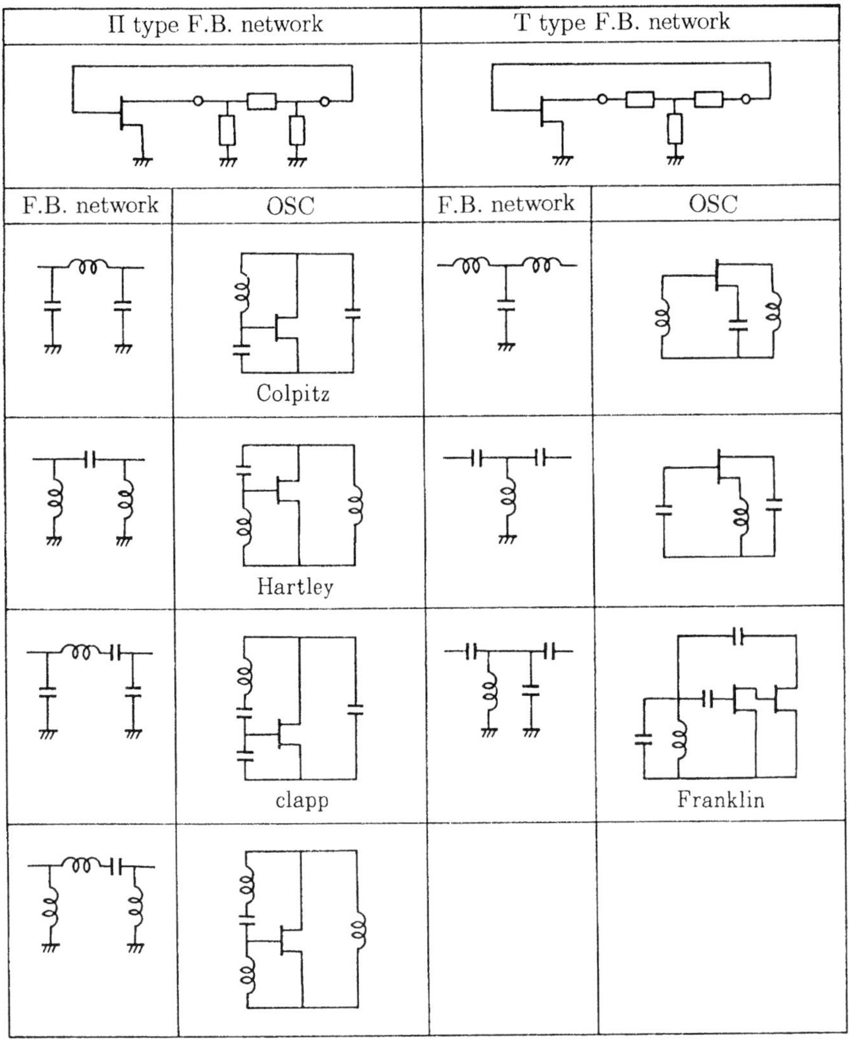

Figure 13. Classification of feedback (F.B.) network of the oscillator (OSC). F.B. feedback; OSC oscillator.

5 OSCILLATOR

We have several kinds of oscillator and the feedback networks can be classified as in Fig. 13. For example, in the case of the colpitz oscillator, it has the circuit in Fig. 14a, the feedback network of Fig. 14b, and the equivalent network including the transistor's impedances as shown in Fig. 14c.

In Fig. 14c, when the constant emitter current flows between E and C of the feedback network, we have the maximum current, I_{max}, flowing in $\dot{Z}_i$, when the network is tuned at f_0. When the frequency bandwidth to satisfy $I \geq I_{max}/\sqrt{2}$ is Δf, the feedback Q_f is

$$Q_f = \frac{f_0}{\Delta f}$$

We, however, have the relation between the noise spectrum of the B region and Q_f as shown in Fig. 15. Then, large values of Q_f are required to get a low-noise oscillator.

We can get the high Q_f by replacing L by a parallel tuned network with low inductance and large capacity. Such a resonator can be obtained by a dielectric resonator. In a higher-frequency region, the $TE^{\circ}_{01\delta}$ mode DR is used (Fig. 4), and in a lower-frequency region, such as the VHF to UHF band, the TEM mode ceramic resonator is available. This is why such ceramic resonators are used in an oscillator such as the VCO (voltage controlled oscillator) and fixed frequency oscillator.

6 CERAMIC INTEGRAL CIRCUIT TECHNOLOGY

Two methods have been used to produce a ceramic multilayer integrated circuit. The first is a paste printing method and the second is a laminating sheet method. These methods are different in the ceramic layer production process before sintering. The paste printing method produces the ceramic multilayer by repeated printing of dielectric material paste. The laminating sheet method produces the layer by laminating nonsintered ceramic sheets, usually called green sheets. Although various improvements have been made to these methods, the laminating sheet method has now been widely used to construct an RF integrated circuit requiring an accuracy in physical dimensions, a design flexibility in connecting conductors between layers, and a shorter production time. An example of the production process of the laminating sheet method is shown in Fig. 16.

The sintering process of the ceramic layer is defined considering a formation method of conductors embedded in the layer. Previously, a sintering of the ceramic using a material fireable lower than the melting point of the conductor has been well known. Recently, several production methods to

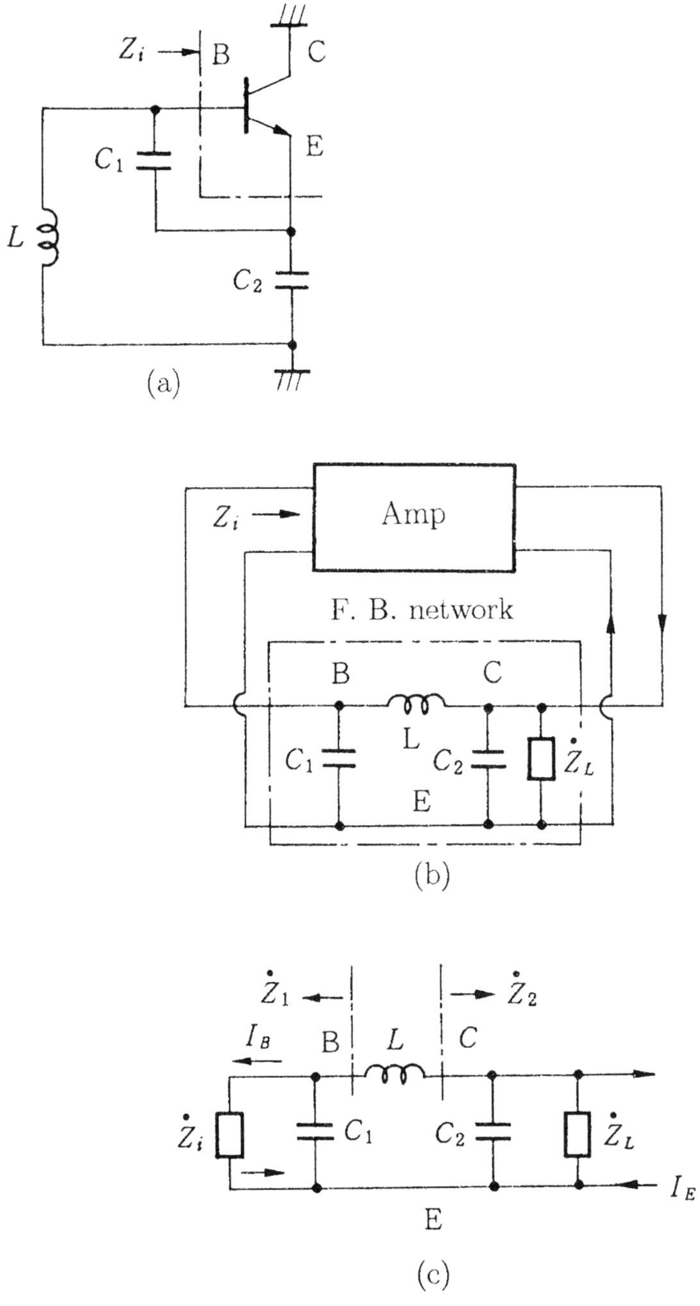

Figure 14. Colpitz oscillator and the feedback network: (a) colpitz oscillator; (b) equivalent network of (a); (c) equivalent network of (b).

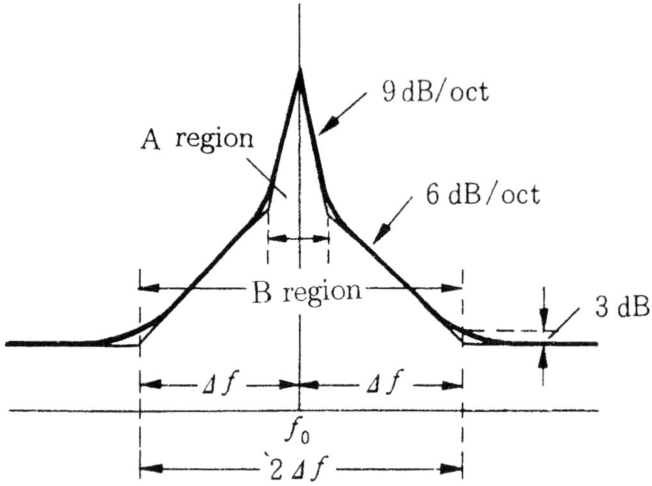

Figure 15. Noise spectrum of oscillator.

sinter the ceramic higher than the melting point of the metal have been developed. Descriptions of these methods will be presented in the following subsections.

6.1 Low-Temperature Firing Method

The idea for this method, called a cofiring method of ceramics and metals, was proposed originally by IBM Corporation at the beginning of 1970. Since then, various kinds of ceramic multilayer integrated RF circuit, such as band-pass filters, duplexers, or VCOs, have been developed using this method.

Usually, ceramic materials having a smaller dielectric loss are sintered at the temperature of 1200°C or higher. The ceramic–metal cofiring technology, however, requires a sintering temperature lower than the conductor melting point. To reduce the sintering temperature of ceramics, mixtures of ceramic powder and glass have been developed. A lower dielectric loss and a lower-temperature coefficient of dielectric constant are also required for the low-temperature fireable ceramic. Typical examples of these materials are listed in the Table 1.

A silver or copper paste is commonly used to construct conductors in the RF circuit. The silver paste shows good conductivity and stable chemical characteristics in the sintering process. It is, however, well known that a migration of the sintered silver causes the serious problem of a short circuit; whereas a copper paste causes a lower diffusion velocity. The copper paste,

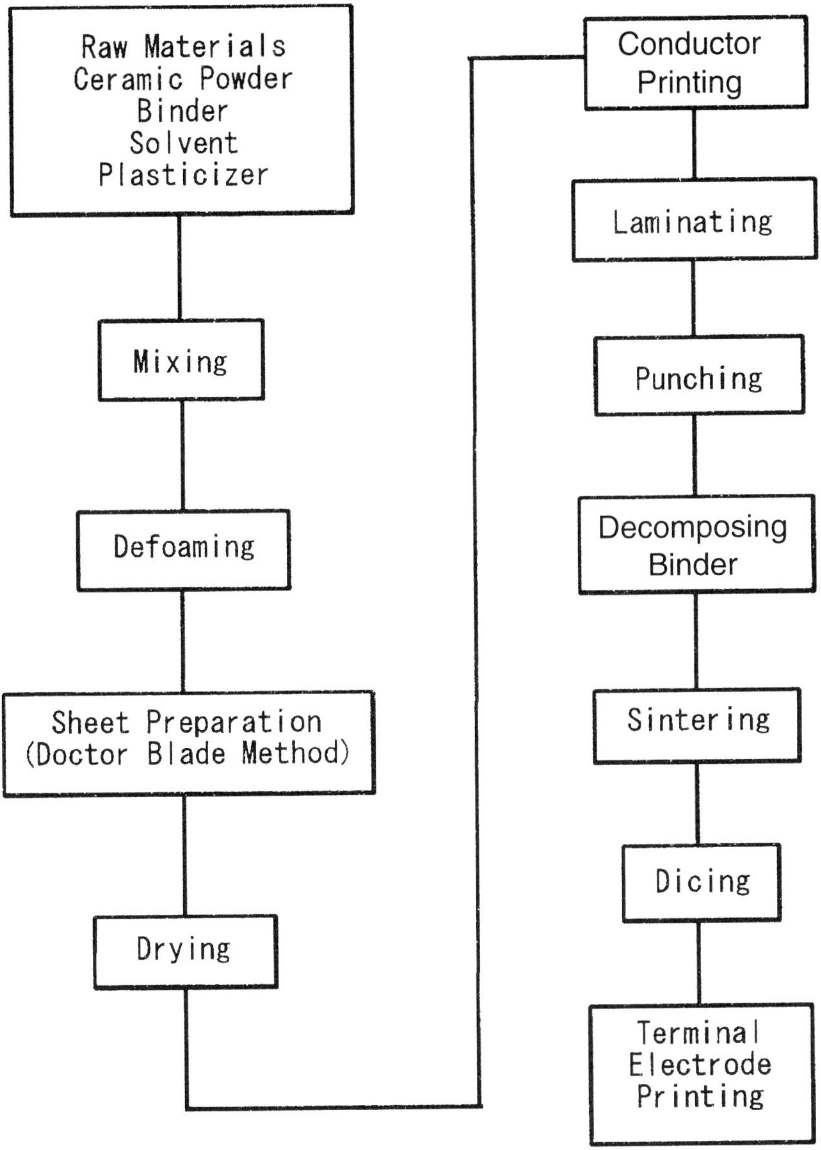

Figure 16. Production process of laminating sheet method.

Table 1. Example of Low-Temperature Fireable Materials

	Dielectric material system			
	$CaZrO_3$–glass	$BaO–TiO_2–Nd_2O_3$–glass	Al_2O_3–glass	$Al_2O_3–TiO_2$–glass
Permittivity	25	75	7.8	10.7
Q Factor	1000[a]	2500[b]	1250	2200[b]
T_{cc} (ppm/°C)	±10		+200	
τ_f (ppm/°C)		±		±0.5
Firing temp. (°C)	980	900	850	1000
	Conductor Material			
	Cu	Ag	Ag	Ag
Resistivity (mΩ/cm²)	1.8	2.5	3–5	1.5

[a]Measured at 1 MHz.
[b]Measured at 1 GHz.

however, should be sintered in an inert atmosphere to avoid oxidation. This fact may complicate the sintering process of the copper.

The migration of the silver paste is caused by a diffusion of the silver ions in the ceramics by a static electric force generated between conductors. Efforts to avoid the migration have been made to produce a fine microstructured ceramics to prevent the diffusion of metal ions. An appropriate design consideration to cut the dc potential is also effective to avoid the migration.

Recently, the silver paste method has been more widely used, applying the countermeasures described above.

6.2 High-Temperature Sintering Method

In spite of various improvements for the low-temperature fireable ceramics, these materials have been shown to have a higher RF loss factor than ordinary sintered ceramics. Previously, it was thought that a high sintering temperature would prevent the embedding of conductors in the ceramics. Two novel methods have overcome this difficulty. The first is the sintering of a conductor embedded within a green sheet, regardless of the melting of the metal. If the conductor if embedded within the sheet, it will remain in the sheet and the conductors will be formed after cooling. A low-insertion-loss bandpass filter and a broadband lumped element circulator with a continuous RF magnetic circuit were developed using this method [2,3].

The other method is a paste-intruding method. The main part of the integrated circuit is fabricated using conventional multilayer technology, except for the printing of dummy conductors in a green sheet using carbon paste. Hollows are constructed after the ceramic sheet is laminated and sintered, as the carbon paste is resolved during the sintering process. A silver paste is intruded into hollows prepared in the ceramics by a cold isostatic pressure method and then fired at a temperature lower than the melting point to fix the conductors in the element. The preparation method of the hollows using the thermal decomposition of a dummy conductor of carbon paste is also well known in ceramic capacitor technology, except that in this case, a silver paste is intruded. An air gap between the ceramics and the conductor is naturally formed in the firing of the silver paste; this is unfavorable for ceramic capacitor technology. However, the fact that the gap has no influence on the generation of magnetic flux made it possible to adopt this method to construct a circulator. A construction of a broadband lumped element circulator with a continuous RF magnetic circuit is reported using this method [3].

REFERENCES

1. Mirua, S., Akanuma, T., Kadoo, S., Funaki, M., Oikawa, K., Honda, H., Koike, T., Kasuga, Y., Takahashi, K., and Kikuchi, T., Development work of transponder for BS-3, IEICE Technical Report SAT87-45, 1987, pp. 13–17.
2. Kobayshi, M., Kawamura, K., and Suzuki, K., High *Q*-factor tri-plate resonator with an inner conductor of melted silver, *Ceram. Trans.*, *41*, 355–362 (1994).
3. Miura, T., Kobayashi, K., and Konishi, Y., Optimization of a lumped element circulator based on eigen inductance evaluation and structural improvement, *IEEE International Microwave Symposium Digest*, pp. 117–120, 1996.

APPENDIX 1

Introduction to Group Velocity

When we consider the amplitude-modulated signal, it has spectra ω, $\omega - \Delta\omega$, and $\omega + \Delta\omega$, where $\Delta\omega$ is the angular frequency of the modulation signal. The spectra are expressed as follows:

$$
\begin{aligned}
\omega: &\quad V_0 e^{j\omega t} e^{-jk_z z} \\
\omega - \Delta\omega: &\quad \frac{m}{2} V_0 e^{j(\omega - \Delta\omega)t} e^{-j(k_z - \Delta k)z} \\
\omega + \Delta\omega: &\quad \frac{m}{2} V_0 e^{j(\omega + \Delta\omega)t} e^{-j(k_z + \Delta k)z}
\end{aligned} \tag{1}
$$

The resultant signal of the above three spectra, $V(t)$ can be expressed by

$$
\begin{aligned}
V(t) &= V_0 e^{j(\omega t - k_z z)} \left[1 + \frac{m}{2} \left(e^{j(\Delta\omega t - \Delta k \cdot z)} + e^{-j(\Delta\omega t - \Delta k \cdot z)} \right) \right] \\
&= V_0 [1 + m \cos(\Delta\omega t - \Delta k z)] e^{j(\omega t - k_z z)} = V_1 e^{j(\omega t - k_z z)}
\end{aligned} \tag{2}
$$

V_1 is the distribution of the amplitude along the z direction, and the amplitude distribution moves toward $z > 0$ with t.

Therefore, the velocity of the moving of the distribution, v_g, is

$$
v_g = \lim_{\Delta\omega \to 0} \frac{\Delta\omega}{\Delta k} = \frac{d\omega}{dk} = \left(\frac{dk}{d\omega} \right)^{-1} \tag{3}
$$

In the case of general signal modulation, we have

$$\psi = A(z, t)e^{j(\omega_0 t - k_z z)} \tag{4}$$

ψ can be expressed by the Fourier integral as

$$\psi = \int a(k)e^{j(\omega t - k_z z)}\, dk \tag{5}$$

Because k exists around k_z, we can get k and ω by

$$k - k_z = \Delta k, \qquad \omega - \omega_0 = \Delta\omega \tag{6}$$

Therefore, $A(z, t)$ of Eq. (4) can be expressed by

$$A(z, t) = \int a(k)e^{j(\Delta\omega \cdot t - \Delta k \cdot \Delta z)}\, dk \tag{7}$$

Because $A(t)$ changes much slowly than $1/\omega_0$, we can approximate $dA/dt = 0$. Therefore, we get the relation

$$\frac{dA}{dt} = \frac{\partial A}{\partial t} + \frac{\partial A}{\partial z} v_g = 0, \qquad v_g = -\frac{\partial A}{\partial t}\left(\frac{\partial A}{\partial z}\right)^{-1} = \frac{\Delta\omega}{\Delta k} = \left(\frac{dk}{d\omega}\right)^{-1} \tag{8}$$

Equation (8) is the same as Eq. (3).

APPENDIX 2

Cavity and Waveguide Perturbation

Cavity Perturbation

The resonant frequency of a cavity changes slightly when the cavity wall or the material inside the cavity is perturbed. Figure 1 shows the perturbation of the cavity wall and Fig. 2 shows that of the material inside the cavity.

The theory of each perturbation is described in the following subsections.

Perturbation of Cavity Wall

As a simple example, we consider the perturbation of the metal wall of a lumped element LC resonant circuit as shown in Fig. 3. In the case of l, w, a, $g \ll \lambda$, the region between two electrodes with an area of S becomes the capacitor and the left region becomes the inductor.

When we push the electrode to narrow the distance between the electrodes, the capacitance C increases by ΔC. Under the condition of a constant displacement current of 1 A, the decrease of the time average of the electric energy, $-\Delta\tilde{W}_e$, can be related to the increase of the capacitance, ΔC, as follows. Because $\tilde{W}_e$ is expressed by C and the displacement current I as

$$\tilde{W}_e = \frac{I^2}{2\omega^2 C}$$

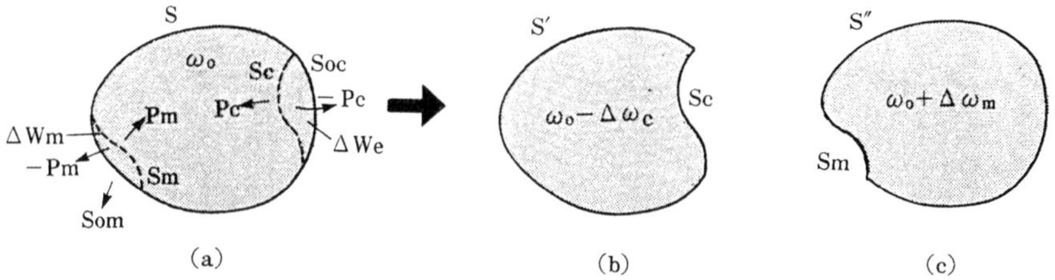

Figure 1. Perturbation of cavity wall: (a) original cavity; (b) cavity perturbed at S_{oc}; (c) cavity perturbed at S_{om}.

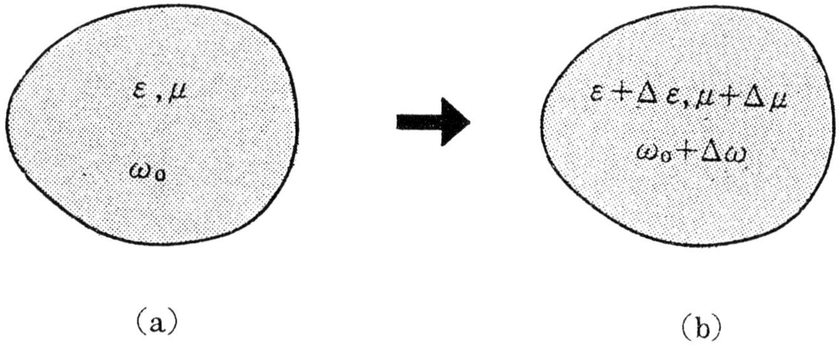

Figure 2. Perturbation of material inside a cavity: (a) original cavity; (b) perturbed cavity.

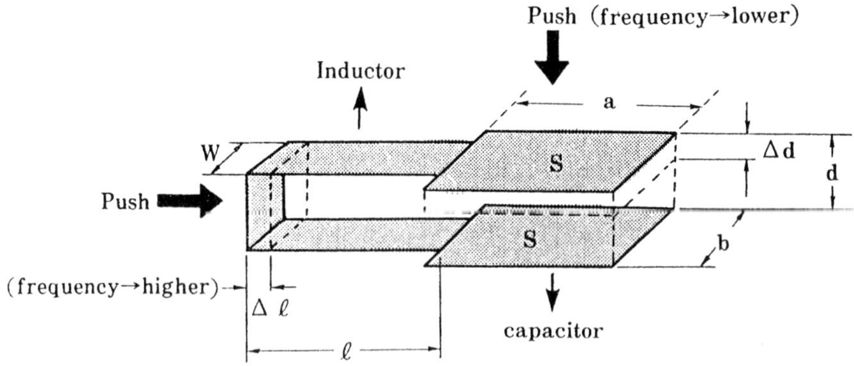

Figure 3. Perturbation of metal wall of a lumped element LC resonant circuit.

we get

$$\Delta W_e \simeq \frac{\Delta C}{2\omega^2 C} I^2$$

and

$$\frac{\Delta \tilde{W}_e}{\tilde{W}_e} \simeq \frac{\Delta C}{C} \tag{1}$$

On the other hand, when we push the shortened end of the inductance as shown in Fig. 3, the inductance L decreases by ΔL. Under the condition of a constant current of 1 A, the decrease of the time average of the magnetic energy, $\Delta \tilde{W}_m$, can be related to the decrease of the inductance, $-\Delta L$, as follows. Because $\tilde{W}_m$ is expressed by L and the current I as

$$\tilde{W}_m = \frac{LI^2}{2}$$

we get

$$-\Delta \tilde{W}_m = \frac{\Delta L}{2} I^2$$

and

$$-\frac{\Delta \tilde{W}_m}{W_m} = \frac{\Delta L}{L} \tag{2}$$

The values of ΔC and ΔL, however, change the resonant angular frequency by $\Delta\omega$ as follows.

Because

$$\omega = \frac{1}{\sqrt{LC}} \tag{3}$$

we get

$$\frac{\Delta\omega}{\omega} \simeq -\frac{1}{2}\left(\frac{\Delta L}{L} + \frac{\Delta C}{C}\right) \tag{4}$$

Using Eqs. (1) and (2) in Eq. (4) and taking into account of the relation $2\tilde{W}_e = 2\tilde{W}_m = \tilde{W}_t$,

$$\frac{\Delta\omega}{\omega} \simeq \frac{\Delta \tilde{W}_m - \Delta \tilde{W}_e}{\tilde{W}_t} \tag{5}$$

Equation (5) shows the following results. When the electric energy decreases by pushing the metal wall, the resonant frequency decreases, and when the magnetic energy decreases by pushing the metal wall, the resonant frequency increases.

Next, we describe more generally the cavity perturbation theory about the construction in Fig. 1. In the cavity surrounded by the metal wall S with a resonant angular frequency ω_0, shown by the solid line of Fig. 1, we consider the complex Poynting power P_c and P_m flowing from the surface S_c and S_m toward inside, respectively.

When the time average of the reactive energy surrounded by S_c and S_{oc} is the electric energy of $\Delta\tilde{W}_e$, P_c should take the values obtained from

$$P_c = 2j\omega\tilde{W}_e \tag{6}$$

This result can be explained from the following important results. The complex Poynting power P_0 on the surface S_0 is generally related to the time average of the reactive power and consumed power in the region of the Poynting power flow by Eq. (7) (see Eq. (41b) of Chapter 1):

$$\dot{P}_0 = P_d + 2j\omega(\tilde{W}_m - \tilde{W}_e) \tag{7}$$

where $\tilde{W}_m$, $\tilde{W}_e$, and P_d are the time-averaged magnetic energy, the electric energy, and the consumed power, respectively. In this case, however, because the reactive energy in the region surrounded by S_c and S_{oc} is only electric energy, we have the condition $P_d = \tilde{W}_m = 0$. Therefore, the Poynting power from the surface of S_c toward S_{oc} should be $-2j\omega\tilde{W}_e$ from Eq. (7).

The Poynting power flowing from the closed surface S toward the inside of the cavity, however, should be zero because of the resonant condition. Therefore, the Poynting power from S_c toward the inside of the cavity, P_c, should be conjugate values of $-2j\omega\tilde{W}_e$. Then, we get Eq. (6). This is shown at point C of Fig. 4.

Next, we have the important relation between the $\partial[\mathrm{Im}(\dot{P})]/\partial\omega$ and the total reactive energy.

Denoting the complex Poynting power, the electric field, and the magnetic field on S_c at ω_0 by P_c, $\boldsymbol{E}_c$, and $\boldsymbol{H}_c$, respectively, P_c is obtained from

$$P_c = \boldsymbol{E}_c \times \boldsymbol{H}_c^* \tag{8}$$

Under the condition of the constant magnetic field, the variation of the total complex Poynting power on S_c toward the inside of the cavity can be obtained from Eq. (8) considering $\delta\boldsymbol{H}_c = 0$ as follows:

$$\delta \iint_{S_c} \boldsymbol{P}_c \cdot \boldsymbol{n}\, dS_c = \delta \iint_{S_c} \boldsymbol{E}_c \times \boldsymbol{H}_c^* \cdot \boldsymbol{n}\, dS_c = \iint_{S_c} \delta\boldsymbol{E}_c \times \boldsymbol{H}_c^* \cdot \boldsymbol{n}\, dS_c \tag{9}$$

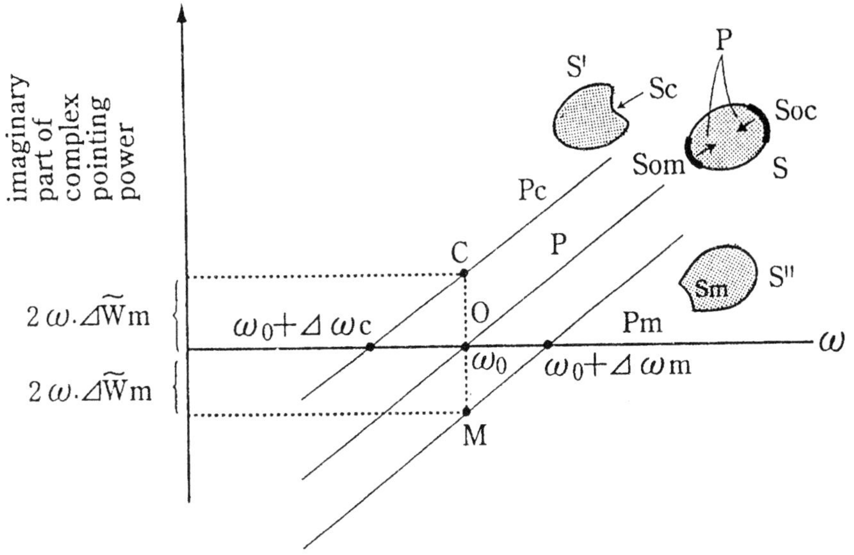

Figure 4. Complex Poynting power P_c, and P_m, and flowing from S_c, S_m, S_{oc}, and S_{om}.

where $\boldsymbol{n}$ is the unit normal vector on S_c.

Comparing Eq. (9) with Eq. (7) of Note 9 in Chapter 2 and considering $\delta \boldsymbol{H}^* = 0$, we get Eq. (10).

Constant magnetic field as Eq. (8).

$$\frac{\partial}{\partial \omega} \int \int_{S_c} \boldsymbol{P}_c \cdot \boldsymbol{n} \, dS_c = 2(\tilde{W}_e + \tilde{W}_m) = 2\tilde{W}_t \tag{10}$$

Now, we will take out the metal wall part of S_c, and consider port T_c at this part. When we consider P_c at port T_c, looking into the cavity, P_c should change corresponding to the frequency change, as shown by the solid line passing through point C of Fig. 4, and this line should satisfy Eq. (8).

Therefore, the values of $\Delta\omega_c$ of Fig. 4 must satisfy

$$-\frac{2\omega_0 \Delta \tilde{W}_e}{\Delta \omega_c} = 2W_t$$

Then, we get

$$-\frac{\Delta \omega_c}{\omega_0} = \frac{\Delta \tilde{W}_e}{W_t} \tag{11}$$

The point of $\omega_0 + \Delta\omega_c$ on the axis ω means the Poynting power is zero. Therefore, we may replace port T_c by the metal plate; that is, the cavity surrounded by the metal surface S', shown in Fig. 1b, is resonant at the angular frequency $\omega_0 - \Delta\omega_c$. It is understood that the frequency decreases by pushing the cavity wall inside in a way that decreases the electric energy.

Next, we will consider the Poynting power P_m. When the time average of the reactive energy surrounded by S_m and S_{om} is the magnetic energy of $\Delta\tilde{W}_m$, P_m should take the values

$$P_m = -2j\omega\tilde{W}_m \tag{12}$$

These results can be explained by the same reasons for the case of P_c.

Because the total reactive energy still takes the values of W_t, the variation of P_m corresponding to the frequency variation can be shown by the solid line passing through point M of Fig. 4, which is parallel to the line showing the performance of P_c. Then, the values of $\Delta\omega_m$ are obtained from

$$\frac{\Delta\omega_m}{\omega} = \frac{\Delta W_m}{W_t} \tag{13}$$

The point $\omega_0 + \Delta\omega_m$ on the axis ω means the Poynting power is zero. We, therefore, can replace the surface S_m by the metal plate, as was done for the angular frequency $\omega_0 + \Delta\omega_m$; that is, the cavity surrounded by the metal surface S'' of Fig. 1c is resonant at an angular frequency of $\omega_0 + \Delta\omega_m$. Therefore, when we push the cavity metal wall inside in such a way as to decrease the magnetic energy, the resonant frequency increases.

As an example, when we push the center of the H plane of the $TE^{\square}_{101}$ cavity as shown in Fig. 5, the resonant frequency decreases. On the other hand, pushing the E plane of the $TE^{\square}_{101}$ cavity as shown in Fig. 5, the resonant frequency increases. It is easily understood as follows. Since the electric fields exist perpendicular to the H plane and they are strongest at the center of the plane, the electric energy decreases by pushing the center of the H plane and the resonant frequency decreases according to Eq. (5). On the other hand, the magnetic fields are strongest at the periphery of the E plane. Then, the resonant frequency increases by pushing the E plane.

Perturbation of Material

When the dielectric constant of the capacitor, ε, and the permeability of the inductor, μ, of Fig. 3 is increased to $\varepsilon + \Delta\varepsilon$ and $\mu + \Delta\mu$, respectively, the capacitance C and the inductance L increased to $C + \Delta C$ and $L + \Delta L$, respectively. The resonant frequency ω_0, therefore, becomes lower in $\Delta\omega$, which is shown in Eq. (4).

Next, we introduce more generally the frequency variation in connection with the electric field $\boldsymbol{E}_0$ and the magnetic field $\boldsymbol{H}_0$ of the unperturbed cavity, together with the change of the material permittivity and permeability.

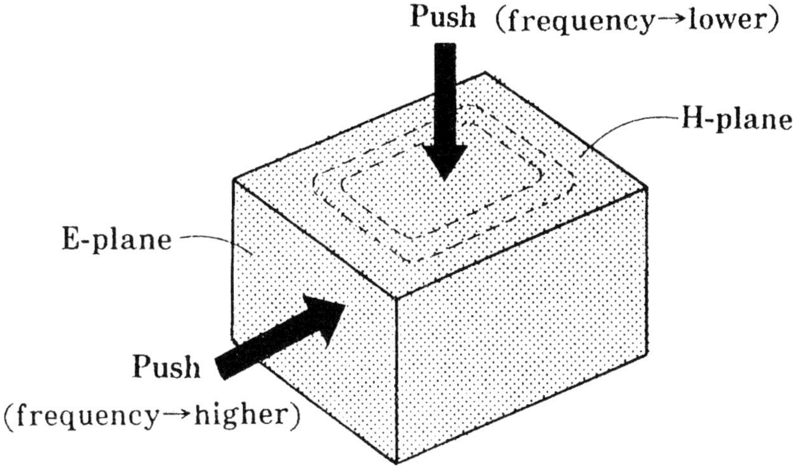

Figure 5. Frequency deviation of $TE^{\square}_{101}$ cavity by pushing the H plane and E plane.

When we denote the electric and magnetic fields in the perturbed cavity by $\boldsymbol{E}$ and $\boldsymbol{H}$, we have the following Maxwell equation:

$$-\nabla \times \boldsymbol{E}_0 = j\omega\mu \boldsymbol{H}_0 \tag{14a}$$

$$\nabla \times \boldsymbol{H}_0 = j\omega\varepsilon \boldsymbol{E}_0 \tag{14b}$$

$$-\nabla \times \boldsymbol{E} = j\omega(\mu + \Delta\mu)\boldsymbol{H} \tag{14c}$$

$$\nabla \times \boldsymbol{H} = j\omega(\varepsilon + \Delta\varepsilon)\boldsymbol{E} \tag{14d}$$

Using $\boldsymbol{E}_0^*$ [Eq. (14d)] + $\boldsymbol{H}_0$ [Eq. (14a)] and $\boldsymbol{E}$ [Eq. (14b)] + $\boldsymbol{H}_0^*$ [Eq. (14c)], we get

$$\nabla \cdot (\boldsymbol{H} \times \boldsymbol{E}_0^*) = j\omega(\varepsilon + \Delta\varepsilon)\boldsymbol{E} \cdot \boldsymbol{E}_0^* - j\omega_0\mu \boldsymbol{H}_0^* \boldsymbol{H} \tag{15a}$$

$$\nabla \cdot (\boldsymbol{H}_0^* \times \boldsymbol{E}) = j\omega(\mu + \Delta\mu)\boldsymbol{H} \cdot \boldsymbol{H}_0^* - j\omega_0 \boldsymbol{E}_0^* \boldsymbol{E} \tag{15b}$$

Calculating $\int\int\int_\tau$ [Eq. (15a) + Eq. (15b)] $d\tau$, the left-hand side becomes

$$\int\int_S (\boldsymbol{H} \times \boldsymbol{E}_0^* + \boldsymbol{H}_0^* \times \boldsymbol{E}) \cdot \boldsymbol{n}\, dS = 0$$

(because $\boldsymbol{E}//\boldsymbol{n}$ and $\boldsymbol{E}_0^*//\boldsymbol{n}$). Therefore,

$$\iiint_\tau [\{\omega(\varepsilon + \Delta\varepsilon) - \omega_0\varepsilon\}\boldsymbol{E}\cdot\boldsymbol{E}_0^* + \{\omega(\mu + \Delta\mu) - \omega_0\mu\}\boldsymbol{H}\cdot\boldsymbol{H}_0^*]d\tau = 0$$

Then,

$$\frac{\Delta\omega}{\omega_0} = -\frac{\iiint_\tau (\Delta\varepsilon\boldsymbol{E}\cdot\boldsymbol{E}_0^* + \Delta\mu\boldsymbol{H}\cdot\boldsymbol{H}_0^*)d\tau}{2\tilde{W}_t} \tag{16}$$

When small samples of a dielectric and a magnetic material are placed inside the cavity and the electric and the magnetic fields $\boldsymbol{E}$ and $\boldsymbol{H}$ are uniform inside the sample,

$$\frac{\Delta\omega}{\omega_0} = -\frac{(\Delta\varepsilon\boldsymbol{E}\cdot\boldsymbol{E}_0^*\Delta V + \Delta\mu\boldsymbol{H}\cdot\boldsymbol{H}_0^*\Delta V')}{2\tilde{W}_t} \tag{17}$$

where ΔV and $\Delta V'$ are the volumes of the dielectric and the magnetic materials, respectively.

In the case of the shape of the sample shown in Fig. 6, $\boldsymbol{E}$ and $\boldsymbol{H}$ can be related to $\boldsymbol{E}_0$ and $\boldsymbol{H}_0$ as the first approximation by

$$\boldsymbol{E} = c\boldsymbol{E}_0, \qquad \boldsymbol{H} = c'\boldsymbol{H}_0 \tag{18}$$

Substituting Eq. (18) into Eq. (17), we get

$$\frac{\Delta\omega}{\omega_0} = -\frac{\Delta\varepsilon C|\boldsymbol{E}_0|^2 + \Delta\mu c'|\boldsymbol{H}_0|^2}{2\tilde{W}_t} \tag{19}$$

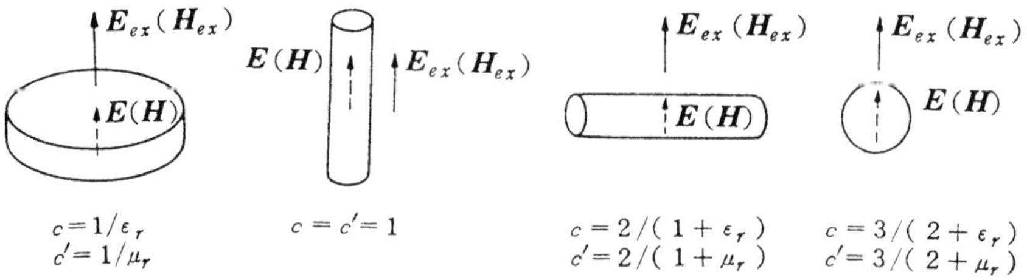

Figure 6. Values of c and c' of Eq. (17) for typical sample shapes.

The insertion of a small sample with an ellipsoidal body inside a cavity is usually used for the measurement of the complex permittivity and the permeability of the sample by the measurement of the complex frequency deviation which is calculated from the frequency deviation and the deteriorator of the Q values of the cavity.

Waveguide Perturbation

Because the waveguide becomes the two-dimensional cavity at the cutoff frequency as shown in Fig. 7, the cutoff angular frequency of the perturbed waveguide, $\omega_c + \Delta\omega_c = \omega_c'$, is obtained by the method of cavity perturbation mentioned earlier. By using the values of ω_c', the phase constant β' and the wave impedance Z_w' of the perturbed waveguide are obtained as follows:

$$\beta' = \beta\sqrt{1 - \left(\frac{\omega_c'}{\omega}\right)^2} \tag{20}$$

$$Z_w' = \frac{Z_w}{\sqrt{1 - (\omega_c'/\omega)^2}} \quad \text{(TE waveguide)} \tag{21a}$$

$$Z_w' = Z_w\sqrt{1 - (\omega_c'/\omega)^2} \quad \text{(TM waveguide)} \tag{21b}$$

where Z_w is the free-space wave impedance. For example, when we push the H plane inward, ω_c decreases, the phase constant increases, and the phase delay increases.

Because the perturbation theory is obtained under the assumption of an unchanged magnetic field at the perturbed place, a change to smooth the shape is required. For example, in the case of the insertion of a sharp edge or a thin metal piece, the evanescent mode generated around the deformed place causes the reactance and makes the error.

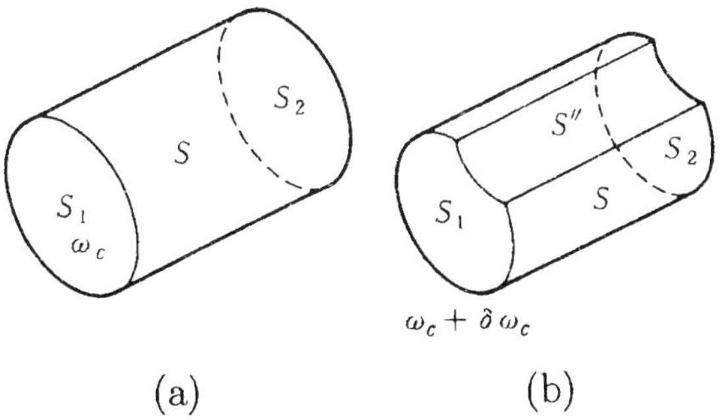

Figure 7. Perturbation of cutoff wavelength: (a) original waveguide; (b) perturbed waveguide.

APPENDIX 3

Relationships Among External Q, Q_e, Coupling Coefficient, k, and g Values in BPF

The values of series inductance L_k and the parallel capacitance C_k of Fig. 3 in Chapter 5 are expressed by g values as

$$L_k = \frac{R_0}{\omega_c} g_k \quad (H)$$

$$C_k = \frac{g_k}{\omega_c R_0} \quad (F) \tag{1}$$

The values of g_k are determined by the characteristics of the bandpass filter BPF as shown in Table 2 of Chapter 5.

The low-pass filter is transformed into the construction in Fig. 7b of Chapter 5, where the constants take the values

$$L_{sk} = \frac{R_0 g_k}{w\omega_0}, \qquad L_{pk} = \frac{wR_0}{\omega_0 g_k}$$

$$C_{sk} = \frac{w}{\omega_0 R_0 g_k}, \qquad C_{pk} = \frac{g_k}{w\omega_0 R_0} \tag{2}$$

where k, s, and p show the kth element, the series resonant network, and the parallel resonant network, respectively.

The external Q_{e1} is the external Q of the L_1C_1 resonant circuit where the impedance of the parallel circuit of L_2C_2 is made zero. The values of Q_{e1}, therefore, take the values $\omega L_1/R$. Using Eq. (2), $L_{s1} = L_1$, $R = R_0$, and $\omega = \omega_0$, together with $Q_{e1} = \omega L_1/R$, we get

$$\frac{\omega_0 L_{s1}}{R_0} = \frac{g_1}{w} \tag{3}$$

In the same way, the external Q of the $L_n C_n$ resonant circuit of Fig. 7b of Chapter 5 is

$$Q_{en} = \omega C_n R_L$$

Substituting Eq. (2) into the above equation and using $C_{nk} = C_n$, $R_0 = R_L$, and $\omega = \omega_0$, we get

$$\omega_0 C_n R_0 = \frac{g_n}{w} \tag{4}$$

Considering the case when all the resonant circuits except the ith and $i + 1$th resonant circuits, we get the circuit of Fig. 1. When the angular frequency ω is

$$\omega = \omega_0 + 2\Delta\omega$$

the admittance of the $L_i C_i$ circuit, Δy_i, is

$$\Delta y_i = 2j\Delta\omega C_i \tag{5}$$

and the impedance of the $L_{i+1} C_{i+1}$ circuit is

$$\Delta z_{i+1} = 2j\Delta\omega L_{i+1} \tag{6}$$

Therefore, the network of Fig. 1 resonates when

$$\frac{1}{\Delta y_i} + \Delta z_{i+1} = 0 \tag{7}$$

Solving Eq. (7), we can get the resonant angular frequency ω_r, which is obtained as follows:

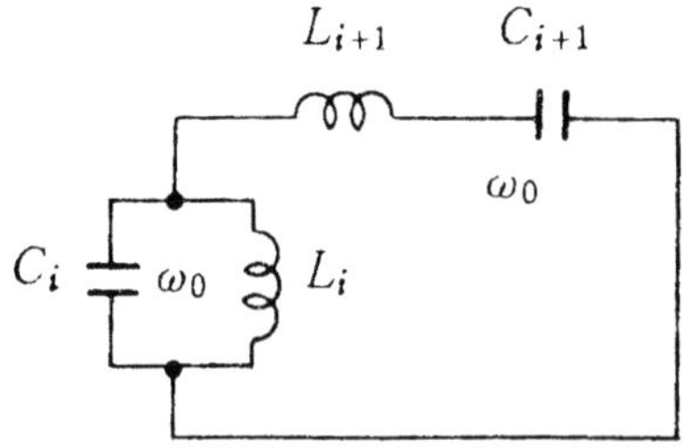

Figure 1

$$\omega_r = \omega_0 + 2\Delta\omega_r$$

Substituting Eqs. (5) and (6) into Eq. (7), we get

$$1 - 4(\Delta\omega_r)^2 C_i L_{i+1} = 0$$

and

$$\Delta\omega_1 = \frac{\pm 1}{2\sqrt{C_i L_{i+1}}} = \frac{\pm w\omega_0}{2\sqrt{g_i g_{i+1}}}$$

Therefore, we get the following two angular resonant frequencies:

$$\omega_+ = \omega_0 + \Delta\omega_r$$
$$\omega_- = \omega_0 - \Delta\omega_r \qquad (8)$$

From Eq. (8), we get

$$\frac{\omega_+ - \omega_-}{\omega_0} = \frac{2\Delta\omega_r}{\omega_0} = \frac{w}{\sqrt{g_i g_{i+1}}} \qquad (9)$$

The values from Eq. (9) are defined as the coupling coefficient $k_{i,i+1}$. If we know the coupling coefficient, we can get the BPF, which is constructed by the neighboring resonators with space electromagnetic coupling through space based on the lumped element construction shown in Fig. 7b of Chapter 5.

In Eq. (9), w denotes the bandwidth ratio of the 3-dB attenuation in the case of the maximum flat BPF and of the equal ripple bandwidth in the case of the Tchebycheff performance.

APPENDIX 4

Determination of Capacitances of the Uniform Coupled Lines in the Anisotropic Inhomogeneous Medium by Resistive Sheet

The Case of Isotropic Medium

When two flat electrodes with area S are placed in parallel at a distance d and the space between the two electrodes is filled by dielectric material with dielectric constant ε, the capacitance C is obtained by

$$C = \frac{\varepsilon S}{d} \quad (F) \tag{1}$$

Next, the dielectric constant is replaced by the resistive material with resistivity ρ (Ω m), the resistance between two electrodes is

$$R = \frac{\rho d}{S} \tag{2}$$

Multiplying Eqs. (1) and (2), we get

$$CR = \varepsilon\rho \tag{3}$$

Such a relation is always kept, whenever the dielectric medium consists of multiple dielectric materials as verified in the following. As shown in Fig. 1, considering the replacement between dielectric media and conductive media, we keep the relation

$$\varepsilon_i = \tau\sigma_i \quad (\tau \text{ is constant}) \tag{4}$$

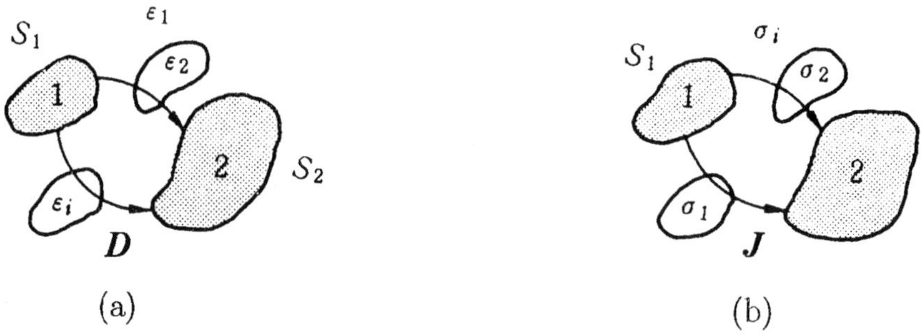

Figure 1. Several dielectrics with ε_i ($i = 1, 2, \ldots$) are replaced by several conductive media with σ_i ($i = 1, 2, \ldots$). (a) Electric flux density between two electrodes in several dielectric materials with dielectric constants ε_i ($i = 1, 2, \ldots$). (b) Current density flowing between two electrodes in several conductive media with conductivity σ_i ($i = 1, 2, \ldots$).

The total charges on electrode I, Q, are related to the electric flux density D and electric field E as follows:

$$Q = \oint\int_{S_1} \boldsymbol{D}\cdot d\boldsymbol{s} = \oint\int_{S_1} \varepsilon_1 \boldsymbol{E}\cdot d\boldsymbol{s} \tag{5a}$$

By the continuity of the electric fields of tangential components and the continuity of the electric flux densities of the normal components on the surface among different media, we have

$$\begin{aligned} E_{t1} &= E_{ti} \\ D_{n1} &= D_{ni}, \qquad \varepsilon_1 E_{n1} = \varepsilon_n E_{ni} \\ E_{t1} &= 0 \quad \text{(on the metal surface)} \end{aligned} \tag{5b}$$

where t and n denote the tangential and normal components, respectively. Inserting the condition of Eq. (4) into Eq. (5b), we get

$$\begin{aligned} E_{t1} &= E_{ti} \\ \sigma_1 E_{n1} &= \sigma_n E_{ni}, \qquad J_1 = J_n \\ E_{t1} &= 0 \quad \text{(on the metal surface)} \end{aligned} \tag{5c}$$

When we replace Fig. 1a by Fig 1b, we have Eq. (5d) under the same values of E in both figures:

$$\boldsymbol{D} \rightarrow \boldsymbol{J}\tau \tag{5d}$$

Therefore, the equation corresponding to Eq. (5a) becomes

$$I = \oint \int_{s_1} \boldsymbol{J} \cdot ds = \oint \int_{s_1} \sigma_1 \boldsymbol{E} \cdot ds \tag{5e}$$

From Eqs. (5a) and (5e), we get

$$\frac{Q|V}{I|V} = CR = \frac{\varepsilon_1}{\sigma_1} = \varepsilon_1 \rho_1 \tag{5f}$$

where ρ_1 is the resistivity (Ω m) and V is the voltage between electrodes 1 and 2.

Because from Eqs. (5b) and (5c) we also get

$$\varepsilon_1 \rho_1 = \varepsilon_n \rho_n \tag{5g}$$

Eq. (5f) can be generalized to

$$CR = \varepsilon_i \rho_i \quad (i = 1, 2, \ldots, n) \tag{5h}$$

When we consider the resistive sheet with thickness δ and resistivity ρ, the surface resistance, ρ_s, is

$$\rho_s = \frac{\rho}{\delta} \tag{5i}$$

Considering the resistive sheet with surface resistance ρ_s for the conductive medium with a 1-m depth, we get

$$C\left(\frac{\mathrm{F}}{\mathrm{m}}\right) = \varepsilon \left(\frac{\mathrm{F}}{\mathrm{m}}\right) \frac{\delta\ (\Omega\ \mathrm{m})}{\delta\ (\mathrm{m})\ R_m\ (\Omega)} = \varepsilon \left(\frac{\mathrm{F}}{\mathrm{m}}\right) \frac{\delta_s\ (\Omega)}{R_m\ (\Omega)} \tag{6}$$

where R_m is the measured values of the resistance on the resistive sheet as shown in Fig. 2.

When ρ_s is unknown, it can be found by measuring the known electrodes, such as the coaxial section shown in Fig. 3. In this measurement, because C is obtained from

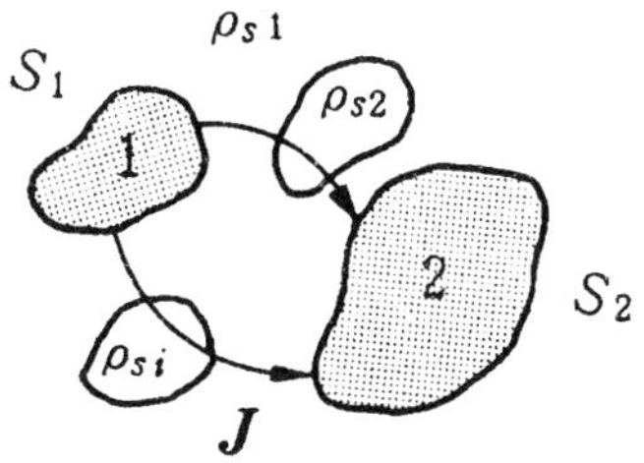

Figure 2. $\rho_i(1/\sigma_i)$ of Fig. 1b is replaced by the surface resistance.

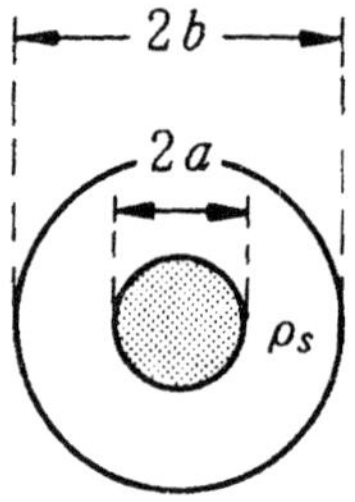

Figure 3. Measurement of ρ_s by coaxial electrode.

$$C\left(\frac{\mathrm{F}}{\mathrm{m}}\right) = \frac{2\pi\varepsilon}{\ln(b/a)} \tag{7a}$$

we get

$$\rho_s = \frac{2\pi}{\ln(b/a)} R_{m0} \quad (\Omega) \tag{7b}$$

The Case of Anisotropic Medium

When ε_i of Fig. 1a is a tensor, the values of E_x, E_y, D_x, and D_y of Fig. 4 can be expressed by

$$\boldsymbol{D} = \hat{\varepsilon}\boldsymbol{E}, \qquad \boldsymbol{D} = \begin{bmatrix} D_x \\ D_y \end{bmatrix}, \qquad \boldsymbol{E} = \begin{bmatrix} E_x \\ E_y \end{bmatrix}$$

$$\hat{\varepsilon} = \begin{bmatrix} \varepsilon_{xx} & \varepsilon_{xy} \\ \varepsilon_{yx} & \varepsilon_{yy} \end{bmatrix} \tag{8a}$$

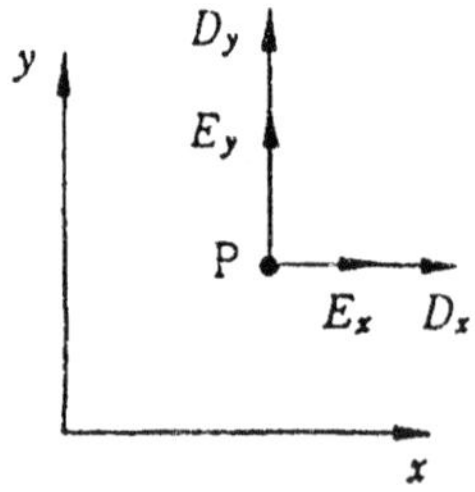

Figure 4. E_x, E_y, D_x, and D_y in the x,y, plane.

In the lossless case, $\hat{\varepsilon}$ becomes the Hermit matrix and is obtained from

$$\hat{\varepsilon} = \hat{\varepsilon}^{*t} \tag{8b}$$

where t is the transpose and the asterisk (*) denotes the conjugate.

When we consider the x,y and x',y' coordinates, we have

$$\begin{bmatrix} x' \\ y' \end{bmatrix} = [\xi] \begin{bmatrix} x \\ y \end{bmatrix}, \qquad [\xi] = \begin{bmatrix} \cos\theta & \sin\theta \\ -\sin\theta & \cos\theta \end{bmatrix} \tag{9a}$$

Denoting electric fields and electric fluxes in the x',y' coordinate by $E_{x'}$, $E_{y'}$, and $D_{x'}$,$D_{y'}$, respectively, we get

$$\begin{bmatrix} D_{x'} \\ D_{y'} \end{bmatrix} = \hat{\varepsilon}' \begin{bmatrix} E_{x'} \\ E_{y'} \end{bmatrix}, \qquad \hat{\varepsilon}' = \begin{bmatrix} \varepsilon_{x'x'} & \varepsilon_{x'y'} \\ \varepsilon_{y'x'} & \varepsilon_{y'y'} \end{bmatrix} \tag{9b}$$

Because we have the relationships

$$\boldsymbol{D} = [\xi]\boldsymbol{D}, \qquad \boldsymbol{E}' = [\xi]\boldsymbol{E}$$

$$\boldsymbol{D}' = \begin{bmatrix} D_{x'} \\ D_{y'} \end{bmatrix}, \qquad \boldsymbol{E}' = \begin{bmatrix} E_{x'} \\ E_{y'} \end{bmatrix} \tag{9c}$$

and considering Eq. (8a), we get

$$\boldsymbol{D}' = \hat{\varepsilon}'\boldsymbol{E}'$$

$$\hat{\varepsilon}' = [\xi]\hat{\varepsilon}[\xi]^{-1} \tag{9d}$$

Because we have the boundary conditions

$$E_{x'1} = E_{x'2} \tag{10a}$$

$$D_{y'1} = D_{y'2} \tag{10b}$$

we get the following from Eq. (10b):

$$\varepsilon_{y'x'1}E_{x'1} + \varepsilon_{y'y'1}E_{y'1} = \varepsilon_{y'x'2} + \varepsilon_{y'y'2}E_{y'2} \tag{10c}$$

If we denote the tensor dielectric constant by $\hat{\varepsilon}_{i'}$ in Fig. 1a and $\sigma_{i'}$, of

$$\hat{\sigma}_{i'} = \frac{1}{\tau} \begin{bmatrix} \varepsilon_{x'x'i} & \varepsilon_{x'y'i} \\ \varepsilon_{y'x'i} & \varepsilon_{y'y'i} \end{bmatrix} \tag{11}$$

is used in Fig. 1b, we have Eq. (12a) corresponding to Eq. (10c):

$$\sigma_{y'x'1}E_{x'1} + \sigma_{y'y'1}E_{y'1} = \sigma_{y'x'2}E_{x'2} + \sigma_{y'y'2}E_{y'2} \tag{12a}$$

Equation (12a) shows the continuity of current at the boundary surface, which means that it is feasible to replace $\hat{\varepsilon}_i$ by σ_i'. Therefore, we get Eq. (12b) using Eq. (2f):

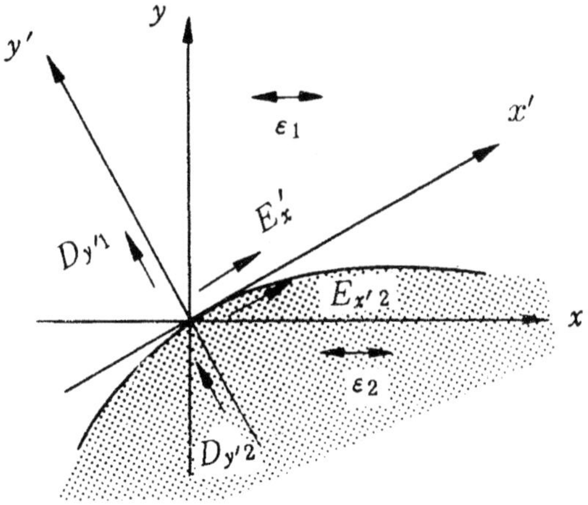

Figure 5. x',y' coordinate is chosen such that x' is tangent on the boundary surface S and y' is normal to S.

$$C\sigma_i' = \frac{\hat{\varepsilon}_i'}{R} \tag{12b}$$

When the medium is lossless and nonreciprocal, $\hat{\varepsilon}$ becomes the real symmetrical matrix. In this case, $\hat{\varepsilon}$ can become the diagonal matrix $\hat{\varepsilon}_d$ as shown in Eq. (13) by using the proper coordinates.

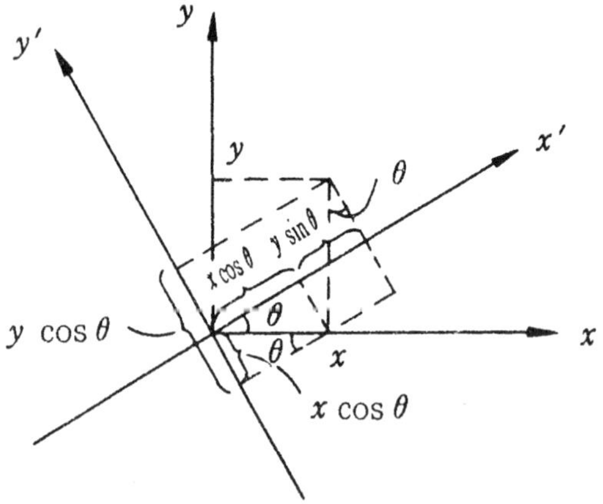

Figure 6. Relative relation between x,y and x',y' coordinates.

$$\hat{\varepsilon}_d = \begin{bmatrix} \varepsilon_i^{(1)} & 0 \\ 0 & \varepsilon_i^{(2)} \end{bmatrix} \tag{13}$$

where $\varepsilon_i^{(1)}$ and $\varepsilon_i^{(2)}$ are positive real numbers. When we denote such a coordinate by x'',y'', the following notation is sometimes used.

$$\begin{aligned} \varepsilon^{(1)} &= \varepsilon_{\parallel} \\ \varepsilon^{(2)} &= \varepsilon_{\perp} \end{aligned} \tag{14}$$

In the coordinate system x'',y'', $\hat{\sigma}_{i'}$ of Eq. (11) becomes

$$\hat{\sigma}_{\text{diag},i} = \begin{bmatrix} \sigma_i^{(1)} & 0 \\ 0 & \sigma_i^{(2)} \end{bmatrix} = \frac{1}{\tau} \begin{bmatrix} \varepsilon_i^{(1)} & 0 \\ 0 & \varepsilon_i^{(2)} \end{bmatrix} \tag{15}$$

When we use the resistive sheet, the surface resistance $\hat{\rho}_{s,\text{diag},i}$ is obtained from Eq. (16) and corresponds to $\hat{\varepsilon}_{\text{diag},i}$ in Eq. (17):

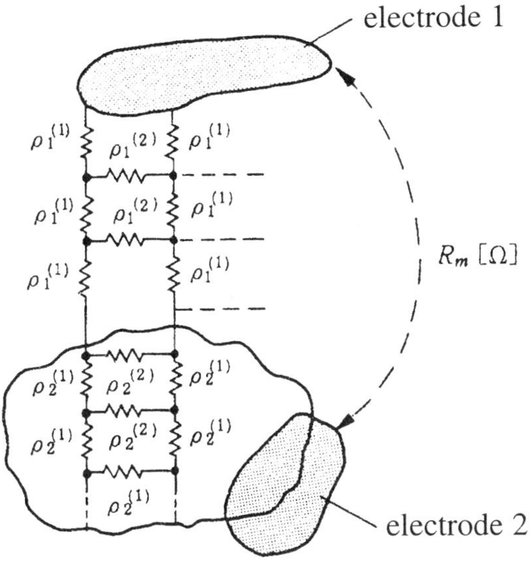

Figure 7. Method to obtain values between electrodes from a resistive network.

$$\hat{\rho}_{s,\mathrm{diag},i} = \begin{bmatrix} \rho_{s,i}^{(1)} & 0 \\ 0 & \rho_{s,i}^{(2)} \end{bmatrix} \tag{16}$$

$$\varepsilon_{\mathrm{diag},i} = \begin{bmatrix} \varepsilon_i^{(1)} & 0 \\ 0 & \varepsilon_i^{(2)} \end{bmatrix} \tag{17}$$

and we have the relation

$$C\left(\frac{\mathrm{F}}{\mathrm{m}}\right) = \varepsilon_i^{(1),(2)}\left(\frac{\mathrm{F}}{\mathrm{m}}\right)\frac{\rho_{si}^{(1)(2)}\ (\Omega)}{R_m\ (\Omega)} \tag{18}$$

The anisotropic resistive sheet can be obtained from the resistive network shown in Fig. 7.

The measurement of the resistance between electrodes can be simulated by a computer program. The theory can be extended to determine the magnetic resistance by using the Babinet theorem, which makes it possible to determine the inductance.

Practical Example

When we measure the capacities of two parallel lines (Fig. 8a), we can measure the values between electrode 1 and the ground as shown in Fig. 8b, R_{m2} between electrode 1 and the ground in Fig. 8c, and R_{m3} between electrode 2 and the ground in Fig. 8d.

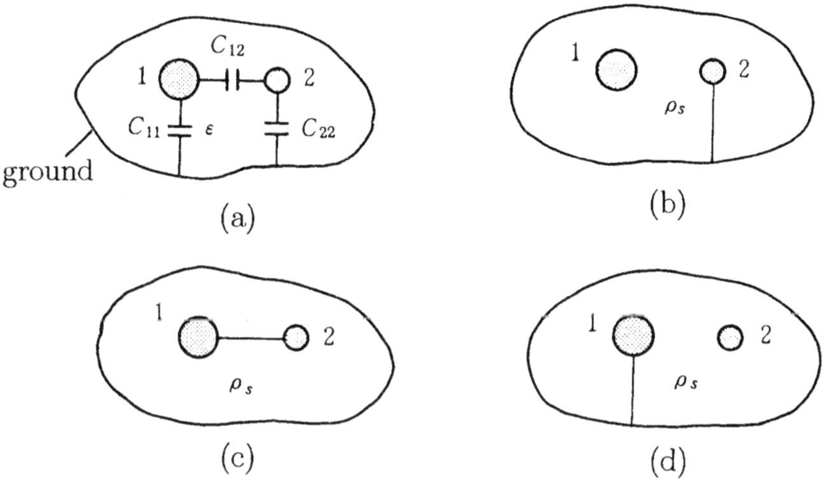

Figure 8. Measurement of the capacities of two parallel lines.

By using the measured values of R_{m1}, R_{m2}, and R_{m3}, C_{11}, C_{22}, and C_{12} can be obtained from

$$C_{12} = \frac{8.855 \times 10^{-12}\varepsilon_r \rho_s}{2}\left(\frac{1}{R_{m1}} + \frac{1}{R_{m2}} - \frac{1}{R_{m3}}\right)$$

$$C_{11} = \frac{8.855 \times {}^{-12}\varepsilon_r \rho_s}{2}\left(\frac{1}{R_{m1}} - \frac{1}{R_{m2}} + \frac{1}{R_{m3}}\right)$$

$$C_{22} = \frac{8.855 \times 10^{-12}\varepsilon_r \rho_s}{2}\left(\frac{1}{R_{m2}} - \frac{1}{R_{m1}} + \frac{1}{R_{m3}}\right) \tag{19}$$

APPENDIX 5

Design of Power Divider with Two Different Output Powers

In Fig. 29 of Chapter 6, the potentials of ports 2′ and 3′ should take the same values when the signal is supplied in port 1. Therefore, the impedances viewed from ports 2′ and 3′ toward the load, the resistances R_2 and R_3, should be calculated from

$$\frac{R_2}{R_3} = K^2 \tag{1}$$

Because the impedance looking at ports 1 and 2 are evaluated as

$$\frac{Z_2^2}{R_2} \quad \text{and} \quad \frac{Z_3^2}{R_3} \tag{2}$$

we get

$$\frac{Z_2}{Z_3} = K^2 \tag{3}$$

From the matching condition of port 1, we get

$$Z_0 = \frac{Z_2^2}{R_2}\frac{Z_3^2}{R_3}\left(\frac{Z_2^2}{R_2} + \frac{Z_3^2}{R_3}\right)^{-1} = \frac{Z_2^2}{R_2}\left(1 + \frac{Z_2^2}{Z_3^2}\frac{R_3}{R_2}\right)^{-1} = \frac{Z_2^2/R_2}{1 + K^2} \tag{4}$$

$$\frac{Z_2^2}{R_2} = (1 + K^2)Z_0$$

On the other hand, the characteristic impedances Z_2' and Z_3' must take the following values:

$$Z_2' = \sqrt{R_2 Z_0} \tag{5}$$

$$Z_3' = \sqrt{R_3 Z_0} \tag{6}$$

In Eqs. (1) and (3)–(6), we have the six unknowns R_2, R_3, Z_2, Z_3, Z_2', and Z_3'. We, therefore, choose one of them arbitrarily. When we choose R_2 as

$$R_2 = KZ_0 \tag{7}$$

we get

$$R_3 = \frac{Z_0}{K}$$

From Eq. (4), we have

$$Z_2 = Z_0\sqrt{K(1 + K^2)}, \qquad Z_2' = Z_0\sqrt{K}$$

$$Z_3 = Z_0 \sqrt{\frac{1 + K^2}{K^3}}, \qquad Z_3' = \frac{Z_0}{\sqrt{K}} \tag{8}$$

Next, we will obtain the values of the absorbing resistance R. When the potential at port 2 is $E(V)$ in the case of the supplied signal at port 2, the potential at port 1 must take the values

$$-j\frac{Z_0}{Z_2}E$$

because the potential of ports 3 and 3′ must be zero. The current flowing into port 3′ through Z_3, therefore, must take the values

$$\frac{-Z_0}{Z_2 Z_3}E \tag{9}$$

On the other hand, the current flown into R must

$$\frac{E}{R} \tag{10}$$

Because the values obtained from Eqs. (9) and (10) should be same, we get

$$R = \frac{Z_2 Z_3}{Z_0} \tag{11}$$

Substituting Eq. (8) into Eq. (11), we get

$$R = Z_0 \frac{1 + K^2}{K} \tag{12}$$

However, the wideband matching condition with a two-stage $\lambda_g/4$ transformer between the input and output impedances of W_1 and W_2 are calculated as follows: that is, the center impedance should be

$$\sqrt{W_1 W_2}$$

Applying such a condition in this case, the following conditions must be satisfied:

$$R_2 = \sqrt{Z_0 \frac{Z_2^2}{R_2}}$$

$$R_3 = \sqrt{Z_0 \frac{Z_3^2}{R_3}} \tag{13}$$

Substituting Eq. (4) into Eq. (13) and using Eq. (1), we get

$$R_2 = Z_0 \sqrt{1 + K^2}$$

$$R_3 = Z_0 \frac{\sqrt{1 + K^2}}{K^2}$$

$$Z_2 = Z_0(1 + K^2)^{3/4}, \qquad Z_2' = Z_0(1 + K^2)^{1/4}$$

$$Z_3 = Z_0 \frac{(1 + K^2)^{3/4}}{K^2}, \qquad Z_3' = Z_0 \frac{(1 + K^2)^{1/4}}{K}$$

$$R = Z_0 \frac{(1 + K^2)^{3/4}}{K} \tag{14}$$

APPENDIX 6

Principle of Directional Coupler with Two Symmetrical Planes

General Theory

The four-port network shown in Fig. 1 has two symmetrical planes S_1 and S_2. Eigenvectors of the network, $\boldsymbol{u}_1$, $\boldsymbol{u}_2$, $\boldsymbol{u}_3$, and $\boldsymbol{u}_4$ are as shown in Table 1, where s_1, s_2, s_3, and s_4 are the corresponding eigenvalues of the scattering matrix. When the signal supplied to port 1 appears at ports 2 and 3 but not at port 4, the scattering matrix can be obtained from the symmetry, as in Eq. (1):

$$[S] = \begin{bmatrix} 0 & \alpha & \beta & 0 \\ \alpha & 0 & 0 & \beta \\ \beta & 0 & 0 & \alpha \\ 0 & \beta & \alpha & 0 \end{bmatrix} \tag{1}$$

From the eigenequation

$$([S - s_i[I])\boldsymbol{u}_i = 0 \tag{2}$$

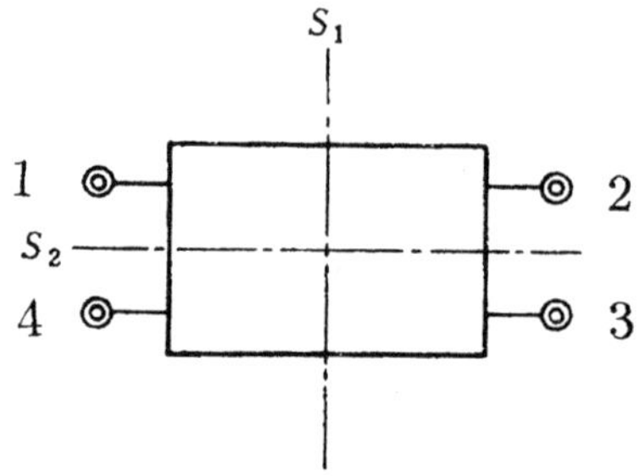

Figure 1. Directional coupler.

we get

$$s_1 = \alpha + \beta$$
$$s_2 = \alpha - \beta$$
$$s_3 = -(\alpha - \beta)$$
$$s_4 = -(\alpha + \beta) \tag{3}$$

From Eq. (3), we get

$$s_1 = -s_4, \qquad s_2 = -s_3 \tag{4}$$

The eigenvalues of the impedance matrix, z_i, is obtained from

$$s_i = \frac{z_i - Z}{z_i + Z} \tag{5}$$

where Z is the impedance of each port using Eq. (4), we get

Table 1. Eigenvectors

Port	u_1	u_2	u_3	u_4
1	+	+	+	+
2	+	+	−	−
3	+	−	+	−
4	+	−	−	+
Eigenvalues	s_1	s_2	s_3	s_4

$$s_2 = \frac{z_2 - Z}{z_2 + Z} = -\frac{z_3 - Z}{z_3 + Z}$$

$$z_2 z_3 = Z^2$$

$$z_1 z_4 = Z^2 \tag{6}$$

α and β can be obtained from Eqs. (3) and (5) as follows:

$$\alpha = \frac{s_1 - s_3}{2} = \frac{(z_1 - z_3)Z}{(z_1 + z_3)Z + (Z^2 + z_1 z_3)}$$

$$\beta = \frac{s_1 + s_3}{2} = \frac{z_1 z_3 - Z^2}{(z_1 + z_3)Z + (Z^2 + z_1 z_3)} \tag{7}$$

Applying the unitary condition to Eq. (1), we get

$$\alpha\beta^* + \alpha^*\beta = 0$$

$$\alpha\beta^* = \pm j|\alpha\beta|$$

Expressing α and β by

$$\alpha = |\alpha|e^{-j\theta_\alpha}$$

$$\beta = |\beta|e^{-j\theta_\beta} \tag{8}$$

we get

$$\theta_\alpha - \theta_\beta = \pm\frac{\pi}{2} \tag{9}$$

By the unitary condition, we also get

$$|\alpha|^2 + |\beta|^2 = 1 \tag{10}$$

Analysis of Distributed Coupling Directional Coupler

In the case of the excitation of $\boldsymbol{u}_1$ and $\boldsymbol{u}_3$, an even mode exists about the S_2 plane, and in the case of $\boldsymbol{u}_2$ and $\boldsymbol{u}_4$, an odd mode exists about the S_2 plane. When we denote the characteristic impedance for even and odd modes by Z_e and Z_0, respectively, we get the eigenvalues of the impedance matrix, z_1, z_2, z_3, and z_4 can be easily obtained:

$$z_1 = -jZ_e \cot\left(\frac{\theta_e}{2}\right)$$

$$z_2 = -jZ_0 \cot\left(\frac{\theta_o}{2}\right)$$

$$z_3 = jZ_e \tan\left(\frac{\theta_e}{2}\right)$$

$$z_4 = jZ_0 \tan\left(\frac{\theta_o}{2}\right) \tag{11}$$

Substituting Eq. (11) into Eq. (6), we get

$$Z_e Z_0 \tan\left(\frac{\theta_e}{2}\right) \cot\left(\frac{\theta_o}{2}\right) = Z^2$$

$$Z_e Z_0 \tan\left(\frac{\theta_o}{2}\right) \cot\left(\frac{\theta_e}{2}\right) = Z^2 \tag{12}$$

As a simple case, we can use

$$\theta_e = \theta_o \tag{13}$$

a condition which is reasonable for the small values of $|\beta|$, that is, the weak coupling.

In Eqs. (11)–(13), θ_e and θ_o are the electric angle of the coupled distributed lines for even and odd modes, respectively.

In the case of Eq. (13), we have

$$Z_e Z_o = Z^2 \tag{14}$$

In this case, substituting Eq. (11) into Eq. (7), we get

$$\alpha = \frac{-jZ}{Z_e + Z_o}\left[\cot\left(\frac{\theta}{2}\right) + \tan\left(\frac{\theta}{2}\right)\right]\left\{1 + j\frac{Z}{Z_e + Z_o}\left[\tan\left(\frac{\theta}{2}\right) - \cot\left(\frac{\theta}{2}\right)\right]\right\}^{-1}$$

$$|\alpha| = \frac{Z}{Z_e + Z_o}\left[\tan\left(\frac{\theta}{2}\right) + \cot\left(\frac{\theta}{2}\right)\right]\left\{1 + \frac{Z^2}{(Z_e + Z_o)^2}\left[\tan\left(\frac{\theta}{2} - \cot\left(\frac{\theta}{2}\right)\right]^2\right\}^{-1/2}$$

$$\theta_\alpha = \frac{\pi}{2} + \tan^{-1}\left\{\frac{Z}{Z_e + Z_o}\left[\tan\left(\frac{\theta}{2}\right) - \cot\left(\frac{\theta}{2}\right)\right]\right\} \tag{15}$$

$$\beta = \frac{Z_e - Z_o}{Z_e + Z_o}\left\{1 + \frac{jZ}{Z_e + Z_o}\left[\tan\left(\frac{\theta}{2}\right) - \cot\left(\frac{\theta}{2}\right)\right]\right\}^{-1}$$

$$|\beta| = \frac{Z_e - Z_o}{Z_e + Z_o}\left\{1 + \frac{Z^2}{(Z_e + Z_o)^2}\left[\tan\left(\frac{\theta}{2}\right) - \cot\left(\frac{\theta}{2}\right)\right]^2\right\}^{-1/2}$$

$$\theta_\beta = \tan^{-1}\left\{\frac{Z}{Z_e + Z_o}\left[\tan\left(\frac{\theta}{2}\right) - \cot\left(\frac{\theta}{2}\right)\right]\right\} \tag{16}$$

Expressing Eq. (16) in decibels, we get Eq. (32) of Chapter 3. From θ_α of Eq. (15) and θ_β of Eq. (16), we get the same results using Eq. (9) which was obtained by the general theory.

At the frequency corresponding $l = \lambda/4$, we get

$$|\alpha| = \frac{2Z}{Z_e + Z_o}, \qquad |\beta| = \frac{Z_e - Z_o}{Z_e + Z_o}, \qquad \theta_\alpha = \frac{\pi}{2}, \qquad \theta_\beta = 0 \tag{17}$$

Analysis of Two-Branch-Lines Directional Coupler

The scattering matrix of Fig. 55 of Chapter 6, $[S]$ should take the form of Eq. (1). When we use the eigenvectors of Table 1, we have Eqs. (3) and (4). By using Eq. (5), the eigenvalues of the impedance matrix, z_1, z_2, z_3, and z_4 are calculated from Eq. (6). The value of β is obtained from Eq. (3) as follows:

$$\beta = \frac{s_1 - s_2}{2} = \frac{s_3 - s_4}{2}$$

and this is expressed using z_i as

$$\begin{aligned}\beta &= \frac{(z_3 - z_4)Z}{(z_3 + z_4)Z + (Z^2 + z_3 z_4)} \\ &= \frac{(z_1 - z_2)Z}{(z_1 + z_2)Z + (Z^2 + z_1 z_2)}\end{aligned} \tag{18}$$

When the two-branch-line directional coupler of Fig. 55 of Chapter 6 is excited by each eigenvector, the symmetrical planes S_1 and S_2 are open or short, which is shown in Table 2.

Based on Table 2, the eigenvalues of the admittance matrix, y_1, y_2, y_3, and y_4 are obtained from

Table 2. Open (Magnetic Wall) or Short (Electric Wall) S_1 and S_2 Corresponding to Each Eigenvector

Eigenvector	S_1	S_2
$\boldsymbol{u}_1$	Open	Open
$\boldsymbol{u}_2$	Open	Short
$\boldsymbol{u}_3$	Short	Short
$\boldsymbol{u}_4$	Short	Open

$$\begin{aligned}
y_1 &= j\frac{1}{Z_1}\tan\left(\frac{\theta}{2}\right) + j\frac{1}{Z_2}\tan\left(\frac{\theta}{2}\right)\\
y_2 &= j\frac{1}{Z_1}\tan\left(\frac{\theta}{2}\right) - j\frac{1}{Z_2}\cot\left(\frac{\theta}{2}\right)\\
y_3 &= j\frac{1}{Z_1}\cot\left(\frac{\theta}{2}\right) - j\frac{1}{Z_2}\cot\left(\frac{\theta}{2}\right)\\
y_4 &= -j\frac{1}{Z_1}\cot\left(\frac{\theta}{2}\right) + j\frac{1}{Z_2}\tan\left(\frac{\theta}{2}\right)
\end{aligned} \tag{19}$$

At the center frequency, substituting $\theta = \pi/2$, we get

$$\begin{aligned}
y_1 &= j(Y_1 + Y_2)\\
y_2 &= j(Y_1 - Y_2)\\
y_3 &= -j(Y_1 + Y_2)\\
y_4 &= -j(Y_1 - Y_2)\\
Y_1 &= \frac{1}{Z_1}, \qquad Y_2 = \frac{1}{Z_2}
\end{aligned} \tag{20}$$

Substituting Eq. (20) into Eq. (6), we get

$$Y_1^2 - Y_2^2 = Y^2, \qquad Y = \frac{1}{Z} \tag{21}$$

Next, β is obtained by substituting Eq. (2) into Eq. (18):

$$\beta = \frac{Y_2}{Y_1}, \qquad |\beta| = \frac{Y_2}{Y_1} \tag{22}$$

From Eqs. (22) and (21), we get

$$Y_1 = \frac{Y}{\sqrt{1 - |\beta|^2}} \tag{23}$$

$$Y_2 = \frac{|\beta| Y}{\sqrt{1 - |\beta|^2}} \tag{24}$$

In the case of a 3-dB coupler, we get

$$Y_1 = \sqrt{2}Y, \qquad Y_2 = Y \tag{25}$$

By using the matrices

$$\boldsymbol{u}_1, \boldsymbol{u}_2, \boldsymbol{u}_3, \boldsymbol{u}_4 = \frac{1}{2}\begin{bmatrix} 1 & 1 & 1 & 1 \\ 1 & 1 & -1 & -1 \\ 1 & -1 & 1 & -1 \\ 1 & -1 & -1 & 1 \end{bmatrix} = [A] \tag{26}$$

and

$$\begin{bmatrix} s_1 & 0 & 0 & 0 \\ 0 & s_2 & 0 & 0 \\ 0 & 0 & s_3 & 0 \\ 0 & 0 & 0 & s_4 \end{bmatrix} = [S_d] \tag{27}$$

we obtain the scattering matrix $[S]$ as follows:

$$[S] = AS_dA^{-1} = AS_d\tilde{A} \tag{28}$$

where A is the orthogonal matrix, A^{-1} is the inverse matrix of A, and $\tilde{A}$ is the transverse matrix of A. Substituting Eqs. (26) and (27) into Eq. (28), we get

$$[S] = \begin{bmatrix} \delta & \alpha & \beta & \gamma \\ \alpha & \delta & \gamma & \beta \\ \beta & \gamma & \delta & \alpha \\ \gamma & \beta & \alpha & \delta \end{bmatrix} \tag{29}$$

where

$$\delta = \frac{s_1 + s_2 + s_3 + s_4}{4}$$

$$\gamma = \frac{s_1 - s_2 - s_3 - s_4}{4}$$

$$\alpha = \frac{s_1 + s_2 - s_3 - s_4}{4}$$

$$\beta = \frac{s_1 - s_2 + s_3 - s_4}{4} \tag{30}$$

The signal supplied at port 1 appears at port 4 with the quantity γ. Substituting

$$s_i = \frac{Y - y_i}{Y + y_i}$$

into Eq. (30), we get

$$\gamma = \frac{1}{4}\left(\frac{2(Y^2 - y_1y_4)}{Y^2 + (y_1 + y_4)Y + y_1y_4} - \frac{2(Y^2 - y_2y_3)}{Y^2 + (y_2 + y_3)Y + y_2y_3}\right)$$

Substituting Eqs. (19), (23), and (24) into the above equation, we get

$$|\gamma| = \frac{1}{4}\sqrt{\frac{1 + |\beta|}{1 - |\beta|}}\left(\frac{\tan^2(\theta/2) + \cot^2(\theta/2)}{2} - 1\right)$$

$$\simeq \sqrt{\frac{1 + |\beta|}{1 - |\beta|}}\,\frac{\Delta\theta}{2} = \sqrt{\frac{1 + |\beta|}{1 - |\beta|}}\,\frac{\pi}{4}\,\frac{\Delta\omega}{\omega} \tag{31}$$

Because the directivity D is defined by

$$D = 20\log_{10}\left(\frac{1}{|\gamma|}\right) \tag{32}$$

$|\gamma|$ is obtained from D, and the bandwidth ratio w is obtained from

$$w = \frac{2\Delta\omega}{\omega} = \frac{8}{\pi}\sqrt{\frac{1 - |\beta|}{1 + |\beta|}}\,|\gamma| \tag{33}$$

Analysis of Tightly-Coupled-Coils-Type Directional Coupler

We will obtain the eigenvalues of the impedance matrix, z_1, z_2, z_3 and z_4, corresponding to eigenvectors $\boldsymbol{u}_1$, $\boldsymbol{u}_2$, $\boldsymbol{u}_3$, and $\boldsymbol{u}_4$, respectively.

For the excitation of $\boldsymbol{u}_1$, the center of the coils and condensers are opened, which results in

$$z_1 = \infty \tag{34}$$

For the excitation of $\boldsymbol{u}_2$, the center of the coils are opened and the center of the condensers are shortened, which results in

$$z_2 = \frac{1}{j\omega C} \tag{35}$$

For the excitation of $\boldsymbol{u}_3$, the center of the coils are shortened and the center of the condensers are opened, which results in

$$z_3 = j\omega\,\frac{L + M}{2} \tag{36}$$

where L and M are the self-inductance and mutual inductance of the coils. For the excitation of $\boldsymbol{u}_4$, the center of the coils are shortened and the center of the condenser are also shortened, which results in

$$z_4 = \frac{1}{j\omega C + [j\omega(L - M)]^{-1}} = \frac{j\omega(L - M)}{1 - \omega^2 C(L - M)} \tag{37}$$

Substituting Eq. (34) into Eq. (5), we get

$$s_1 = 1 \tag{38}$$

Substituting Eq. (37) into Eq. (4), we have $s_4 = -1$, which results in $z_4 = 0$, that is,

$$L = M \tag{39}$$

This means that the coil should be coupled completely.

Substituting Eqs. (39) and (35) into Eq. (6), and substituting $Z = R$ in Eq. (6), we get

$$\sqrt{\frac{L}{C}} = R \tag{40}$$

From Eq. (5) and with z_3 of Eq. (36), we get

$$s_3 = \frac{j\omega L - R}{j\omega L + R} = -s_2 \tag{41}$$

Substituting Eq. (41) and the relation $s_1 = -s_4 = 1$ into Eq. (7), we get

$$\alpha = \frac{1}{2}\left(1 - \frac{j\omega L - R}{j\omega L + R}\right) = \frac{1}{1 + j\omega L/R} \tag{42}$$

$$\beta = \frac{1}{2}\left(1 + \frac{j\omega L - R}{j\omega L + R}\right) = \frac{1}{1 - jR/\omega L} \tag{43}$$

APPENDIX 7

Faraday Rotation in an Infinite Medium

We consider the linearly polarized wave propagating in the z direction in the infinite medium of ferrite under the dc magnetic field in the z direction.

The linearly polarized wave with an amplitude of 1 can be separated into positive and the negative circularly polarized waves with an amplitude of $\frac{1}{2}$ as shown in the below side of Fig. 1.

Positive and the negative polarized waves can be expressed by

$$(\boldsymbol{i}_x - j\boldsymbol{i}_y)e^{-j\beta_+ z}e^{j\omega t}, \qquad \beta_+ = \omega\sqrt{\mu_+\varepsilon} \tag{1a}$$

$$(\boldsymbol{i}_x + j\boldsymbol{i}_y)e^{-j\beta_- z}e^{j\omega t}, \qquad \beta_- = \omega\sqrt{\mu_-\varepsilon} \tag{1b}$$

thus, we can represent the electric fields at $t = 0$ as in Fig. 1.

For the dc magnetic field, H_0, shown in Fig. 2, we have the following values:

$$\beta_+ < \beta_-, \quad \lambda_+ = \frac{2\pi}{\beta_+} > \lambda_- = \frac{2\pi}{\beta_-}$$

because

$$0 < \mu_+ < \mu_-$$

Therefore, the twisted angle corresponding to the same values of z, is smaller for the positive circularly polarized wave than for the negative circularly polarized wave, as shown in Fig. 1.

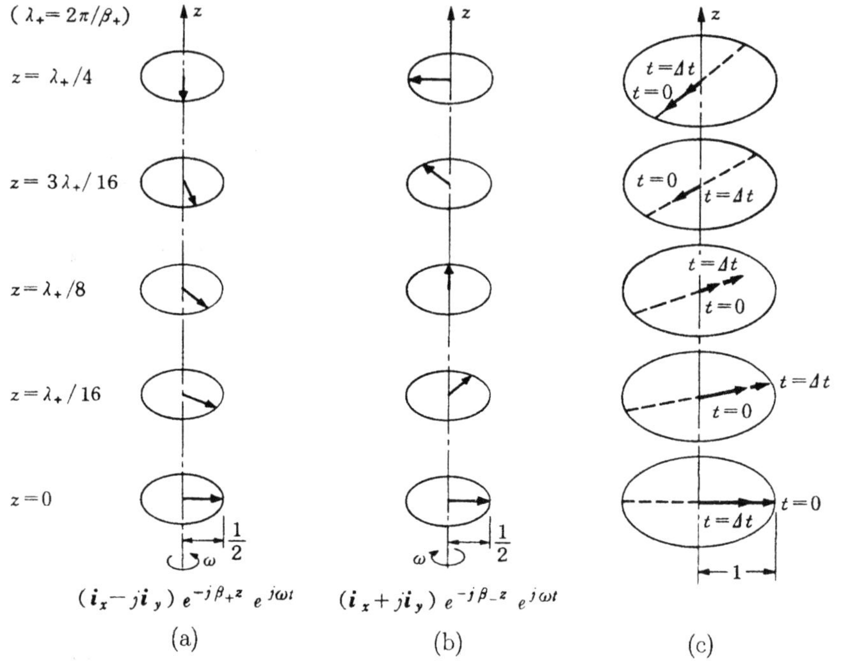

Figure 1. Explanation of Faraday rotation: (a) decomposed positive circularly polarized wave; (b) decomposed negative circularly polarized wave; (c) resultant field of (a) and (b).

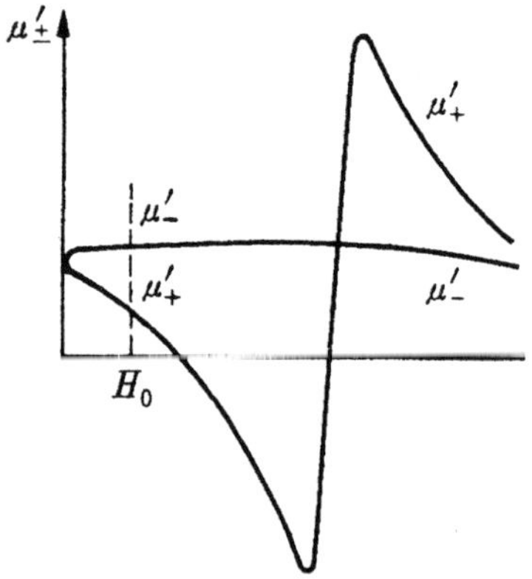

Figure 2. μ'_+ and μ'_- corresponding to the dc magnetic field H_0.

Therefore, the direction of the resultant field is twisted to the right as z increases. The twisted angle θ, therefore, can be obtained from Fig. 3 by

$$\theta = \frac{\beta_- - \beta_+}{2} l = \frac{\omega}{2c} (\sqrt{\mu_{-r}} - \sqrt{\mu_{+r}}) l \quad \text{(rad)} \tag{2}$$

where $\mu_{-r} > \mu_{+r}$ and c is the velocity of light.

Substituting $\mu_{\pm r}$ of Eq. (58d) of Chapter 6 into Eq. (2) and assuming $\alpha = 0$, we get

$$\theta = \frac{\omega\sqrt{\varepsilon} l}{2c} \left(\sqrt{1 + \frac{\omega_M}{\omega_0 + \omega}} - \sqrt{1 + \frac{\omega_M}{\omega_0 - \omega}} \right) \tag{3}$$

As understood from Fig. 2, H_0 is generally much smaller than H_r; using the definition

$$H_r = \frac{\omega}{|\gamma|}, \qquad H_0 = \frac{\omega_0}{|\gamma|}$$

we get $\omega \gg \omega_0$.

When M is small and ω is large, we have the relation $\omega_M \ll \omega$ from Eq. (58d) of Chapter 6. In such a case, Eq. (3) is simplified as

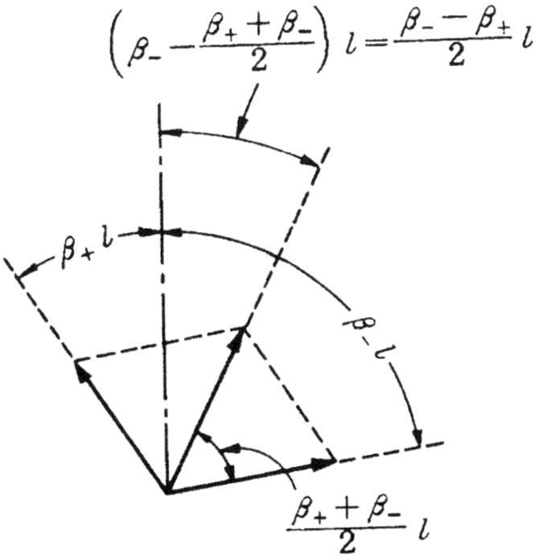

Figure 3. Rotated angle by Faraday rotation.

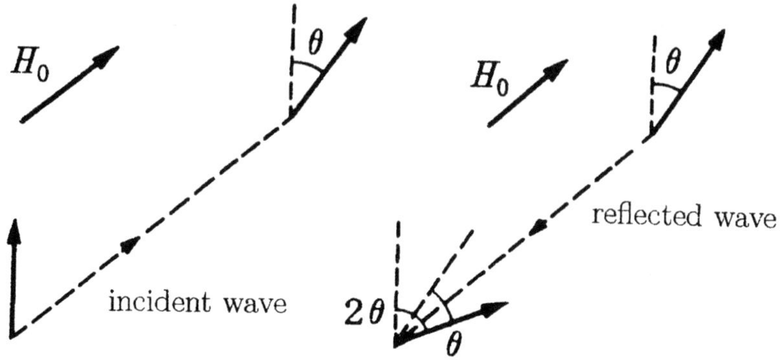

Figure 4. Faraday rotation angle θ of the incident and reflected waves.

$$\theta = \frac{\sqrt{\varepsilon}\omega_M l}{2c} \quad \text{(rad)} \tag{4}$$

Thus far, we explained the incident wave.

In the case of the reflected wave, the wave travels from $z > 0$ toward $z = 0$. Therefore, we can consider the case when the dc magnetic field is supplied in reverse in Figs. 1a and 1b. In this case, we can consider the positive and negative circular polarized waves in Figs. 1a and 1b. Then the direction of the resultant linearly polarized wave field is twisted to the left as z decreases, which is the direction of the reflective wave. This means that the twisted direction is the same toward the z direction when the values of the θ are the same as for the incident wave.

The statement can be represented by Fig. 4. Therefore, the twisted angle and the direction is the same between the incident and reflected waves as for looking the same direction of $z > 0$.

Index